Lecture Notes in Physics

Edited by J. Ehlers, München, K. Hepp, Zürich
R. Kippenhahn, München, H. A. Weidenmüller, Heidelberg
and J. Zittartz, Köln
Managing Editor: W. Beiglböck, Heidelberg

128

Neutron Spin Echo

Proceedings of a
Laue-Langevin Institut Workshop
Grenoble, October 15–16, 1979

Edited by F. Mezei

Springer-Verlag
Berlin Heidelberg New York 1980

Editor

Ferenc Mezei
Central Research Institute for Physics, P.O.B. 49
H-1525 Budapest

ISBN 3-540-10004-0 Springer-Verlag Berlin Heidelberg New York
ISBN 0-387-10004-0 Springer-Verlag New York Heidelberg Berlin

Printing and binding: Beltz Offsetdruck, Hemsbach/Bergstr.
2153/3140-543210

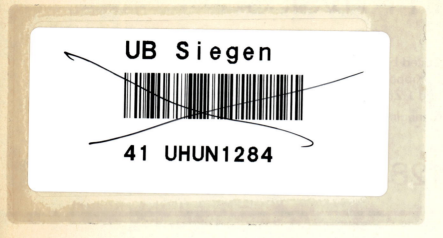

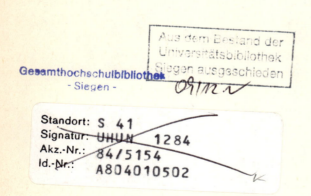

P R E F A C E

The idea of Neutron Spin Echo was born in April 1972 at a red traffic light at the corner of Alagút street in Budapest. Within two weeks the basic points were experimentally verified at the reactor of the Budapest Central Research Institute for Physics. By the end of the year I was also able to demonstrate (this time at the Institut Laue-Langevin in Grenoble) that by this method one can really observe very small velocity changes of a neutron beam, independently of the velocity spread. Soon after, in January 1973, the ILL Council approved the construction of a proposed spin echo spectrometer (later to become known as IN11) for high resolution inelastic neutron scattering experiments.

The design of the new project was kept technically fairly simple and inexpensive, with a maximum of flexibility for tests. Nevertheless, the novel machine presented the ILL technical services with a number of unusual problems. By the end of the year 1973 John Hayter joined the project, and it was not until early 1977, after a large amount of work by many people, that real-life tests could begin. Routine user operation started in mid-1978 and the demand for beam-time has been increasing ever since.

There existed one uncertain point when the project was launched: how to produce a good polarized neutron flux. Fortunately, during 1976, Paul Dagleish and I succeeded in developing a new type of neutron polarizer, the "supermirror", which provides several times higher flux on IN11 than the next best solution. Without this development we would have had a lot fewer experimental results to talk about today. The problem of neutron flux in polarization analysis was, perhaps fortunately, not fully appreciated back in 1972 when I made the proposal to build IN11, and it remains the main experimental difficulty.

The present first workshop on Neutron Spin Echo, the organization of which was initiated by John White, took place at a time when there was enough "battlefield" experience with NSE to make a summary, especially intended to help future users who seem to be steadily increasing in number at the ILL. These proceedings were therefore made with the ambition of providing as complete as possible a reference manual for readers with just a little background in neutron scattering research, and also for specialists from other areas of neutron physics. For their convenience, the volume includes an appendix of facsimile reproductions of a number of original publications, which appear by the kind permission of the Copyright holders: Springer-Verlag (Appendices A and B), International Atomic Energy Agency (Appendix C), The Institute of Physics (Appendix D), The American Physical Society (Appendix E) and North-Holland Publishing Company (Appendices F and G), whose courtesy is gratefully acknowledged.

As the editor of this volume, I am indebted to the authors who accepted the burden of writing up their talks, often in extended form. The book has no monolithic structure. The papers appear, however, in the same logical order as they were presented at the workshop. They are essentially self-contained articles, in which the authors have freely expressed their own points of view and understanding of the subject. I also wish to thank Mmes. Wegener, Parisot and Volino in Grenoble and Miss Polgàr in Budapest for their careful typing of nearly all of the contributions and MM. Paul Dagleish and Harvey Shenker for their part in the proofreading.

To conclude I would like to express my very personal gratitude to all those whose help and collaboration marks the way from the original idea to the reality of this volume. They include many many people from the technical services at the ILL, numerous colleagues, particularly those who contributed to this book, and the successive directors of both the Institut Laue-Langevin and the Central Research Institute for Physics in Budapest, MM. Dreyfus, Jacrot, Joffrin, Lomer, Pàl, Springer, Szabô, White and especially Prof. Mössbauer. It has been the continuous interest, support and generous approach by these directors that have smoothed my way from that red traffic light onwards.

Budapest, March 1980 F. Mezei

TABLE OF CONTENTS

CHAPTER I:

The Neutron Spin Echo Method

THE PRINCIPLES OF NEUTRON SPIN ECHO

F. MEZEI

Institut Laue-Langevin
156X, 38042 Grenoble Cédex, France

and

Central Research Institute for Physics
H-1525 Budapest 114, P.O.Box 49, Hungary

INTRODUCTION

Neutron Spin Echo (NSE) is a particular experimental technique in inelastic neutron scattering. It is substantially different from the other, the "classical", methods both conceptually and technically. Conventionally, an inelastic neutron scattering experiment consists of two steps, viz. preparation of the incoming monochromatic beam and analysis of the scattered beam. The values of the measured energy and momentum transfer are then determined by taking the appropriate differences between the incoming and outgoing parameters measured in the two above steps. In NSE, both the incoming and outgoing velocity of a neutron (more precisely given components of these) are measured by making use of the Larmor precession of the neutron's spin. This kind of measurement could be called "internal" for each neutron, since the Larmor precession "spin clock" attached to each neutron produces a result stored on each neutron as the position of the spin vector serving like the hand of a clock. This is in contrast to the classical monochromatization or analysis, in which cases neutrons within a given velocity band are singled out "externally", i.e. by a selecting action measuring equipment. This difference is the technical one. In addition, since the Larmor precession information on the incoming velocity (component) of each neutron is stored on the neutron itself, it can be compared with the outgoing velocity (component) of one and the same neutron. Thus in NSE the velocity (component) change of the neutrons can be measured directly, in a single step, which is its conceptual novelty.

In this introductory paper the principles and the different types of applications of NSE are described. Although the presentation is self-contained, most technical and mathematical details are omitted here. These are extensively dealt with in the subsequent contributions and in the original papers reproduced in the Appendix of this volume, and the reader will be provided with ample references to these. In the first section the basic facts about Larmor precession in a polarized beam and the notion of the spin echo action are discussed. The second section is devoted to the introduction of the simplified principle of Neutron Spin Echo as a method of inelastic neutron scattering spectroscopy, applicable to quasi-elastic and non-dispersive inelastic scattering processes. The following section gives the generalization of the NSE principle for the study of dispersive elementary excitations; the final one describes the effect of sample magnetism introducing the notions of Paramagnetic, Ferromagnetic and Antiferromagnetic NSE.

1. LARMOR PRECESSION AND SPIN ECHO

To the best of my knowledge Larmor precession in a neutron beam traversing a magnetic field region was first observed by Drabkin et al.[1] as early as 1969. Unfortunately this work was not known to me until recently; it was in 1972 that I started to work on Larmor precession by introducing a simple new technique for turning the neutron spin direction in any desired direction with respect to the magnetic field direction[2] (see also the Appendix). This technique is described in the following paper by Otto Schärpf, together with more details about Larmor precessions. For the moment it is sufficient to recall that in a neutron beam travelling through a homogeneous magnetic field H_o and polarized originally parallel to the magnetic field direction $z||H_o$, one can initiate Larmor precession by turning the polarization direction $\vec{P}$ perpendicular to the z axis, say into the x direction at a given point (surface) A along the trajectory (Fig. 1).

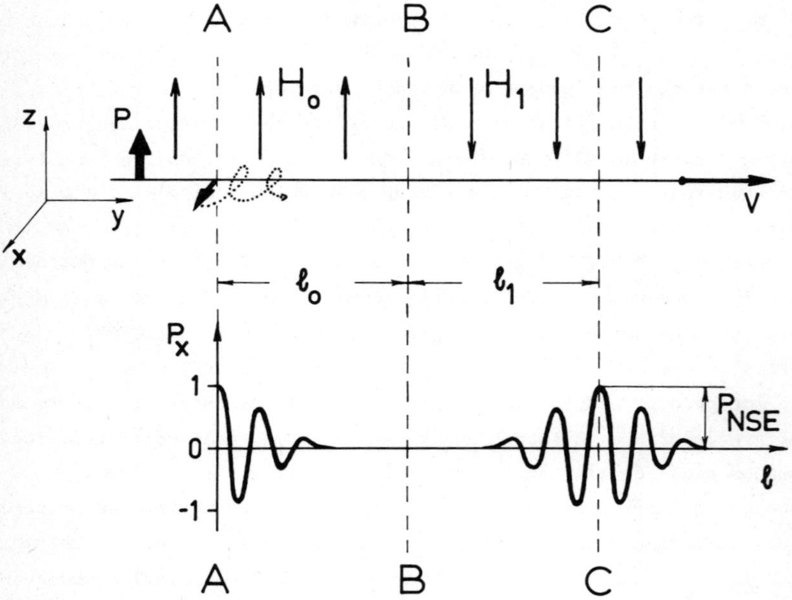

Fig. 1. Larmor spin precession of neutrons in a beam and the simple spin echo effect.

This $\frac{\pi}{2}$ turn initiates the Larmor precession which can be physically characterized, for example, by the x component of the neutron polarization - which obviously has a value of 1 at A. The basic fact about Larmor precessions in spin $\frac{1}{2}$ particle beams is that they can be exactly described classically except in situations where the Stern--Gerlach effect is appreciable, which only happens in very extreme cases with neutrons. This means that the particle beam will be described by a classical velocity distribution function f(v), and for each point-like particle the "classical" spin vector $\vec{S}$ follows the $d\vec{S}/dt=\gamma_L[\vec{S}\times\vec{H}]$ classical equation of motion. A rigorous quantum mechanical proof of this theorem has been described recently by the author[3]. Thus the Larmor precession angle φ for a given neutron at a distance ℓ from A (Fig. 1) will be given as

$$\varphi = \gamma_L \frac{\ell H_o}{v} \quad , \tag{1}$$

where γ_L=2.916 kHz/Øe. Since we measure φ with respect to the initial direction x, the polarization component P_x for the beam is given by the beam average

$$P_x = \langle\cos\varphi\rangle = \int f(v)\cos\left(\frac{\gamma_L \ell H_o}{v}\right)dv \tag{2}$$

(Notice that here P_x is given as the Fourier transform of the distribution function for $\frac{1}{v}$, viz. $F(\frac{1}{v})=v^2 f(v)$, which is in fact the wavelength spectrum. This point is discussed in detail in the contribution of John Hayter; and also in the Appendix[4].) The behaviour of P_x with ℓ is easily seen from Eq.(2). As ℓ increases, the differences between φ's for different v's become bigger and bigger, i.e. the Larmor precessions for different neutrons become more and more out of phase. Consequently, the average $\langle\cos\varphi\rangle$ will tend to zero, and we obtain the characteristic behaviour of P_x shown in the lower part of Fig. 1 between A and B; this behaviour was observed by Drabkin et al. in 1969. The period of the damped oscillation is obviously related to the average beam velocity. Thus the observation of Larmor precessions is a simple way of measuring neutron velocities though it tends to be somewhat over-sensitive except for special high precision problems such as the one described by W. Weirauch et al. later in this volume. This sensitivity is illustrated by the large value of φ=1832 rad for H_o=100 Øe, ℓ=1 m and v=1000 m/sec (λ=4 Å).

In order to make more general use of the high sensitivity of Larmor precessions we have to eliminate this dephasing effect arising from the velocity distribution f(v). This is where the echo principle, common to various physical phenomena (one of which is described in the contribution of Badurek, Rauch and Zeilinger), becomes instrumental. In the present case it is realized by making the neutrons precess in the opposite sense after a certain time. This happens in section BC in Fig. 1, where field H_1 is opposite H_o. At point C

$$\varphi = \varphi_{AB} - \varphi_{BC} = \gamma_L (H_o \ell_o - H_1 \ell_1)/v \qquad (3)$$

and if the configuration is "symmetric", that is, $H_o\ell_o = H_1\ell_1$, φ will be zero for all velocities v and thus $P_x = \langle\cos\varphi\rangle = 1$. Obviously, as is also illustrated in Fig. 1, P_x will show the same damped oscillation behaviour on both sides of C as that described for point A, since differences in φ build up in exactly the same way on moving away from C. It is clear from Eq.(3) that only the difference $H_o\ell_o - H_1\ell_1$ is important, and in view of this the number of both the forward and the backward precessions, φ_{AB} and φ_{BC}, respectively, can be arbitrarily big (assuming that the fields H_o and H_1 are sufficiently stable and homogeneous). We will call this behaviour of the polarization P_x a "spin echo group" and the amplitude of the P_x oscillation at the symmetry position C will be called "spin echo signal", P_{NSE}. As has been pointed out, the spin echo group is the Fourier transform of the $\frac{1}{v}$ distribution function, $v^2 f(v)$, thus the narrower this distribution, the more oscillations are contained in the group, as shown by the measured curves in Fig. 2. Note that in practice one would change H_1 rather than ℓ_1; furthermore, H_o and H_1 will be parallel and the neutron spins are flipped at B instead, as in NMR spin echo (cf. Otto Schärpf's paper for details).

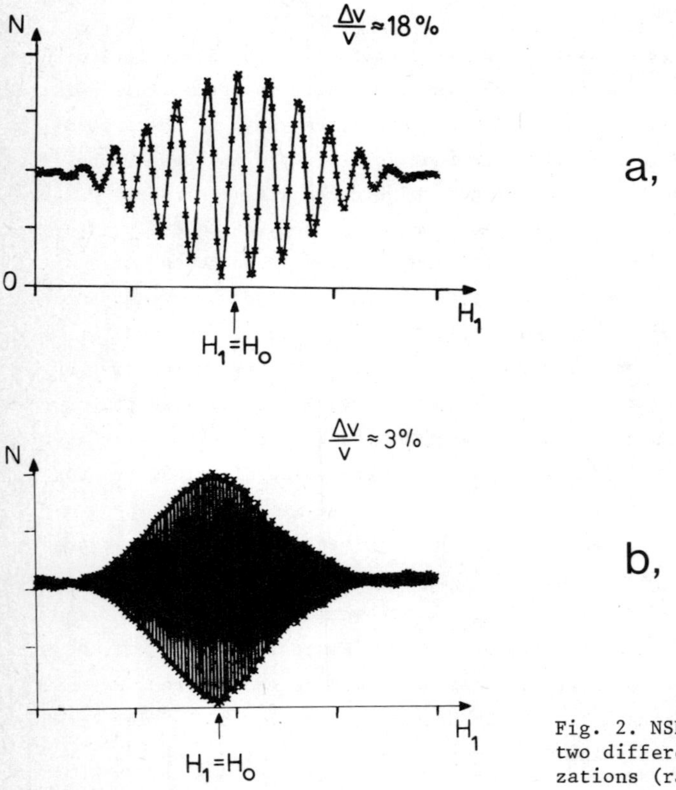

Fig. 2. NSE groups measured with two different beam monochromatizations (raw data).

The way P_x can be measured at C consists of applying another $\frac{\pi}{2}$ turn at this point, thus turning P_x into $\pm P_z$ depending on the sense of the $\frac{\pi}{2}$ turn. Then a usual spin analyser placed before the detector allows the determination of P_z by showing a maximum transmission for $P_z=1$ and a minimum for $P_z=-1$. Depending on the sense of the $\frac{\pi}{2}$ turn at C the $P_x=1$ echo maximum will be observed either as a maximum, or as a minimum of the detector counting rate N. (A comparison of these two values is one way of measuring P_{NSE}.) The data in Fig. 2 were obtained by choosing a $P_x \to P_z$ turn.

For those readers familiar with classical polarized neutron work, there is a general remark to be made here about the calculation of polarization figures in NSE. The usual definition of $P=(N_+-N_-)/(N_++N_-)$ has one inconvenience: it is nonlinear in N_+ and N_-, thus it is not additive for different contributions, and the comparison of different polarization figures for the same beam, e.g. P_x and P_z, becomes tedious too. Since the polarizer and analyser efficiencies are constant for the whole experiment, it is generally more convenient to take the polarization as being just proportional to the modulation: $P=(N_+-N_-)/N_o$, where the constant denominator N_o has been determined in the usual way from just one of the polarization measurements. This gives a linear definition for P, and since polarization figures will only be considered relative to each other (e.g. P_{NSE} with respect to the incoming beam polarization P_o or P_{NSE} signals measured for different samples), the polarization efficiency factors for the different spectrometer components need not be known. For example in Fig. 2 we can define the spin echo signal P_{NSE} as $(N_{max}-N_{min})/2N_{ave}$, where N_{max} and N_{min} correspond to the highest maximum and the lowest minimum of the curve, and N_{ave} is the counting rate outside the spin echo group.

To conclude this section let us remark that the essential physical reason for the appearance of the NSE group is that at C the Larmor precession angle φ is stationary over the beam:

$$\left(\frac{\partial\varphi}{\partial v}\right)_{beam} = 0 \tag{4}$$

and not that φ is identically zero. The P_x precessing polarization is produced by all neutrons in the beam having the same spin direction φ at a given point, no matter how this common direction comes about. The shape of the NSE group is then determined by the dephasing of the spin precessions "locally", around the echo point, thus it depends only on the f(v) distribution of the detected neutron beam. Equation (4) is thus the echo condition to be used in what follows.

2. SPIN ECHO WITH SCATTERED NEUTRON BEAMS

Let us consider the configuration shown in Fig. 3. This configuration only differs from the previous one by the neutron beam being scattered on a sample between the forward and backward precessions. For simplicity the precessions in the well controlled small field around the sample, between H_o and H_1, are neglected. (In practice

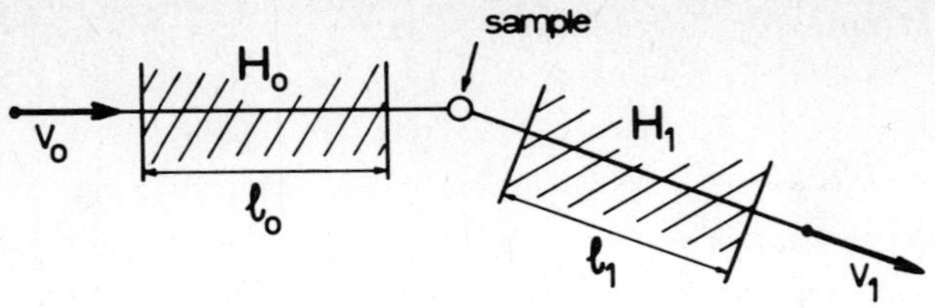

Fig. 3. Application of NSE to inelastic neutron scattering.

these count either for H_o or H_1, the two precession regions being separated by a spin flipper coil, cf. the paper by Otto Schärpf.) The total Larmor precession angle now reads, instead of Eq.(3),

$$\varphi = \varphi_{in} - \varphi_{out} = \gamma_L \left(\frac{\ell_o H_o}{v_o} - \frac{\ell_1 H_1}{v_1} \right) = \varphi(v_o, v_1) \quad , \tag{5}$$

where v_o and v_1 are the incoming and outgoing neutron velocities, respectively. If there is only elastic scattering $v_o = v_1 = v$, in which case the situation is exactly what it was before. In inelastic scattering the neutron energy change is the relevant quantity in which we are interested, viz.

$$\hbar\omega = E_1 - E_o = \tfrac{1}{2} m v_o^2 - \tfrac{1}{2} m v_1^2 = \hbar\omega(v_o, v_1) \quad . \tag{6}$$

The basic idea of NSE inelastic spectroscopy is to use φ as given by Eq.(5) to measure ω. Since $\varphi(v_o, v_1)$ and $\omega(v_o, v_1)$ are different functions, this is only possible locally with respect to an arbitrarily chosen value ω_o. To achieve this, we will use neutron beams such that the average neutron velocities $\bar{v}_o$ and $\bar{v}_1$ correspond to ω_o: $\omega(\bar{v}_o, \bar{v}_1) = \omega_o$. Then we require

$$\varphi - \bar{\varphi} = t(\omega - \omega_o) \tag{7}$$

where $\bar{\varphi} = \varphi(\bar{v}_o, \bar{v}_1)$ and t is a proportionality constant. This is the fundamental equation of NSE spectroscopy. It postulates that locally around "ideal" values given by $\bar{v}_o$, $\bar{v}_1$ the Larmor precession angle φ becomes a measure of the neutron energy transfer ω. Obviously this equation can only be satisfied in first order in $\delta v_o = v_o - \bar{v}_o$ and

$\delta v_1 = v_1 - \bar{v}_1$. Namely, from Eq.(5):

$$\varphi - \bar{\varphi} = -\gamma_L \frac{\ell_o H_o}{\bar{v}_o^2} \delta v_o + \gamma_L \frac{\ell_1 H_1}{\bar{v}_1^2} \delta v_1 \tag{8}$$

and from Eq.(6)

$$\omega - \bar{\omega}_o = \frac{m}{\hbar} v_1 \delta \bar{v}_1 - \frac{m}{\hbar} \bar{v}_o \delta v_o \quad . \tag{9}$$

Thus the NSE equation (7) is satisfied if the coefficients of the independent variables δv_o and δv_1 agree on both sides:

$$\gamma_L \ell_o \frac{H_o}{\bar{v}_o^2} = t \frac{m}{\hbar} \bar{v}_o \ , \qquad \gamma_L \ell_1 \frac{H_1}{\bar{v}_1^2} = t \frac{m}{\hbar} \bar{v}_1 \tag{10}$$

Since these equations are identical for both indices 0 and 1, in what follows i stands for both. Equations (10) are called the NSE conditions. They can be given in the more practical form

$$\frac{\ell_o H_o}{\ell_1 H_1} = \frac{\bar{v}_o^3}{\bar{v}_1^3} \quad , \qquad t = \frac{\hbar \gamma_L \ell_i H_i}{m \bar{v}_i^3} = \hbar \frac{\bar{\varphi}_i}{2\bar{E}_i} \quad . \tag{11}$$

Equations (11) are used in practice to choose the ratio of the magnetic fields to obtain the echo signal and to calculate the proportionality parameter t for Eq.(7). It is obvious that for elastic scattering or for the simple straight beam echo experiment shown in Fig. 1, we get what we had before, namely: $\ell_1 H_1 = \ell_o H_o$, since $\bar{v}_1 = \bar{v}_o$.

It can be shown quite generally that the NSE conditions in Eqs.(10) or (11) describe the centre of the NSE group for a scattering process with energy change ω_o. The essential thing Eq.(7) implies is that φ depends only on the energy transfer ω, which is the relevant parameter for the sample scattering, and it does not explicitly depend on v_o and v_1 separately. Thus the NSE equation (7) is equivalent to the condition

$$\left(\frac{\partial \varphi}{\partial v_i}\right)_{\omega=\omega_o} = 0 \tag{12}$$

which has just the form of Eq.(4), and it states that phase φ is stationary over the scattered beam, i.e. the neutrons having suffered an energy change ω_o in the scattering will produce much the same NSE signal as in Fig. 1, but centred at $\ell_1 H_1 / \ell_o H_o \neq 1$ [cf. Eq.(11)].

The sample scattering is characterized by the scattering function $S(\vec{\kappa},\omega)$, which will be assumed at this point to depend only on ω, since for the moment we are looking at the energy transfers only. This function describes the probability that the neutrons are scattered with the energy change ω, and thus it will give, via Eq.(7), the distribution function of φ's in the scattered beam. Consequently, the NSE signal will be given as

$$P_{NSE} = P_S \langle \cos(\varphi-\bar{\varphi}) \rangle = P_S \frac{\int S(\vec{\kappa},\omega)\cos[t(\omega-\omega_o)]d\omega}{\int S(\vec{\kappa},\omega)d\omega} , \tag{13}$$

where P_S takes into account the eventual change of the neutron polarization by the scattering action itself which will be dealt with in detail in Section 4. If there is no spin scattering and high magnetic field involved, $P_S=1$. We have to remember, however, that the proportionality (7) between $\varphi-\bar{\varphi}$ and $\omega-\omega_o$, only holds to first order around $\bar{v}_o$ and $\bar{v}_1$, i.e. within a restricted range of δv_o and δv_1. Thus the integrations in Eq.(13) have to be restricted to this range, which means that in practice the incoming beam has to be roughly monochromatic and eventually an analyser has to be used for the outgoing beam, too. This amounts to using NSE in combination with a classical background spectrometer. This is a very general feature and, in addition to the ω resolution obtained by the spin echo, this background spectrometer provides the momentum $\vec{\kappa}$ resolution. Thus Eq.(13) shows that the NSE signal P_{NSE} corresponds to the ω Fourier transform of a given part of $S(\vec{\kappa},\omega)$ as singled out by the transmission function ("resolution ellipsoid" in the usual terminology) of the background spectrometer. This situation is illustrated in Fig. 4 where the shaded areas correspond to this transmission function, and it is also shown that by measuring only $\omega-\omega_o$ one can study quasi-elastic scattering problems ($\omega_o=0$) and optical-like, flat sections of elementary excitation branches ($\omega_o\neq0$). Experimental examples for both cases are described in various contributions to this volume.

In order to obtain full information about the scattering function in Eq.(13), P_{NSE} has to be measured at several values of the Fourier parameter t, i.e. at several values of H_o at constant H_o/H_1 [cf. Eqs.(11)]. For example, if the studied part of $S(\vec{\kappa},\omega)$ corresponds to a Lorentzian line, which is narrow compared with the background spectrometer transmission function (Fig. 4) and is centred at ω_o:

$$S(\vec{\kappa},\omega) \propto \frac{\gamma}{\gamma^2+(\omega-\omega_o)^2} , \tag{14}$$

the integration in Eq.(13) can be taken from $-\infty$ to ∞, and it gives

$$P_{NSE}(t) = P_S \frac{\int_{-\infty}^{\infty}[\gamma^2+(\omega-\omega_o)^2]^{-1}\cos[t(\omega-\omega_o)]d\omega}{\int_{-\infty}^{\infty}[\gamma^2+(\omega-\omega_o)^2]^{-1}d\omega} = e^{-\gamma t} . \tag{15}$$

Thus, from the t dependence of P_{NSE} one can check if the line is really Lorentzian and one can then determine the linewidth parameter γ.

In practice, the measurement of P_{NSE} occurs in two steps. First the spectrometer has to be calibrated by measuring the NSE signal for a standard sample with

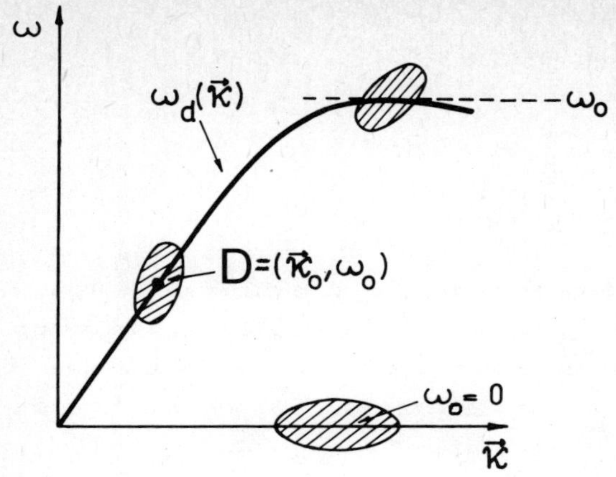

Fig. 4. The role of the transmission function of the host ("background") spectrometer in providing momentum resolution and selection of the scattering process in different types of NSE experiments.

γ=0 (e.g. elastic scatterer), and P_S=1, of course. The obtained $P^o_{NSE}(t)$ function can be considered as the instrumental resolution function since it differs from 1 due only to the finite overall polarization efficiency of the spectrometer and, what is more important, due to the residual dephasing of the Larmor precessions brought about by the inhomogeneities of the magnetic precession fields H_o and H_1. The absolute value of these inhomogeneities increases with increasing H_o and H_1, thus $P^o_{NSE}(t)$ itself depends strongly on t, as shown by the experimental curve in Fig. 5.

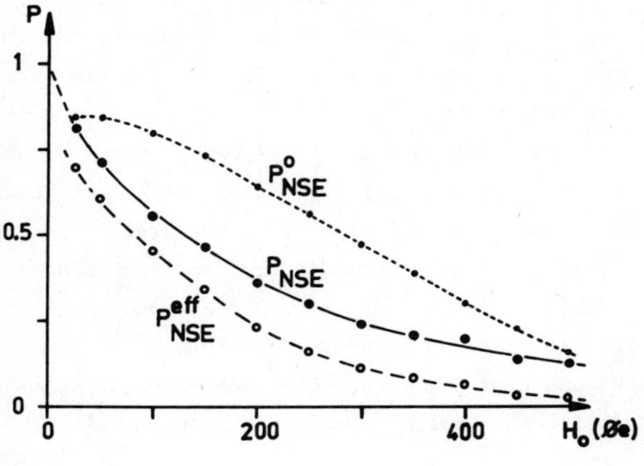

Fig. 5. An example for NSE spectrometer calibration and data reduction (see text). The corrected spectrun P_{NSE} corresponds to a quasielastic scattering line with 0.24 μeV half-width at half maximum. H_o is proportional to the time parameter t, cf. Eq. (11). (IN11 data)

This instrumental broadening of the φ distribution will then convolute with the one that corresponds to $S(\vec{\kappa},\omega)$ of the sample, which convolution will be translated by the Fourier transformation in Eq.(13) to multiplication, viz. the NSE signal P_{NSE}^{eff} directly observed for the sample will be given as

$$P_{NSE}^{eff}(t) = P_{NSE}(t) \cdot P_{NSE}^{o}(t) \quad , \tag{16}$$

where $P_{NSE}(t)$ is the function we have been considering above and the one we wish to obtain from the NSE experiment. Thus, in contrast to the usual methods, the instrumental resolution effects are taken into account by simple division instead of an often ambiguous and tedious deconvolution. This feature of NSE is fundamental for line-shape studies.

Let us note that for quasi-elastic scattering with $\omega_o=0$, $P_{NSE}(t)$ can be given a concrete interpretation. Since the scattering function of the sample $S(\vec{\kappa},\omega)$ is given as the Fourier transform of the space - time correlation function $S(\vec{r},t)$, the cosine Fourier transform in Eq.(13) will lead back to the so-called intermediate scattering function, viz.

$$P_{NSE}(t) = P_S \cdot \mathrm{Re}\ S(\vec{\kappa},t) \quad , \tag{17}$$

where t has the meaning of real, physical time. For isotropic systems $S(\vec{\kappa},t)$ is purely real, thus $P_{NSE}(t)$ is precisely the time dependent correlation function for fluctuations with the selected wave vector $\vec{\kappa}$.

Before turning to the more general formulation of the NSE method, let us consider the conceptual difference between NSE and conventional inelastic neutron scattering techniques. Conventionally, the energy transfer $\hbar\omega$ is measured by taking the difference of the outgoing and incoming energies, i.e. as a difference of two beam averages: $\frac{1}{2}m\langle v_1^2\rangle - \frac{1}{2}m\langle v_o^2\rangle$. The resolution in ω is thus limited by the scatter of both v_1 and v_o. In view of this, high resolution implies correspondingly good monochromatization, i.e. low neutron intensity, which relation comes directly from the well-known Liouville theorem. In NSE the beam average of $\cos\varphi$ is measured in a single step, i.e. of a quantity which is directly related to the energy change ω. Consequently, the ω resolution becomes independent of the incoming and outgoing beam monochromatization, and the Liouville relation between intensity and resolution does not apply. This feature and the high inherent sensitivity of the Larmor precession technique are the clues to the new possibilities offered by NSE in high resolution inelastic neutron scattering spectroscopy.

3. THE GENERAL PRINCIPLE OF NSE

In the previous section we dealt with scattering functions which locally depend only on ω and not on $\vec{\kappa}$. The NSE method in its most general form, as introduced in

Ref. 5, (see also the Appendix) can be applied to $\vec{\kappa}$ dependent processes too, like elementary excitations with a general dispersion relation (cf. Fig. 4). In Ref. 5 this generalization was obtained from the following basic idea. The scattering of a given neutron is described by the incoming and outgoing velocity or momentum vectors $\vec{\kappa}_o = \frac{m}{\hbar}\vec{v}_o$ and $\vec{\kappa}_1 = \frac{m}{\hbar}\vec{v}_1$, respectively. The sample scattering functions on the other hand depend on the transfer parameters

$$\vec{\kappa} = \frac{m}{\hbar}(\vec{v}_1 - \vec{v}_o)$$
$$\hbar\omega = \frac{1}{2}m(v_1^2 - v_o^2) \qquad (18)$$

(Note that throughout this paper the absolute value of a vector is denoted by its symbol without the arrow, i.e. $v_1 = |\vec{v}_1|$, etc.) Thus the four parameters $(\vec{\kappa},\omega)$ are the relevant ones, whereas in conventional neutron scattering spectroscopy the six irrelevant parameters $(\vec{v}_o,\vec{v}_1)$ are measured separately. The redundant 6-4=2 free parameters are at the origin of the complexity of the resolution calculations, and since $\vec{\kappa}$ and ω are both obtained as differences of separately measured quantities $\vec{\kappa}_o$ and $\vec{\kappa}_1$ [cf. Eqs.(18)], the Liouville relation between intensity and resolution applies and it alone sets the limits for resolution. (In the classical type backscattering method basically designed for $\vec{\kappa}$ independent scattering effects, very much intensity can be gained back by the use of huge detector solid angles, cf. the paper by A. Heidemann et al.) The quantity involved in general NSE experiments, the total Larmor precession angle φ – which most generally will be given as $\varphi=\varphi(\vec{v}_o,\vec{v}_1)$ – depends on both the incoming and the outgoing neutron parameters. It is then conceivable that this function φ has special symmetry in that it depends (locally) on only four relevant combinations (18) of six irrelevant parameters $(\vec{v}_o,\vec{v}_1)$. Thus for the local variation we require that $\varphi=\varphi(\vec{\kappa},\omega)$, that is,

$$\delta\varphi = \vec{\alpha}\delta\vec{\kappa} + \beta\delta\omega \quad , \qquad (19)$$

where $\vec{\alpha}$ and β are constants. This equation is the most general formulation of NSE. It implies that φ becomes locally, around a point in the $(\vec{\kappa},\omega)$ space, exactly the same kind of four parameter function as $S(\vec{\kappa},\omega)$, and thus, if properly matched, it can effectively probe $S(\vec{\kappa},\omega)$ directly in a single step measurement, and it is not affected by the Liouville intensity-resolution relation.

Here I will present a more obvious though less general reasoning which, however, leads to the same result. The aim of doing this is to give insight to the significance of the above argument, which in turn made sure that nothing gets omitted.

Let us consider how the fundamental NSE equation (7) could be generalized for the study of dispersive elementary excitation branches. What we have to know in such a case is the behaviour of $S(\vec{\kappa},\omega)$ as a function of the distance from the dispersion relation $\omega_d(\vec{\kappa})$. Indeed, we expect that $S(\vec{\kappa},\omega)$ changes little along the $\omega_d(\vec{\kappa})$ surface, i.e. in going along the "ridge" of the excitation branch, but it changes rapidly when going across. This amounts to the assumption that locally, around the point $D=(\vec{\kappa}_o,\vec{\omega}_o)$ (cf. Fig. 4), the scattering function can be approximated by the single variable function S_D as

$$S(\vec{\kappa},\omega) = S_D[\omega-\omega_d(\vec{\kappa})] \tag{20}$$

Note that in effect $S(\vec{\kappa},\omega)$ as it appears in the experiment is already modified by the background spectrometer transmission function $T(\vec{\kappa},\omega)$. Thus, $S_D[\omega-\omega_d(\vec{\kappa})]$ actually corresponds to the behaviour of $S(\vec{\kappa},\omega)T(\vec{\kappa},\omega)$ averaged over the shaded area where $T(\vec{\kappa},\omega)\neq 0$.

If there are scattering contributions inside the transmission area other than the ω_d dispersion relation we are interested in, these will also be included in S_D but they will tend to give only a flat background below the peak at $\omega-\omega_d=0$.

In order to probe function S_D using the NSE Larmor precession angle φ, we require, instead of Eq.(7),

$$\varphi-\bar{\varphi}=t[\omega-\omega_d(\vec{\kappa})] \quad , \tag{21}$$

where $\vec{\kappa}=m(\vec{v}_1-\vec{v}_o)/\hbar$, furthermore $\varphi=\varphi(\vec{v}_o,\vec{v}_1)$, and $\bar{\varphi}=\varphi(\bar{\vec{v}}_o,\bar{\vec{v}}_1)$ with the average velocities $\bar{\vec{v}}_o$ and $\bar{\vec{v}}_1$ being assumed to correspond to the point D on the dispersion relation around which it is being looked at, that is,

$$\hbar\vec{\kappa}_o = m(\bar{\vec{v}}_1-\bar{\vec{v}}_o) \qquad \hbar\omega_o = \tfrac{1}{2}m(\bar{\vec{v}}_1^{\,2}-\bar{\vec{v}}_o^{\,2})$$
$$\omega_d(\vec{\kappa}_o) = \omega_o \tag{22}$$

Equation (21) is the final, general NSE equation which can again be satisfied to first order in $\delta\vec{v}_i=\vec{v}_i-\bar{\vec{v}}_i$ (i=0,1). Note that in order to tackle $\vec{\kappa}$ dependent scattering functions we have to consider the full vector-character of the incoming and outgoing neutron velocities, whereas above it was sufficient to look at their absolute values.

The right hand side of our fundamental equation (21) can, in view of Eqs.(22), be written in the differential form

$$\omega-\omega_d(\vec{\kappa}) \simeq \omega-[(\vec{\kappa}-\vec{\kappa}_o)\cdot\mathrm{grad}\,\omega_d(\vec{\kappa}_o)+\omega_d(\vec{\kappa}_o)] =$$
$$= \frac{m}{\hbar}(\bar{\vec{v}}_1-\mathrm{grad}\,\omega_d)\delta\vec{v}_1 - \frac{m}{\hbar}(\bar{\vec{v}}_o-\mathrm{grad}\,\omega_d)\delta\vec{v}_o \quad , \tag{23}$$

where $\mathrm{grad}\,\omega_d=(\frac{\partial\omega_d}{\partial\kappa_x}, \frac{\partial\omega_d}{\partial\kappa_y}, \frac{\partial\omega_d}{\partial\kappa_z})$.

It is seen that $\omega-\omega_d(\vec{\kappa})$ is a vectorial function of $\delta\vec{v}_i$'s, and obviously the matching $\varphi(\vec{v}_o,\vec{v}_1)$ function has to have the same character i.e. φ has to depend on the direction of v_i's too, not only on their absolute values as in Eq.(5). This can be achieved, as suggested in Ref. 5 and shown in Fig. 6, by using precession field regions tilted with respect to the main beam directions $\vec{\bar{v}}_i$. (Another solution has already been tried out, and it is described in this volume in the report on ^{4}He excitations.)

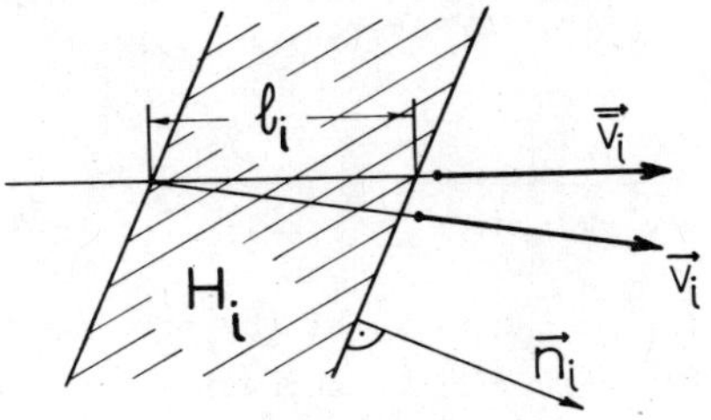

Fig. 6. Magnetic field configuration for producing Larmor precessions which depend on the direction of the neutron velocity.

It easy to see in Fig. 6 that the Larmor precession angles φ_i can be given in the differential form, [cf. Eq.(1)]:

$$\varphi_i(\vec{v}_i) = \gamma_L \frac{\ell_i H_i}{\bar{v}_i} - \gamma_L \frac{\ell_i H_i}{\bar{v}_i^2}(\vec{n}_i \delta\vec{v}_i) \quad , \tag{24}$$

where $\vec{n}_i$ is normalized so that $(\vec{n}_i\vec{\bar{v}}_i)=\bar{v}_i$. Thus the left hand side of the NSE equation (21) reads $[\bar{\varphi}_i=\varphi_i(\vec{\bar{v}}_i)]$

$$\varphi-\bar{\varphi} = (\varphi_o-\bar{\varphi}_o)-(\varphi_1-\bar{\varphi}_1) = -\gamma_L \frac{\ell_o H_o}{\bar{v}_1^2}(\vec{n}_o\delta\vec{v}_o)+\gamma_L \frac{\ell_1 H_1}{\bar{v}_1^2}(\vec{n}_1\delta\vec{v}_1) \quad . \tag{25}$$

Equation (21) is satisfied if and only if the vector coefficients of the $\delta\vec{v}_i$'s agree on both sides, i.e. by Eqs.(23) and (25) (i=0,1)

$$\gamma_L \frac{\ell_i H_i}{\bar{v}_i^2}\vec{n}_i = t\frac{m}{\hbar}[\vec{\bar{v}}_i-\mathrm{grad}\,\omega_d(\vec{\kappa}_o)] \quad . \tag{26}$$

This equation immediately defines the direction of the vectors $\vec{n}_i$, i.e. the field tilt angles ϑ_i contained by $\vec{n}_i$ and $\vec{\bar{v}}_i$:

$$\vec{n}_i \,||\, [\vec{\bar{v}}_i-\mathrm{grad}\,\omega_d] \quad , \tag{27}$$

and by multiplying both sides of Eq.(25) with $\vec{\bar{v}}_i$, we get

$$\gamma_L \frac{\ell_i H_i}{\bar{v}_i} = t\frac{m}{\hbar}[\bar{v}_i^2 - (\vec{\bar{v}}_i \cdot \mathrm{grad}\omega_d)] \quad ,$$

which for convenience can be rewritten (with $\bar{E}_i = \frac{1}{2}m\bar{v}_i^2$) as

$$\frac{\ell_o H_o}{\ell_1 H_1} = \frac{\bar{v}_o^3 - \bar{v}_o(\vec{\bar{v}}_o \cdot \mathrm{grad}\omega_d)}{\bar{v}_1^3 - \bar{v}_1(\vec{\bar{v}}_1 \cdot \mathrm{grad}\omega_d)} \quad , \tag{28}$$

$$t = \hbar \frac{\gamma_L \ell_i H_i}{\bar{v}_i m(\bar{v}_i^2 - \vec{\bar{v}}_i \cdot \mathrm{grad}\omega_d)} = \hbar \frac{\bar{\varphi}_i}{2\bar{E}_i - m(\vec{\bar{v}}_i \cdot \mathrm{grad}\omega_d)} \quad . \tag{29}$$

Equations (28) and (29) determine the ratio of the precession field strengths and the time-like constant t in Eq.(21). It will be shown later how t is related to the real, physical time.

The distribution of $\omega - \omega_d(\vec{\kappa})$ for the investigated scattering process being given by the S_D function [cf. Eq.(20)], by virtue of Eq.(21) one gets for the NSE signal

$$P_{NSE}(t) = P_S \frac{\int_{-\infty}^{\infty} S_D(\omega)\cos(t\omega)d\omega}{\int_{-\infty}^{\infty} S_D(\omega)d\omega} \quad . \tag{30}$$

Note that since S_D is a single parameter function vanishing outside the width of the transmission function T, in Eq.(30) ω is just the energy integration parameter. In the experiment, as already discussed above, $P_{NSE}(t)$ will be measured using the normalization to the instrumental effects contained in $P^o_{NSE}(t)$, and S_D can then be obtained by Fourier transformation if necessary.

The setting up of the NSE experiment for the study of a given excitation consists of tuning the NSE parameters: $\vec{n}_o$, $\vec{n}_1$ (i.e. ϑ_o, ϑ_1) and H_o/H_1 to the values given by Eqs.(27) and (28). It can be seen that these depend not only on $(\vec{\kappa}_o, \omega_o)$ via $\vec{\bar{v}}_o$ and $\vec{\bar{v}}_1$, but also on the slope of the dispersion relation $\mathrm{grad}\omega_d(\vec{\kappa}_o)$. This special feature of NSE can be useful in selecting out a given excitation branch at points where branches overlap or hybridize. Obviously for the special case $\mathrm{grad}\omega_d = 0$ this general formulation reduces to what we have seen in the previous section.

The transmission areas in Fig. 4, as pointed out above, are defined by the background spectrometer with which the NSE setup is combined. For the study of elementary excitations, where a reasonable $\vec{\kappa}$ resolution is required (few %), the most obvious choice for the background spectrometer is the triple axis spectrometer. This particular combination is discussed in detail by Roger Pynn later in this book, and in the Appendix[(6)].

Let us keep in mind that all the conditions for matching the NSE to a particular elementary excitation have been shown to be satisfiable to first order only. Higher order terms will introduce instrumental resolution limitations, especially since in many cases they will not be directly measureable but only calculable, similarly to usual spectrometer resolution functions. To assess their importance we performed Monte Carlo simulation calculations with Laci Mihály[7], which have shown that this fundamental limitation is typically not worse than 1-5 μeV for usual phonons and magnons. This is illustrated in Fig. 7 by the sample simulation result obtained for an NSE-triple axis configuration with typical collimations and monochromator-analyser mosaicities. It is seen that the distribution of $\varphi-\bar{\varphi}$, the $p(\varphi-\bar{\varphi})$ function (dots and dashed lines) reproduces with minimal (~2 μeV) broadening the distribution of $\omega-\omega_d(\vec{\kappa})$, which for the assumed three parallel phonon branches (insert to Fig. 7) corresponds to an S_D function consisting of three δ-functions separated from each other by only 20 μeV.

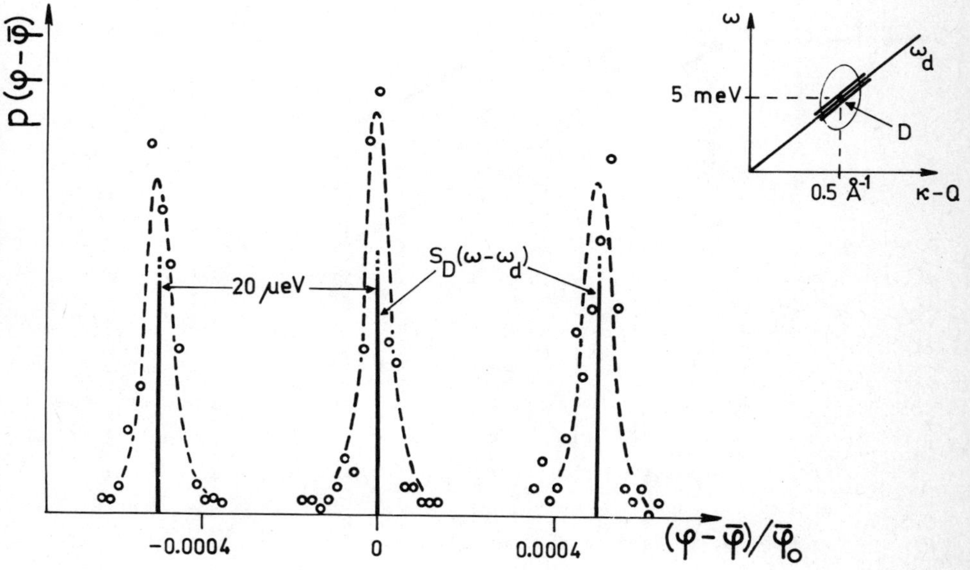

Fig. 7. Distribution of the total Larmor precession angle showing the general NSE focussing effect for inelastic scattering on dispersive elementary excitations (see text). The open circles are the data points calculated by Monte-Carlo simulation method, the dashed lines are guides to the eye. The vertical lines represent the $\omega-\omega_d$ energy distribution for the model longitudinal phonon branches with zero linewidth (see insert). The parameters of the assumed host triple-axis spectrometer configuration are: 2.5 Å incoming neutron wavelength; 40' graphite monochromator and analyser; energy loss scattering; $Q=3$ Å^{-1} reciprocal lattice vector; 30'x30'x30'x30' collimation.

4. EFFECT OF SAMPLE SCATTERING

In this section we will consider the direct effect of the sample scattering process on the neutron spin polarization which is taken into account by the factor P_S [cf. Eq.(12)]. As mentioned above $P_S=1$ if we investigate non-magnetic scattering effects and, in addition, there is but a small magnetic field on the sample. This latter condition is essential because the various possible neutron paths in the sample differ considerably in length. Thus there is always a spurious dephasing of the Larmor precession angles, which is of the order of $\Delta\varphi=2\gamma_L H_s d\cdot\sin\frac{\vartheta}{2}/v$ where H_s is the magnetic field at the sample, d the relevant dimension of the sample, ϑ the scattering angle, and v the neutron velocity (Fig. 8). As is discussed in the paper by

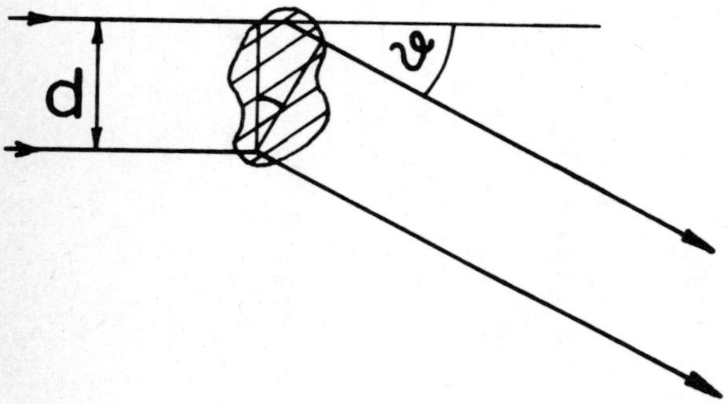

Fig. 8. Neutron path length differences in scattering on extended samples.

Otto Schärpf, some minimum guiding magnetic field is always required at the sample, so H_s cannot be arbitrarily small. For example, for $H_s=0.5$ Øe we get for a typical maximum sample size d=3 cm and $\lambda=5$ Å neutron wavelength $\Delta\varphi = 40^o\cdot\sin\frac{\vartheta}{2}$, which means just half of this amount as deviation from the average value of $\bar{\varphi}$. Table I below shows the effect of dephasing on the precessing polarization, assuming uniform distribution of precession angles between $\bar{\varphi}-\frac{1}{2}\Delta\varphi$ and $\bar{\varphi}+\frac{1}{2}\Delta\varphi$. It is seen that H_s fields of the order of 0.5 Øe, already sufficient as guide fields, produce negligible polarization losses. On the other hand fields with values of H_s greater than 5 Øe destroy the echo for large samples and scattering angles above 90°, and for higher fields this limiting scattering angle decreases. Strong H_s fields of the order of 1 kØe and above at the sample would therefore be excluded by this effect alone and, in addition, such high fields inevitably display considerable inhomogeneities too. (However, by the use of a trick which will be introduced below as the Ferromagnetic Neutron Spin Echo, this difficulty can be circumvented at the expense of a 50 % reduction of the echo signal.) At this point we conclude that in ordinary NSE the field at the sample has to be kept at a low value, which depends on the geometry of the experiment. This condition, usually corresponding to an upper limit of few Øe, is generally easily met.

Table I.

Reduction factor η for the precessing polarization produced by uniform distribution of precession angle differences between 0 and Δφ

Δφ	η	Δφ	η
20°	0.995	180°	0.637
40°	0.980	200°	0.564
60°	0.955	220°	0.489
80°	0.921	240°	0.413
100°	0.878	260°	0.338
120°	0.827	280°	0.263
140°	0.769	300°	0.191
160°	0.705	320°	0.121

It is interesting to observe that the very reason that NSE, an essentially time--of-flight method, can be of very high resolution is that it is not simply the path lengths that count here, but those weighted by the magnetic field values $S_L=\int Hds$. Thus, a high precision of the flight path definition is only required for the precession field regions H_o and H_1 (where it means stringent requirements, equivalent to better than 0.1 mm geometrical precision, which is to be met by the homogeneity of the fields), whereas differences of several centimetres can be tolerated in the low field regions, viz. at the sample, where, for high scattering angles, these differences are bound to be comparable to the sample dimensions.

In order to study the effect of magnetic sample scattering, we need to have a look at the action of the spin flipper device separating the two precession fields H_o and H_1 near to the sample (Fig. 9). As mentioned above, in practice the opposite sign of the φ_o and φ_1 precessions is produced by this device, also known as a π-coil, and not by a theoretically equivalent 180° flip of the field direction between H_o and H_1. This point will be discussed at length in the next contribution, here we merely anticipate part of it. If we consider the spin precession plane (x,y) we can envisage the flipper action as a 180° turn of the neutron spin around an arbitrary axis in this plane, e.g. around the x axis. Thus a neutron spin polarization $\vec{P}=(P_x,P_y,P_z)$ will be flipped into $\vec{P}'=(P_x,-P_y,-P_z)$. The precession angle φ_o, measured for convenience with respect to the x axis (Fig. 9a) and given as $\varphi_o=\arctan(P_y/P_x)$, will correspondingly be transformed in $\varphi_o'=\arctan(-P_y/P_x)=-\varphi_o$. This is precisely the key action to the echo: neutrons arriving at the π-coil with a precession angle φ_o, will appear to leave it with $-\varphi_o$.

At this point, if not earlier, the reader will probably ask himself a question about the significance of an obviously arbitrary ±2πn (n=integer) term which can be added at any point to the phase φ. The question is completely justifiable and in

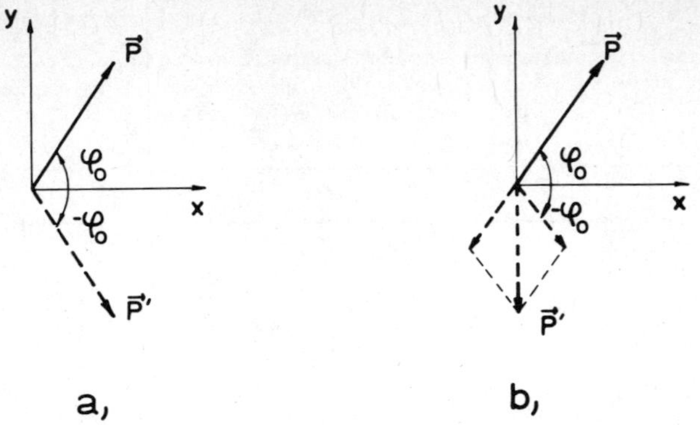

Fig. 9. Inversion of the sign of the Larmor precession angle by the application of a π spin flip coil (a) and in paramagnetic scattering (b).

effect this phase ambiguity drops out only as a consequence of a very natural assumption about the f(v) velocity distribution function, which was not mentioned before. This assumption is that f(v) is not too narrow, or has no too sharp structures so that after having gone through the first precession field H_o, the precessing polarization is completely dephased $<\exp(i\varphi_o)>=\int f(v)\exp[i\varphi_o(v)]dv=0$. In this case the $\partial\varphi/\partial v=0$ sufficient echo condition, obviously unaffected by an extra $\pm 2\pi n$ term, becomes a necessary one, and no precessing polarization can be found outside the neutron spin echo group. To see a conspicuous example to the contrary, imagine that $f(v)=\delta(v-v_A)+\delta(v-v_B)$, with v_A and v_B being constants. In this case precessing polarization could occur around any values of H_o and H_1, due to the beating between the two velocity components v_A and v_B, and subsequent beating maxima would indeed be separated by a 2π phase difference. The NSE polarization signal measured at the values of the NSE parameters as given by Eqs.(11) or (28) above, will obviously be correct, i.e. correspond to the scattering function as required by Eqs.(13) or (30). But by inspection of the spectrometer response alone we would not be able to find out which of the beating maxima is the right NSE group. In practice such problems can occur only when the incoming beam is too monochromatic with respect to the number of Larmor precessions, that is $<\exp(i\varphi_o)>\neq 0$. This corresponds to the illogical situation in which the NSE ω resolution is not really better than that of the background spectrometer. In such a case the maximum of the NSE group might not correspond to the ideal NSE conditions, examples for which are shown in Roger Pynn's article in this volume. Normally such a situation does not arise if NSE is used as it should be, i.e. to improve the ω resolution of the background configuration.

In what follows we will consider how NSE can work for various magnetic sample scattering processes in which the neutron spin direction changes. This change can be checked by simple neutron spin polarization analysis, without the echo. Its effect on the echo will be twofold: on the one hand it calls for an appropriate spin flipping scheme at the sample, and on the other, it gives rise to NSE sample scattering polarization factors P_S [introduced in connection with Eq.(13)] which are less than unity. Note that if there are several scattering processes contributing to the echo signal, the NSE spectrum $P_{NSE}(t)$ will be given by the weighted average. The knowledge of the different polarization behaviours of the different components might be helpful in their decomposition.

4.1 NSE with nuclear spin incoherent scattering

This process is characterized by the $\vec{P}' = -\frac{1}{3}\vec{P}$ equation, which relates the polarization $\vec{P}$ of the beam impinging on the sample with that of the scattered one $\vec{P}'$. Thus, in addition, to the slow Larmor precessions in the H_s field at the sample there is a $\varphi_o' = \varphi_o + \pi$ transformation of the precession angle. This, having no effect on $\partial\varphi/\partial v$, the echo signal appears at the same position, but with opposite sense (minima instead of maxima) and with amplitude reduced to $\frac{1}{3}$. Of course, the π-coil is necessary to produce the negative sign, and the total transformation between H_o and H_1 reads as $\varphi_o' = -\varphi_o \pm \pi$ (the sign of π being immaterial). This process obviously corresponds to $P_S = -\frac{1}{3}$.

4.2 Paramagnetic Neutron Spin Echo[8] (PNSE)

Paramagnetic scattering, which can be defined most generally as scattering on a macroscopically isotropic magnetic sample (i.e. with no strong magnetic field H_s and with isotropically distributed orientations for any kind of eventual locally ordered domains), is characterized by the relation

$$\vec{P}' = -\vec{\kappa}(\vec{P}\cdot\vec{\kappa})/\kappa^2 \qquad (31)$$

between the incoming and scattered beam polarization. This polarization behaviour is well-known for the scattering on spin paramagnets, but it will be shown in Sub-section 4.4 below, that it is valid more generally for isotropic magnetic systems with both orbital and spin magnetism.

It is clear from Eq.(31) that if $H_s || \vec{\kappa}$, i.e. the precession plane (x,y) is perpendicular to $\vec{\kappa}$, $\vec{P}'$ will be 0 for any $\vec{P}$ in this plane, and thus no spin echo can occur in this geometry. We will therefore assume that the $\vec{\kappa}$ scattering vector lies in the precession plane and, for convenience, its direction will be taken as the y axis with φ_o measured with respect to the x axis. (If $\vec{\kappa}$ is at an angle Θ with respect to the (x,y) plane, its projection is the y axis, and the final polarization will be reduced by a factor of $\cos\Theta$). Equation (31) then corresponds to the transformation:

$$\vec{P}' = (0,-P_y,0) = (\tfrac{1}{2}P_x,-\tfrac{1}{2}P_x,0)+(-\tfrac{1}{2}P_x,-\tfrac{1}{2}P_y,0) \qquad (32)$$

The first term on the right hand side corresponds to the $\varphi_o' = -\varphi_o$ flip; the second to

$\varphi'_o=\varphi_o+\pi$. Thus without any further π-coil action the first term with the key negative sign for φ_o will give an echo signal of an amplitude reduced by $\frac{1}{2}$, that is, $P_S=\frac{1}{2}$ (cf. Fig. 9b) or if we use a flipper at the sample, an echo signal of similar amplitude but opposite phase will be observed, due to the second term ($P_S=-\frac{1}{2}$), just as for case 4.1. The first possibility is of particular importance since without a flipper there will be no confusing contributions to the echo from non-magnetic or nuclear spin scattering effects.

In order to be able to normalize the PNSE signal, we have to know the denominator in Eq.(13), viz. the total paramagnetic scattering intensity separated from the other contributions. This can easily be achieved by the following, novel procedure(8). Let us take classical polarization analysis "spin up" (no-spin-flip) and "spin down" (spin-flip) counts, with H_s parallel to the x direction of a coordinate system, and repeat this with $H_s||y$ and $H_s||z$. Since in this case $\vec{P}||H_s$ to start with, and $\vec{P}'$ is also analysed parallel to $H_s||\vec{P}$, the paramagnetic contribution to the $N_\uparrow-N_\downarrow$ counting rate modulation is given as [cf. Eq.(32)]

$$N_\uparrow^P-N_\downarrow^P = N_P(\vec{P}\cdot\vec{P}') = -N_P(\vec{P}\cdot\vec{\kappa})^2/\kappa^2 \quad , \tag{33}$$

where N_P is the total paramagnetic intensity we want to determine. Adding together the modulations for the three subsequent $\vec{H}_s$ orientations, x, y and z, we get

$$\sum_{j=x,y,z} (N_\uparrow^P-N_\downarrow^P)_j = -N_P|\vec{P}|^2(\kappa_x^2+\kappa_y^2+\kappa_z^2)/\kappa^2 = -N_P P^2 \quad , \tag{34}$$

where in effect P^2 stands for the efficiency of the polarizer-flipper-analyser system. In addition we make sure that one of the axes, say the y axis, is perpendicular to $\vec{\kappa}$ (e.g. by being perpendicular to the scattering plane) in which case we get $\kappa_y=0$, i.e. the paramagnetic modulation $(N_\uparrow^P-N_\downarrow^P)_y=0$, and we can observe the pure contribution of the non-paramagnetic (NP) effects, for which $\vec{P}'=\vec{P}$ always holds:

$$(N_\uparrow^{NP}-N_\downarrow^{NP})_y=N_{NP}P^2 \quad . \tag{35}$$

Since this contribution is the same for x and z, we find for the combination below of the total modulations with $N=N^P+N^{NP}$,

$$\sum_{j=x,y,z} (N_\uparrow-N_\downarrow)_j-3(N_\uparrow-N_\downarrow)_y=(N_\uparrow-N_\downarrow)_x+(N_\uparrow-N_\downarrow)_z-2(N_\uparrow-N_\downarrow)_y=-N_P P^2 \tag{36}$$

An example of PNSE experiment is described in the contribution of Amir Murani and in Ref. 8 (see the Appendix).

4.3 Ferromagnetic Neutron Spin Echo (FNSE)

If we wish to study a ferromagnetic sample by NSE, we have to apply a strong magnetic field to saturate it, in order to avoid the well-known complete depolarization of the beam. Thus, as pointed out above, the Larmor precession will be completely dephased around the sample, and all information contained in φ_o's concerning the incoming beam polarization are lost, unless we do something about it. The situation is obviously the same for all samples which have to be studied in a "high" magnetic field, independently of ferromagnetism.

We will utilize the simple fact that the non-precessing $z||H_s$ component of the polarization is maintained in a proper adiabatic guide field, as it is well-known in polarized neutron work. Thus, applying a 90° turn to the neutron spins after H_o, we can turn one component of the (x,y) precessing polarization, say x, into the z direction. This can then be turned back by another $\frac{\pi}{2}$ coil to the relevant x direction after the scattering, when entering H_1. The other two components are assumed to be completely dephased in the sample region, i.e. the spin history between H_o and H_1 is the following

$$(P_x,P_y,P_z) \underset{\frac{\pi}{2}\text{ coil}}{\rightarrow} (P_z,-P_y,P_x) \underset{H_s\text{ field}}{\rightarrow} (0,0,P_x) \underset{\frac{\pi}{2}\text{ coil}}{\rightarrow} (P_x,0,0)$$

The final polarization $\vec{P}'=(P_x,0,0)$ can be considered again as a sum of two terms

$$(P_x,0,0) = (\tfrac{1}{2}P_x,\tfrac{1}{2}P_y,0)+(\tfrac{1}{2}P_x,-\tfrac{1}{2}P_y,0) \quad , \qquad (37)$$

where the first corresponds to $\varphi_o'=\varphi_o$ and the second to $\varphi_o'=-\varphi_o$. Thus the second term gives an echo signal of 50 % amplitude ($P_S=\frac{1}{2}$).

Note that here the complete elimination of P_y, a necessary condition for FNSE to work correctly, corresponds to "forgetting" the sense of the rotation for φ_o, i.e. mathematically it corresponds to going from $e^{i\varphi_o}$ to $\cos\varphi_o=\frac{1}{2}e^{i\varphi_o}+\frac{1}{2}e^{-i\varphi_o}$. This method has been tested but not yet used in experiments. Its applications can include, besides experiments in high fields, in particular the study of magnons for which an additional spin-flip occurs at the scattering, $P_x' \rightarrow -P_x$, which produces a change in the sign of the echo signal described by $P_S=-\frac{1}{2}$.

4.4 Antiferromagnetic Neutron Spin Echo (AFNSE)

In the previous section we have seen that the introduction of a preferred magnetization direction in a ferromagnet necessarily leads to the primary dephasing of the Larmor precessions around the sample, due to the saturation moment and the high magnetizing field. On the other hand, in an antiferromagnetic single crystal the crystalline symmetry might produce a preferred direction, or a few preferred directions for the atomic magnetic moments, without any strong field to be applied. We will therefore consider NSE experiments on magnetically anisotropic samples in low magnetic fields. The cardinal point is that the polarization behaviour of the neutron beam scattered from a macroscopically isotropic magnetic sample is exactly the same as for paramagnetic samples, i.e. $\vec{P}'=-\vec{\kappa}(\vec{P}\cdot\vec{\kappa})/\kappa^2$. (Because this theorem does not seem to have been recognized before, it will be proved below.) Thus the paramagnetic NSE situation will apply to any sample, paramagnetic or not, in which the eventual local anisotropies (e.g. preferred direction of the moments) are averaged out on the macroscopic scale. Similarly, the antiferromagnetic NSE will be defined as the situation that applies to samples with zero saturation moment and with macroscopically anisotropic magnetic correlations, even if they are not actually antiferromagnetic; an example of such a material would be a tetragonal single crystal displaying critical fluctuations in the paramagnetic phase.

Let us recall here that magnetic neutron scattering is related to the magnetization-magnetization correlation functions, which include both spin and orbital contributions[9]. For an isotropic system we have (i,j=x,y,z)

$$\langle M_i \rangle = 0$$

$$\langle M_i^* M_j \rangle = 0 \quad i \neq j \tag{38}$$

$$\langle M_i^* M_i \rangle = \langle M_j^* M_j \rangle \quad .$$

The first two lines follow from the fact that the symmetry operation of 180° rotation, e.g. around the y axis, would transform M_x to $M_{-x}=-M_x$ and $\langle M_x M_y \rangle$ to $\langle M_{-x} M_y \rangle$.

Identifying the $\vec{\alpha}$ interaction potential in equations (10.37) and (10.41) on pp. 330 and 331 of Ref. 9 with the component of the magnetization vector perpendicular to the momentum transfer $\vec{\kappa}$, viz.

$$\vec{M}_\perp(\vec{\kappa}) = \vec{M}(\vec{\kappa}) - \vec{\kappa}[\vec{M}(\vec{\kappa})\cdot\vec{\kappa}]/\kappa^2 \quad ,$$

we are left with the following very general expression, after suppression of the terms vanishing by virtue of Eqs.(38), for the scattered beam polarization:

$$\vec{P}' = \frac{\langle \vec{M}_\perp^* (\vec{M}_\perp \vec{P}) \rangle + \langle (\vec{M}_\perp^* \vec{P}) \vec{M}_\perp \rangle - \vec{P} \langle (\vec{M}_\perp^* \vec{M}_\perp) \rangle}{\langle (\vec{M}_\perp^* \vec{M}_\perp) \rangle} \quad , \tag{39}$$

where the < > brakets stand for the averaging over the sample states as given in detail in equation (10.33) in Ref. 9. For clarity and simplicity we will consider this equation in a coordinate system for which $\vec{\kappa}/\kappa$ is one of the base unit vectors, say $\vec{x}$. Thus $\vec{M}_\perp = (0, M_y, M_z)$ and

$$\vec{P}' = \frac{\langle \vec{M}_\perp^* (P_y M_y + P_z M_z) \rangle + \langle (P_y M_y^* + P_z M_z^*) \vec{M}_\perp \rangle}{\langle M_y^* M_y + M_z^* M_z \rangle} - \vec{P} =$$

$$= \frac{2\vec{y} P_y \langle M_y^* M_y \rangle + 2\vec{z} P_y \langle M_z^* M_z \rangle}{\langle M_y^* M_y + M_z^* M_z \rangle} - \vec{P} \quad . \tag{40}$$

In view of the last of the equations (38), finally we get

$$\vec{P}' = (0, P_y, P_z) - \vec{P} = -\vec{x}\cdot P_x = -\vec{\kappa}(\vec{P}\cdot\vec{\kappa})/\kappa^2 \quad , \tag{41}$$

which is the required result.

Note that the premises we used, i.e. Eqs.(38), mean somewhat less than total isotropy; they will be satisfied, for example, by a cubic single crystal too if it contains equal volume fractions of opposite antiphase domains (if they apply), which is the condition for the first two equations in (38) to hold.

The simplest case of antiferromagnetic NSE is the example of a tetragonal single crystal with $0 \neq \langle M_y^* M_y\rangle - \langle M_z^* M_z\rangle = \langle M_y^* M_y + M_z^* M_z\rangle A$, where A describes the anisotropy, and there is no antiphase domain asymmetry. Thus from Eq.(40) we readily find that

$$\vec{P}' = (-P_x, AP_y, -AP_z) \quad .$$

If the sample guide field H_s is in the z direction we get for the precessing polarization, if (x,y) is the precession plane [P_z=0, cf. Eq.(32)]

$$\vec{P}' = \frac{1+A}{2}(-P_x, P_y) + \frac{1-A}{2}(-P_x, -P_y) \quad , \tag{42}$$

where the first term shows the familiar $\varphi_o \rightarrow -\varphi_o$ flip, producing an NSE signal with the amplitude $P_S = \frac{1+A}{2}$. (The paramagnetic case corresponds to A=0.) The anisotropy also makes P_S depend on the crystal orientation, showing a maximum when the preferred magnetization direction is parallel to the y axis. For a tetragonal collinear antiferromagnet this maximum is unity which was in fact observed in an early test in 1974.

Finally, if the sample contains only a single AF antiphase domain, which is often difficult to achieve anyway, none of the above trivial symmetry arguments holds, and one has to go through the whole complicated formalism of polarization analysis in order to interpret the NSE results. This is the most specific case of AFNSE but, unfortunately, nothing simple and general can be said about it.

The main conclusions in this section are summarized in Table II below, which gives a list of the sample environments and the NSE signal amplitudes for the different types of echoes considered

Table II.

Different types of NSE configurations

Type	Sample environment: flipper coils	Sample environment: H_s field	NSE signal P_S
Normal NSE			
non-magnetic scattering	π	small	1
nuclear spin incoherent scattering	π	small	$-\frac{1}{3}$
Paramagnetic NSE (isotropic samples)	none	small	$\frac{1}{2}$
Ferromagnetic NSE	$\frac{\pi}{2} - \frac{\pi}{2}$	high	$\frac{1}{2}$
Antiferromagnetic NSE (anisotropic samples)	none	small	$\frac{1}{2} < P_S^{max} \leq 1$

Conclusion

We have seen that the NSE technique can be applied for a variety of inelastic scattering processes and for both magnetic or non-magnetic samples. But the real conclusion of this introductory paper is in fact contained in the contributions which follow. At this point I should just like to express my gratitude for many stimulating discussions to a large number of colleagues, first of all to the authors of this volume and to many others at the ILL together with those I have already thanked in the preface. I should particularly like to thank the audiences who attended the talks I have given at various places, whose questions and sometimes difficulties with these ideas were certainly invaluable for my own understanding.

REFERENCES

(1) G.M. Drabkin, E.I. Zabidarov, Ya.A. Kasman, and A.I. Okorokov, Zh'ETF 56, 478 (1969); Sov. Phys. JETP, 29, 261 (1969)

(2) F. Mezei, Z. Physik 255, 146 (1972)

(3) F. Mezei, in "Imaging Processes and Coherence in Physics", edited by M. Schlenker et al. (Lecture Notes Series, Springer Verlag, Heidelberg, 1980) p. 282

(4) J.B. Hayter and J. Penfold, Z. Physik B 35, 199 (1979)

(5) F. Mezei, in Neutron Inelastic Scattering 1977 (IAEA, Vienna, 1978) p. 125

(6) R. Pynn, J. Phys. E 11, 1133 (1978)

(7) L. Mihâly, Thesis, Central Research Institute of Physics, Budapest, (1977) unpublished

(8) F. Mezei, and A.P. Murani, Proc. Neutron Scattering Conf., Jülich, August 1979, J. Mag. Mag. Mat. 14, 211 (1979)

(9) W. Marshall, and S.W. Lovesey, Theory of Thermal Neutron Scattering (Clarendon Press, Oxford, 1971)

THE POLARISED NEUTRON TECHNIQUE OF NEUTRON SPIN ECHO

O.Schärpf

Institut Laue-Langevin, 156X
38042 Grenoble Cedex, France

INTRODUCTION

All spin echo properties described in the following result from neutron precession. Neutron precession can most easily be visualized through a classical treatment. The reason for this is, that even the quantum mechanical treatment by introducing Pauli spin matrices into the Schrödinger equation is in effect a classical treatment if one considers the origin of those matrices. They result from the problem of mapping three dimensions on two dimensions introducing one complex component and were treated by Cayley and Klein 1897 [1] in connection with problems of the spinning top, a really classical problem. There exists a correspondence of the classical description and the spinor description and because the classical description is more conspicuous and more suited to pictorial representation, the classical treatment only will be used. We shall discuss: 1.Precession of polarised neutrons in a homogeneous magnetic field, 2.Effect of a guide field on polarised neutrons, 3.Mezei π-coils, 4.Mezei $\pi/2$-coils, 5.The combination of guide fields and Mezei coils to obtain the neutron spin echo, 6.Some properties of the spin echo.

1. PRECESSION OF POLARISED NEUTRONS IN A HOMOGENEOUS MAGNETIC FIELD

Classical mechanics shows that a torque exerted on a magnetic moment $\underline{\mu}$ by a magnetic field $\underline{H}$ inclined at an angle θ relative to the magnetic moment (fig.1) causes the magnetic moment of the neutron to precess about the direction of the field with a frequency ω_L, the Larmor frequency.

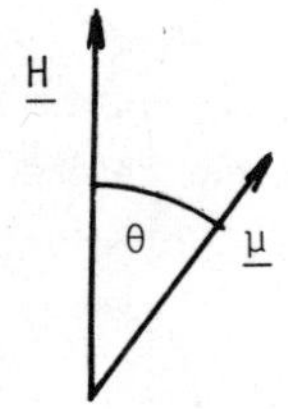

Fig.1 Definition of angle θ

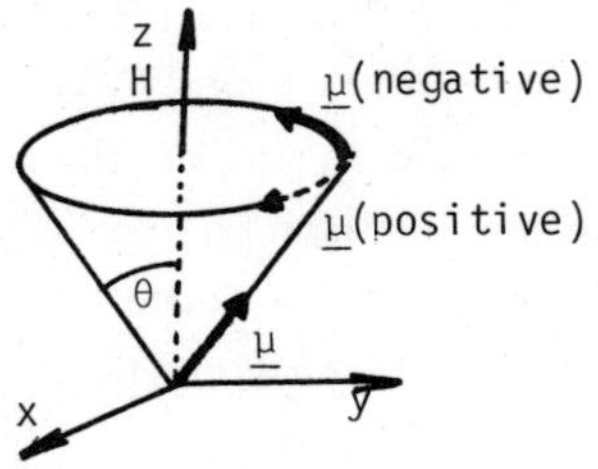

Fig.2 Direction of the precession for positive and negative $\underline{\mu}$

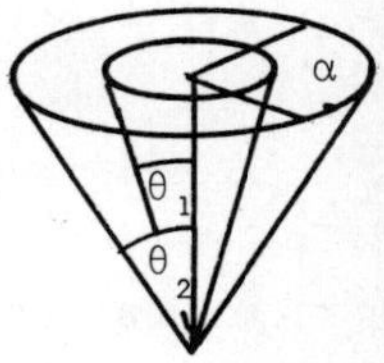

Fig.3 Geometrical significance of the precession angle $\alpha = \omega_L t$ for different θ

The direction of this precession is given in fig.2. As ω_L is independent of θ (see below) the precession angle $\alpha=\omega_L \cdot t$ is independent of θ and in a given field only a function of the time of flight in this field (fig.3).

Neutron precession in a magnetic field is easier to understand than the precession of a spinning top under the influence of the gravity. Similar to the neutron precession in the case of the spinning top one has also an angular momentum, a torque resulting from the gravity acting on the center of mass about the point of contact with the floor, and one can observe a precession, a motion at right angles to the angular momentum and the force producing the torque. But for describing this motion one needs the momentum of inertia. The resulting motion is in general no mere precession but contains a nutation and can in general be very complicated depending on the initial conditions and the relation of the components of inertia [2] (fig.4).

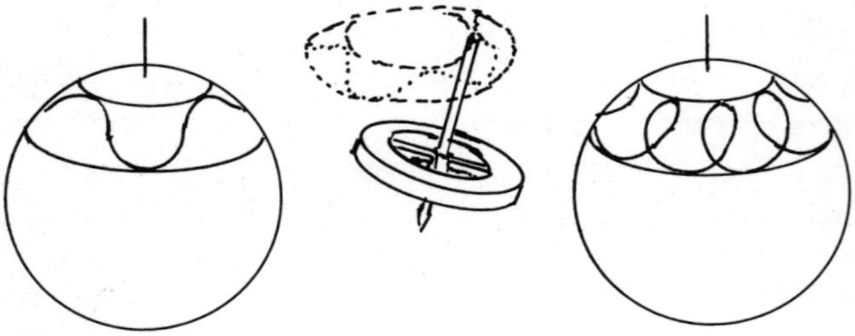

Fig.4 Spinning top and some of its general motions given as the projection of the axis of the top on a sphere (with loops and without loops). These nutations are not possible for neutrons

For a neutron in a magnetic field the behaviour is easier to understand and to describe. The neutron has always the same angular momentum $\underline{L}=\underline{J}\hbar$, a constant, so that the energy, resulting from this angular momentum, is a constant that can be omitted in the equation of motion, the Schrödinger equation. J = the spin quantum number of the neutron = 1/2 and $\hbar$ is the Planck constant divided by 2π. There exists a fixed relationship between angular momentum $\underline{L}$ and magnetic moment $\underline{\mu}$

$$\underline{\mu} = \gamma \cdot \underline{L}$$

where γ is called the "gyromagnetic ratio" of the neutron. An applied magnetic field tries to align this magnetic moment. Thereby it exerts a torque and if the field is homogeneous it does not exert a force on the magnetic moment. The resulting equation of motion in this case is simply

$$\frac{d\,\underline{L}}{dt} = -\gamma\,\underline{L} \times \underline{H} = \underline{L} \times \underline{\omega_L}$$

with $$-\gamma\,\underline{H} = \underline{\omega_L} \quad \text{and} \quad \gamma_{neutron}/2\pi = -2916.4\ \text{Hz/Oe}$$

This equation of motion gives a change of $\underline{L}$ in time that is normal to $\underline{L}$ and to $\underline{H}$ and is just the precession.

We summarize: if the field $\underline{H}$ and the magnetic moment $\underline{\mu}$ of the neutron have different directions, one observes a precession of $\underline{\mu}$ about $\underline{H}$ without nutations, with the same angular frequency of precession ω_L for all angles between $\underline{\mu}$ and $\underline{H}$.

2.EFFECT OF A GUIDE FIELD ON POLARISED NEUTRONS

A magnetic guide field is normally necessary to maintain the spin direction $\underline{S}$ or the polarisation $\underline{P}$ of the neutron beam in a direction parallel to this field. Therefore the guide field usually has the direction parallel or antiparallel to $\underline{S}$ or $\underline{P}$ (see fig.5).

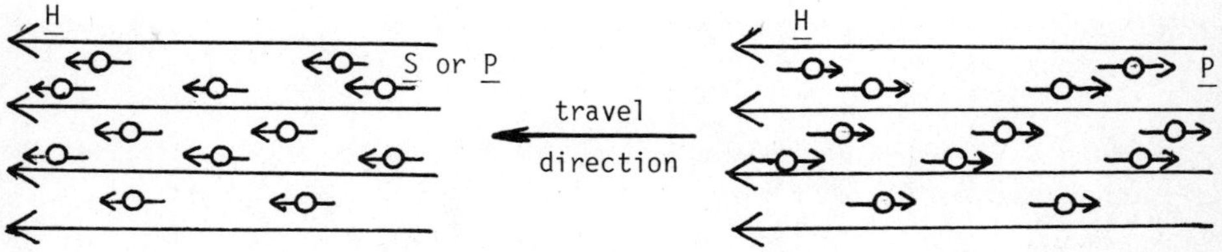

Fig.5a Guide field for longitudinal polarisation, $\underline{P} \parallel \underline{H}$, the small arrow is the spin of the neutrons

Fig.5b Guide field for longitudinal polarisation, $\underline{P}$ antiparallel $\underline{H}$

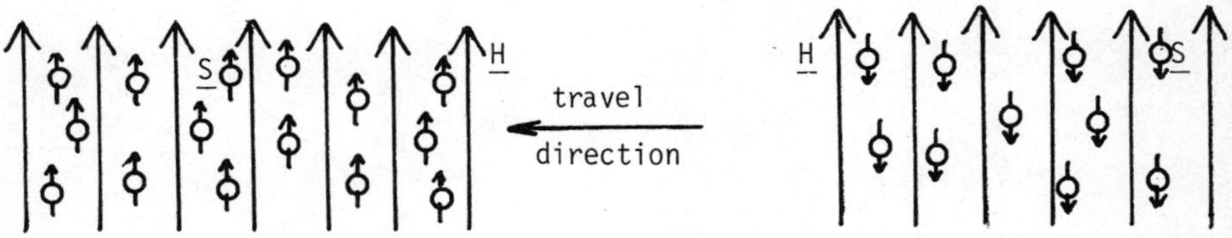

Fig.5c Guide field for transversal polarisation, $\underline{P}$ parallel $\underline{H}$

Fig.5d Guide field for transversal polarisation, $\underline{P}$ antiparallel $\underline{H}$

If θ,ϕ are spherical coordinates with the z-direction in the direction $\theta=0$ then the spin $\underline{S}$ is given by the spinor

$$\begin{pmatrix} S_1 \\ S_2 \end{pmatrix} = \begin{pmatrix} \cos\theta/2 \ \exp(-i\phi/2) \\ \sin\theta/2 \ \exp(i\phi/2) \end{pmatrix}$$

equivalent to a three dimensional vector $\underline{P} = (\cos\phi \ \sin\theta, \ \sin\phi \ \sin\theta, \ \cos\theta)$.The modulus of this vector is always one, the polarisation of one neutron being always 100%. For a polarised beam containing neutrons of different spin directions an averaging process must be used, which is best described by the spin density matrix. In the following we speak always of the behaviour of the spin of the single neutrons and therefore we use polaisation $\underline{P}$ and spin $\underline{S}$ without special distinction.

The guide field should have a constant direction and magnitude in the space between polariser and the position where the polarised neutrons are used. It can be set up by permanent magnets or by current carrying coils. We ask first for the normal case where the neutron spin is parallel to the guide field: If the guide field changes its direction, what is the effect of this change on the spin direction of the neutrons? In answering this question we have to distinguish between a) slow change, b) sudden change and c) general case. We then consider the less normal case d) where the neutron spin includes a certain angle with the guide field. Here, we describe briefly the precession of the spin and e) the effect of a change of the direction of the guide field on this precession.

a) Neutron spin parallel or antiparallel to guide field; effect of a slow change of field direction

In the case of a slow change of field direction the spin of the neutrons follows the field direction (see fig.6). What means "slow"? Slow means: $|\omega| << |\omega_L|$ with $\omega_L = -\gamma H$ = Larmor frequency and $\omega = d(\underline{H}/|\underline{H}|)/dt$. For the definition of ω see also fig.6, where we give an example.

$$\omega = \frac{\pi}{2}\frac{v}{s} = \frac{\pi}{2}\cdot\frac{3956}{0.8\lambda[\text{Å}]}$$

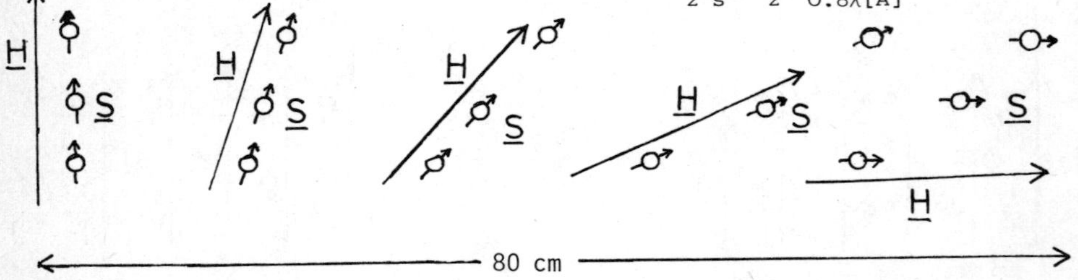

Fig.6 Polarisation of neutrons follows field direction in the adiabatic case: $|\omega| << |\omega_L|$
v = velocity of the neutrons, s = distance for the rotation of $\pi/2$

This is the so called adiabatic case in the sense of Ehrenfest or in the sense of quantum mechanics, not in the sense of thermodynamics. If e.g. the direction of $\underline{H}$ changes by 90^0 in a distance of 80 cm as in fig.6; this yields an ω that can be obtained by $\omega = (2\pi\cdot 3956)/(4\text{x}0.8\text{x}\lambda[\text{Å}])$ for λ=5.5Å as ω = 1412 s^{-1}. A Larmor frequency of this magnitude would be produced by a field of 0.08 Oe. For λ=8Å one obtains ω = 970 s^{-1} corresponding to a field strength of 0.05 Oe. In the case of fig.6 even the earth field fulfills the condition of adiabaticity, and if there is no other guide field, changes in the direction of the earth field by pieces of iron or similar magnetic objects in the neighbourhood could change the direction of the polarisation by adiabatic change of field direction. This is the reason why one normally needs a guide field.

b) Neutron spin parallel or antiparallel to guide field; effect of a sudden change of field direction

If the field direction changes suddenly, the polarisation of the neutron cannot follow the field direction. The spin direction remains the same as before, but its direction relative to the field is changed. This is the so called "sudden" approximation for $\omega \gg \omega_L$. It can be used to flip the field and by this the spin relative to the field (see fig.7). Majorana flippers, Dabbs foils, Kjeller eight, and kryo flippers all use this principle.

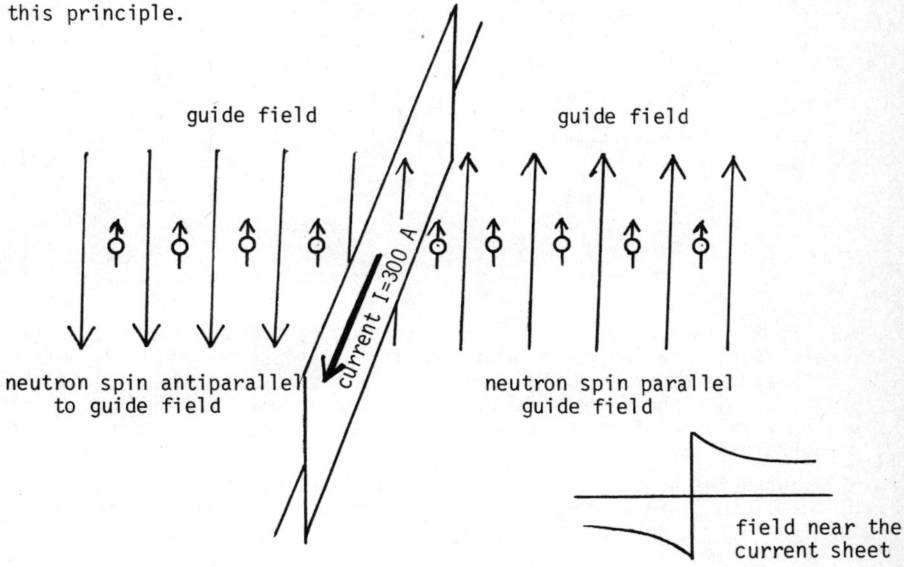

Fig.7 Use of a current sheet with a sudden change of field direction to flip the neutron spin relative to the field direction and field near the current sheet

This sudden approximation is also used to obtain a defined direction between field and spin in the Mezei coils (see section 3 and 4).

c) Neutron spin parallel or antiparallel to the guide field; effect of change of direction of the guide field neither slow nor sudden. General case including $\omega \simeq \omega_L$

In the general case one expects that the changes of the directions of the field and of the neutron spin are different. We can construct the essential features of the motion of the spin direction, by fixing our attention on the intersection of the field direction with a unit sphere having its center at M (see fig.8) . It is interesting to compare this motion of the field direction with the motion of the intersection of the spin direction with the same unit sphere in the general case. It is given by the dotted line in figure 8. Its origin can be described by a cone with its apex in the center of the sphere, which is unrolling on the area in which the field direction

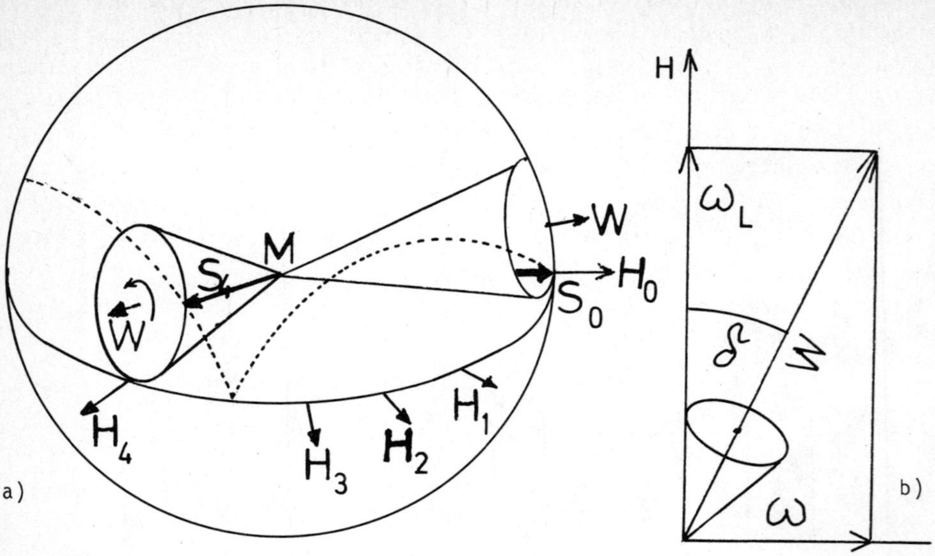

a) b)

Fig.8 a) Behaviour of the neutron spin $\underline{S}$ relative to the field directions $\underline{H}_i$ in the general case. The intersection of the field directions with the unit sphere describes the curve H_0, H_1, H_2, H_3, H_4. The intersection of the spin direction with the unit sphere desribes the dotted curve. This dotted curve originates by the unrolling of the cone with axis MW and with its apex in M. The apex angle is 2δ

b) Relation between δ, ω_L, ω and $W = \sqrt{\omega_L^2 + \omega^2}$ for the case where the rotation of the field is in a plane (axis of rotation normal to the field $\underline{H}$)

changes, in fig.8 the plane through H_0, H_1, H_2, H_3, H_4. In this case $\underline{\omega}$ and $\underline{\omega}_L$ are normal to each other as in fig. 8b. This is not necessary [3]. The line of the cone shaped shell touching the plane $H_0, \ldots H_4$ must always be parallel to the momentary field direction e.g. the line MH_4 in fig.8 parallel to $\underline{H}_4$ for the moment when the spin direction is $\underline{S}_4$. The behaviour of the spin is then the behaviour of a line of the cone shell fixed to this unrolling cone given by the arrow $\underline{S}_0$ or $\underline{S}_4$ on the shell of the cone of fig.8. This line must start unrolling at the original direction of the spin (e.g. $\underline{S}_0$ in fig.8 starts at the same point as the field $\underline{H}_0$). The spin direction intersects the surface of the sphere in a sort of cycloide of the base circle of the cone on the line that describes the movement of the field direction on the surface of the sphere, as given for example by the dotted line in fig.8 [3].

From this one can immediately see that the maximum deviation of the spin direction from the field direction is just the angle at the top of the cone $2\delta = 2\ \mathrm{arctg}(\omega/\omega_L)$. If $\omega = \omega_L$ this maximum deviation is just 90^0.

Now using this description for the general behaviour one can also understand the adiabatic and the sudden case and has a scale for the accuracy with which they are realized. In the adiabatic case $\omega \ll \omega_L$ and the unrolling cone is very small. The maximum deviation of spin direction from field direction is then $2\mathrm{arctg}(\omega/\omega_L)$ which is as small as the ratio of ω to ω_L (fig.9). In the sudden case $\omega \gg \omega_L$ the apex angle of the unrolling cone is large, in the limit 180°. In this case there is no unrolling possible, the plane of the cone touching everywhere the plane of the field change. Therefore the spin direction remains the same as before.

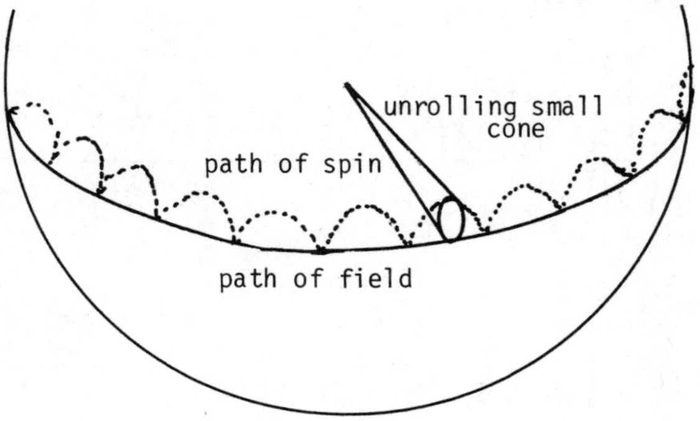

This general behaviour can also be applied to the behaviour of the spin inside the wires of the Mezei coils (see section 3).

Fig.9 Field and spin behaviour in the adiabatic case $\omega \ll \omega_L$: small unrolling cone and therefore small deviations of spin from field path of field ———, path of spin ....

d) Precession of the neutron spin if it includes an angle with the guide field

If the spin of the neutron includes an angle with the guide field, the neutron spin precesses about the field direction (fig.10). The precession angle $\alpha = \omega_L t$ in the field H depends only on H via $\omega_L = -\gamma H$ and on the time t, during which the neutron is in the field region, given by $t = l/v$ with l = length of the path and v = velocity of the neutron. Therefore the precession angle is $\varphi = \gamma H \frac{l}{v}$. Using this relation one finds the necessary field integral lH for one precession. If the field integral is in Oe·cm and the wavelength in Å, this relation is

$$\frac{135.65}{\lambda[\text{Å}]} \text{ Oe cm} \quad \text{is needed for one full precession}$$

That gives for λ = 8 Å a field integral of 17 Oecm corresponding to 0.34 mm for a field of 500 Oe. For λ = 5.5 Å the field integral must be 24.7 Oecm or .5 mm with 500 Oe.

If the neutrons have different wavelengths the precession in traversing the same distance is different for different neutrons. If a given beam of neutrons has a range of wavelengths of $0.5\lambda_o<\lambda<\lambda_o$, then, after one full precession of neutrons of the shortest wavelength, the directions of the spins of all neutrons will be distributed over the whole circle (fig.11) because neutrons with λ_o have then made a precession angle of 4π and all other neutrons have precession angles between these two i.e. between 2π and 4π.

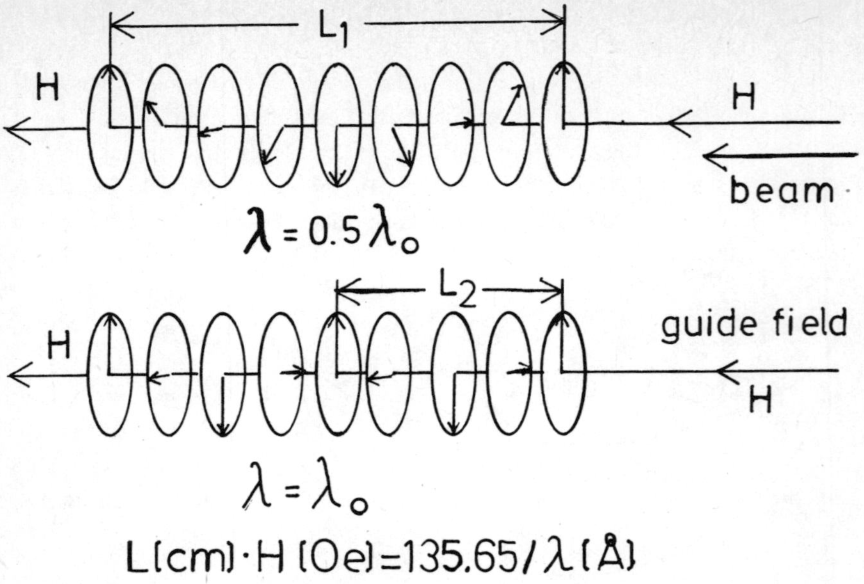

Fig.10 If the spin includes an angle of 90° with the guide field, the neutron precesses about the field direction in the plane normal to the field; one needs for one full precession $135.65/\lambda$[Å] Oe·cm . The figure shows the spin direction after traversing certain distances in the guide field H for two different λ

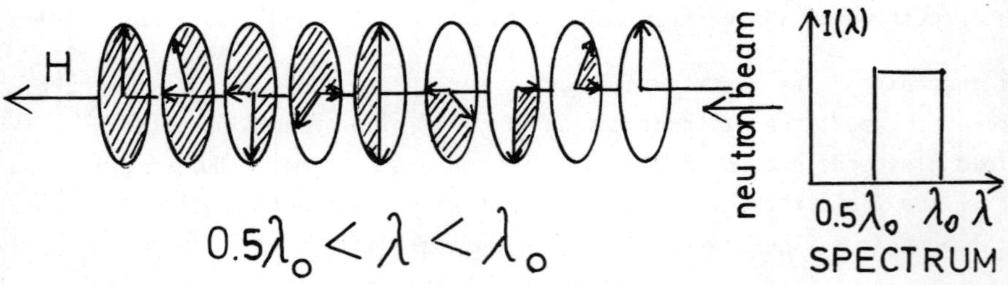

Fig.11 For a wavelength spectrum of $0.5\lambda_o < \lambda < \lambda_o$ after one precession of the spin of the neutrons with $0.5\lambda_o$ the direction of the spins of this wavelength range ist distributed over the whole circle 2π. The hatched part gives the range of spin directions after traversing a certain distance in the guide field. For the limiting wavelengths see fig.10

Fig.10 and fig.11 show the behaviour for the neutron spin, if it is initially at 90° to the guide field. If the spin includes another angle, the precession of the neutrons is described by a respective cone, where the cone angle only depends on the angle between $\underline{S}$ and $\underline{H}$ (or $\underline{P}$ and $\underline{H}$).

e) Effect of the precession if the guide field changes direction

It is easy to see what happens in the adiabatic case and in the sudden case. In the adiabatic case the axis of the precession cone follows the field direction (see e.g. fig.24). In the sudden case the precession,after the sudden change,takes place about the new field direction,with the spin direction before the change as initial condition. In the general case the movement of the spin is described by a more general cycloide. One obtains then curves that remind one of nutations of a spinning top in a field of gravity (fig.4).

3. MEZEI π-COILS

Mezei coils apply the principles that we have described for guide fields. They consist of long rectangular coils, an example of which is given in fig.12. The advantage of such a long rectangular coil is, there is a sudden transition from one direction of the field outside the coil (horizontal, H_z) to the other field direction inside the coil (vertical, H_π). The transition takes place in the range of the wire thickness. This renders possible a predetermined precession mode (described below) and thus e.g. a flipping of the spin.

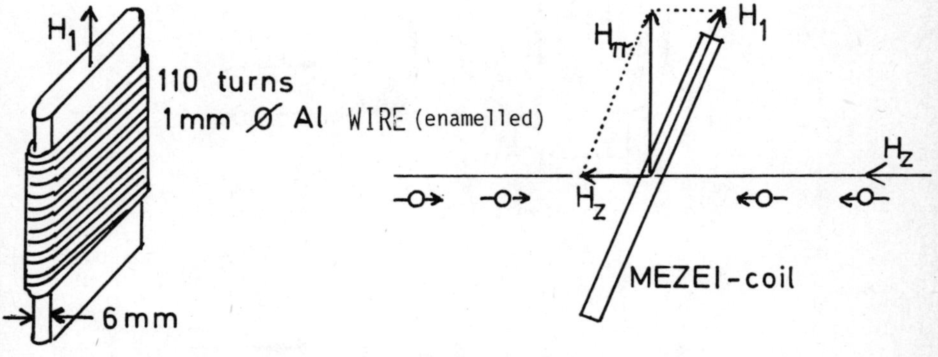

Fig.12 Mezei coil and its application as a spin flip coil

The application of the Mezei π-coil as a flipper for monochromatic neutrons is visualized in fig.13. The neutrons fly from right to left, the guide field H_z is in the flight direction and the neutrons are polarised parallel to the field direction. Inside the coil the field direction is $\underline{H}_\pi$ normal to the field $\underline{H}_z$ outside the coil. Inside the coil the neutrons begin to precess in a plane normal to $\underline{H}_\pi$. A spin flip is obtained if the neutron spin undergoes half a precession on its way d through the coil. $\underline{H}_z$ outside is necessary as a guide field to avoid depolarisation by the earth field and other undefined stray fields (see sect.2). H_π is determined by the relation $\gamma H_\pi \cdot d/v = \pi$. One can also use the relation of fig.10. For half a precession one needs for a width d of the coil the field

$$H_\pi = \frac{67.825}{\lambda[\text{Å}]d[\text{cm}]} \text{ Oe}$$

For example for d = 5 mm, λ = 8 Å one needs H_π = 17 Oe corresponding to 1.35 A for the coil of fig.12. For d = 7 mm, λ = 8 Å, H_π = 12.1 Oe or 0.96 A. To achieve the correct H_π in the presence of H_z correction coils are required (see fig.14).

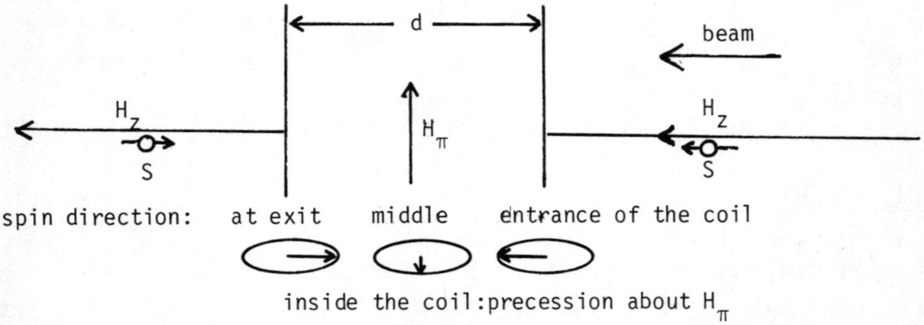

Fig.13 Application of the Mezei coil as a flipper for monochromatic neutrons.
d = range inside the coil where the resulting field direction is H_π

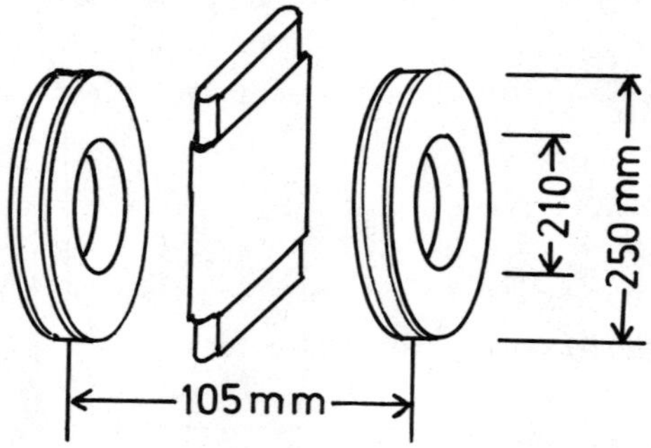

Fig.14 Correction coils for the Mezei coil

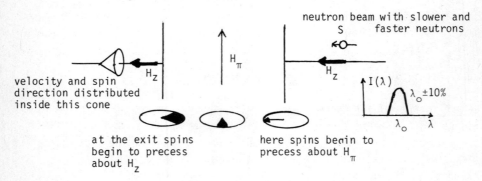

Fig.15 Application of Mezei coil as a flipper for a polychromatic beam. Inside the Mezei coil the spins spread over a certain range of angles

The application of the Mezei coil as a flipper for a polychromatic beam is shown in fig.15. Assume a perfectly polarised beam with a wavelength distribution of say $\lambda = \lambda_0 \pm 10\%$. If this beam enters the π-coil, the spins begin to precess about the H_π-direction as shown in fig.15. As they have different velocities, some of them fulfill more then π and some less than π precession angle. Thus the spin directions at the exit of the coil will be distributed over the area shown in black in fig.15, representing an angular range of $180^0 \pm 18^0$. Averaging over the whole wavelength distribution with a maximum at 180^0 gives a polarisation in the z-direction, P_z, of better than 95% (corresponding to $\cos \pm 18^0$).

If we wish to describe the behaviour of this flipper with one number, the flipper efficiency f defined as $P_z = 1 - 2f$, then for the above example f would be 97.5%, as P_z after flipping is -95%. The flipping ratio, given by $R = (1+P_1P_2)/(1+P_1P_2P_z)=2/(1-0.95)$ with P_1 and P_2 the polarising efficiency of the polariser and analyser, would be R = 40 in our example assuming $P_1 = P_2 = 1$. At the exit of the π-coil the different spins begin to precess about the H_z-direction. Behind the π-coil the spin directions are therefore distributed inside the cone shown in fig.15. It is in first approximation sufficient to look only for the component parallel or antiparallel to H_z, which is equivalent to using only the flipper efficiency (no account is taken by this method of the precession angle).

The method of the adjustment of the π-coil: In order to obtain H_π with the desired magnitude and direction it can be seen from fig.12 that H_1 and H_z must both be adjusted so as to give the correct resultant. One way to set up a π-flipper is to use the following iterative method which converges very rapidly. We start with a polarised neutron beam passing through the non-energised flipper and an analyser arranged, as in fig.16, to give a maximum count rate at the detector. The flipper is tilted at a small angle (θ) to the normal of the z-direction (i.e. beam direction). This determines the direction of H_1. Starting with a value of H_1 calculated to give the required H_π together with H_z, H_1 is varied to give a minimum count rate. H_z is then varied by altering the current in the correction coils to again minimise the count rate. H_1 and H_z are thus alternately adjusted until no further minimum can be reached.

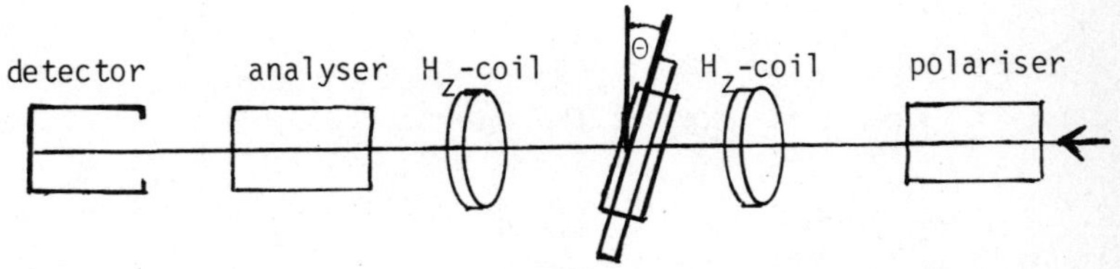

Fig.16 Arrangement for setting up a Mezei π-flipper coil. The $H_z$coils form a Helmholtz pair. (If the field is asymmetric as in fig.19 one uses only one correction coil)

So far we have assumed that the transition through the wires of the Mezei coils is a sudden transition. In fig.6 of section 2 we developped a criterium, whether this is really the case. We now want to apply this criterium to the Mezei coils of fig.12. There we have a change of field direction of 90^0 inside the wire of 1 mm thickness. This corresponds, for a wavelength of 8 Å, to an angular frequency of ω = (2πx3956)/(4x0.001x8) =776760 Hz. An ω_L of this magnitude would correspond to a field of 42.4 Oe. As we have seen that H_π is only 17 Oe the condition for a sudden change is just realised as $\omega \approx 2.5\omega_L$. Deviations of the spin by this poor realisation of the sudden case are partially eliminated by the above experimental process of seeking the optimum.

4. MEZEI π/2-COILS

The same rectangular coil (fig.12) can also be applied to rotate the spin only 90^0. In this case one has a Mezei $\frac{\pi}{2}$ - coil. Fig.17 shows how this can be achieved. Again the neutrons fly from right to left, the guide field H_z is in the flight direction and the neutrons are polarised parallel to the field direction. Inside the coil the field direction is $\underline{H}$, inclined 45^0 to the direction of the guide field. This is achieved by a field in the Mezei coil, that is $H/\sqrt{2}$ and a guide field H_z adjusted also to $H/\sqrt{2}$. Then the neutrons fulfill half a precession, on the way d through the coil, about the resulting $\underline{H}$ inclined 45^0 to the polarisation direction of the incident neutrons.(lower half of fig.17). After a precession of 180^0 about $\underline{H}$, the neutron spin at the exit of the coil is at 90^0 to H_z. In the space after the Mezei $\frac{\pi}{2}$ - coil they begin to precess in a plane normal to H_z.

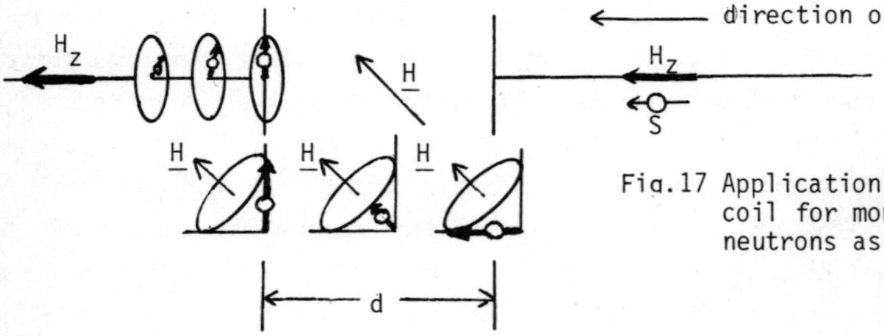

Fig.17 Application of the Mezei coil for monochromatic neutrons as a π/2-coil

For a non monochromatic beam the behaviour is shown in fig.18. As a first approximation to understand the principle of the spin-echo, we take account only of the component in the xy-plane, z-component neglected. In the same way one can describe the behaviour using only flipper efficiency, as for the π-coil.

The method of adjustment is easy once the π-coil has been set up. The current in the π-coil needs to be known. The coil field for the π/2-coil must be $H_\pi/\sqrt{2}$, H_π as determined for the π-coil before. Then using the correction coils, the field H_z is changed so that a 90^0 rotation of the spins is obtained. This can be found using the fact that $\underline{P}$, after a 90^0 rotation, appears to be totally depolarised at the analyser. Therefore one has to select the correction coil field so that the counting rate is not changed if the spin is flipped using the π-coil (switching the current on or off).

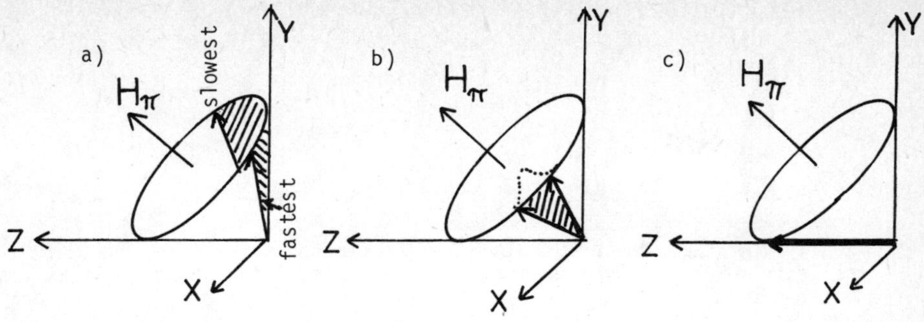

Fig.18 Different wavelengths in the $\pi/2$ coil have different end positions of spin directions, not all in the xy-plane.
a) distribution of spin directions at coil exit , b) distribution of spin directions in the middle of the coil (dotted curve indicates the intensity distribution), c) direction of spins of the incident beam

5. NEUTRON SPIN ECHO

The total apparatus consisting of a helical velocity selector, polariser, $\frac{\pi}{2}$ - coil with correction coils, precession coils, sample on a rotatable table, a second arm rotatable about the sample for different scattering directions, a π - coil with corrections coils, a second precession coil, a second $\frac{\pi}{2}$ - coil with correction coils, an analyser (super-mirrors) and a detector, is shown in fig.19.

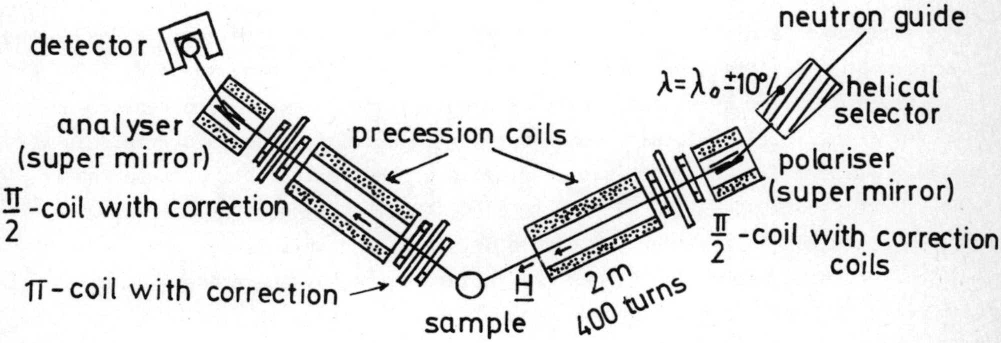

Fig.19 Spin echo spectrometer, IN11.

The neutrons are first monochromatized by a helical selector giving a spectrum of $\lambda = \lambda_0 \pm 10\%$. After this helical selector the neutrons are polarised by a super mirror arrangement of two parallel mirrors of 60 cm length, 1 cm apart. This gives a polarised beam of 2 cm width and a very high polarisation of 97-99%. For a change of wavelength the angle between the super-mirrors and the beam can be changed. The super-mirrors are mounted inside a solenoid 66 cm long. Thus one obtains a

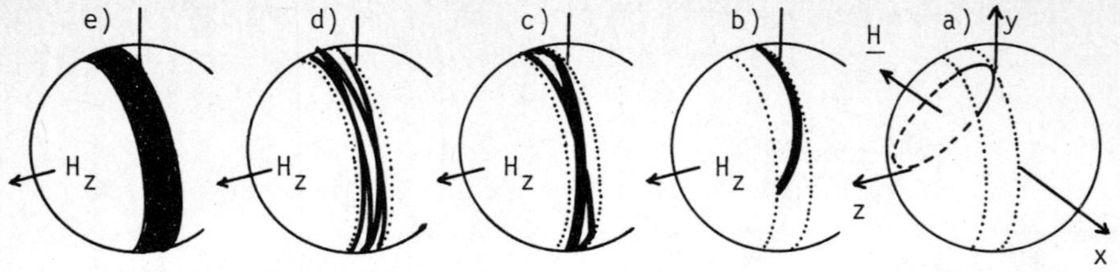
e)
d)
c)
b)
a)
y
H
H_z
H_z
H_z
H_z
z
x
finally the full zone of the sphere is covered
the broadening and elongation is stronger and stronger after more full precessions
after a full precession about H_z this line is elongated and broadened (see text)
line of initial distribution of spin directions c.f. fig.18

Fig.20 Effect of the precession coil. In a) the effect of the $\pi/2$-coil is indicated which results in the line of initial distribution of spin directions, shown as the full line. This line spreads more and more with the precessions about the guide field H_z and finally fills the whole respective zone of the sphere

longitudinally polarised beam. This beam is converted in the first $\pi/2$-coil to a transversally polarised beam as shown in fig.18 and fig.20a. In fig.20a the line of initial distribution of spin directions after the $\pi/2$-coil on the surface of a sphere is shown. It is essential that the transition to transversal polarisation is achieved with a sudden transition to the following longitudinal guide field. Fig.20b-e shows the effect of the guide field produced by the first precession coil. In this coil the line of the initial distribution of fig.20a spreads more and more and finally covers the whole zone of the sphere as shown in fig.20e.

In every precession a difference in time of flight due to any wavelength spread gives a difference in the precession angle, and this difference is then, for e.g. 6000 precessions, multiplied by even this factor 6000, giving rise to a broadening and elongation of the initial distribution shown in fig.20. This precession field is very essential in determining the possible resolution of the instrument, because the angle of precession is the time of flight "clock" that every neutron carries with itself. If a neutron looses energy its velocity falls and this is measured as an increase in precession angle. The higher the factor reached by the precession coils the higher is the precision of the measurement of this change of time of flight. As we shall see later, for the measurement of the line shape it is desirable that one can change the precession field from very low to very high fields without depolarisation. For this we use a solenoid with inner radius 10 cm, outer radius 15 cm, length 2 m and 400 turns. This gives a field path integral of 10^5 Oe cm at 200 A, which corresponds to 4070 full precessions for $\lambda = 5.5$ Å and 6000 full precessions for $\lambda = 8$ Å (the maximum coil current is 600 A).

One difficulty of the solenoid field is that it is not totally homogeneous. The field near the wires is higher than on the coil axis, giving a difference in the line integral for different paths through the coil (see fig.21), resulting in a difference in the number of precessions. To avoid this, one can use a coil of larger cross section, or a smaller beam cross section (3 cm diameter is used for IN11), or a correction coil (Fresnel coil, see F.Mezei, last paper).

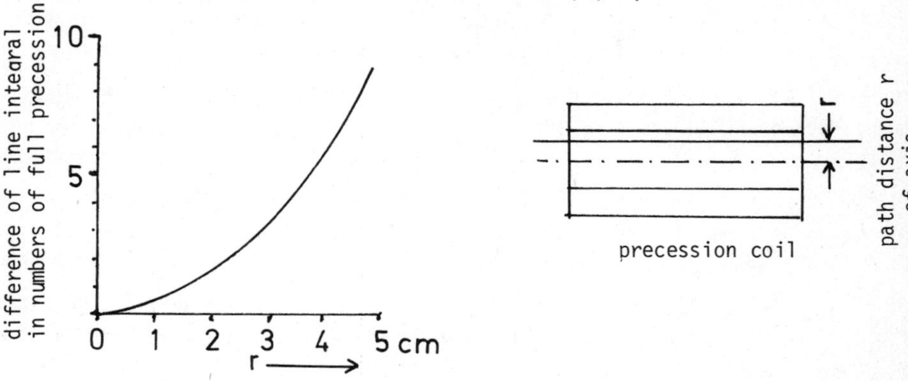

Fig.21 Difference in the line integral for paths of distance r from the coil axis for λ = 5.5 Å, I = 200 A for a coil of 2 m length with inner diameter 20 cm, outer diameter 30 cm, 400 turns.

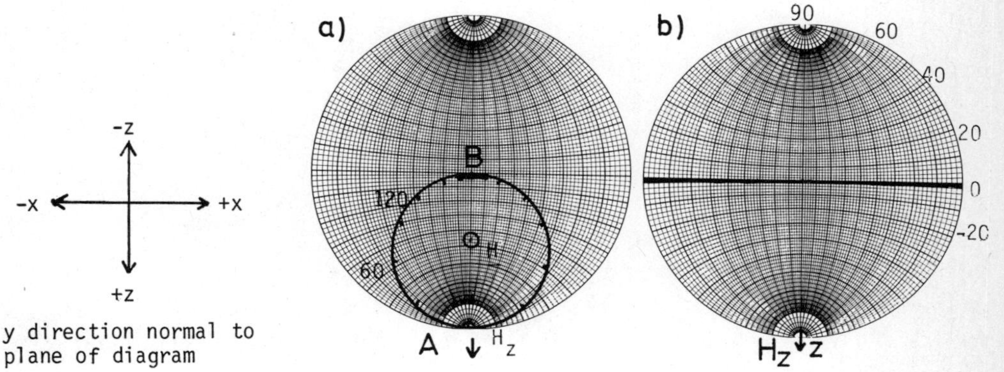

Fig.22 Effect of the $\pi/2$ - coil and the guide field in stereographic projection
a) Effect of the $\pi/2$ - coil: change of the initial direction at A into the range of directions at B with $180^0 \pm 18^0$. Each division represents 30 deg. of precession.
b) Effect of the first guide field: smearing out of the spin directions over black shadowed range (note that this range extends only in a negative direction)

To see quantitatively the effect of the $\frac{\pi}{2}$ - coil and the first guide field, these two steps can be visualized in stereographic projection by the Wulff net (fig.22). It can be seen that the smearing out of all spins by the first guide field is very flat and so it is possible to consider, for simplicity, only the components in the xy-plane.

The next step is the π-coil or the flipper coil which is placed as near as possible to the scattering sample. The effect of the flipper coil is shown in fig.23. As the direction of the guide field is normally the same in both arms of the spectrometer (but see fig.25) the direction of the precession in both guide fields remains the same, given by the arrows A and B in fig.23. Since the π-coil cannot directly distinguish integral 2π differences in the precession angles of individual neutrons, only differences less than 360^0 are considered here. Assume, that before being flipped, the spins have the directions 1, 2, 3, 4, and 5 (1 is the most advanced, 5 the least). Then the π-coil flips these spins over a precession cone (fig.23b) e.g. from 5 to 5'. This is just a precession angle of 180^0. (For the definition of the precession angle in the case of a precession cone see fig.3). Simultaneously 1 is flipped to 1' and so on and at the end 1' is the most retarded and 5' is the most advanced. The different spins now have the reversed order compared to the order before the flipping. The reason for this difference were their different velocities when they traversed the incoming guide field. If they now traverse a second guide field equal to the first, each with is original velocity, these differences will just cancel, and all the spins will be realigned when they reach the exit (see fig.26).

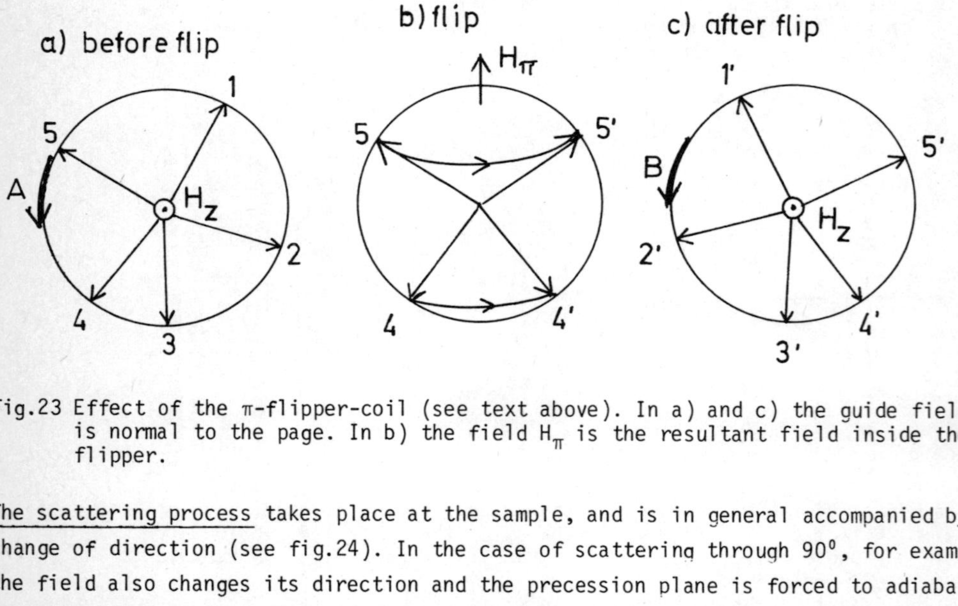

Fig.23 Effect of the π-flipper-coil (see text above). In a) and c) the guide field H_z is normal to the page. In b) the field H_π is the resultant field inside the flipper.

The scattering process takes place at the sample, and is in general accompanied by a change of direction (see fig.24). In the case of scattering through 90^0, for example, the field also changes its direction and the precession plane is forced to adiabatically follow this field as indicated in fig.24. As we saw in section 2, the field at the sample should, in this case, be greater than 0.05 Oe. On IN11 a field of 6 Oe is obtained with 200 A guide field current. If the current is 100 times less, i.e. 2 A, one could expect a strong depolarising effect, because then we have only 0.06 Oe. This depolarisation is observed by the disappearance of the spin-echo.

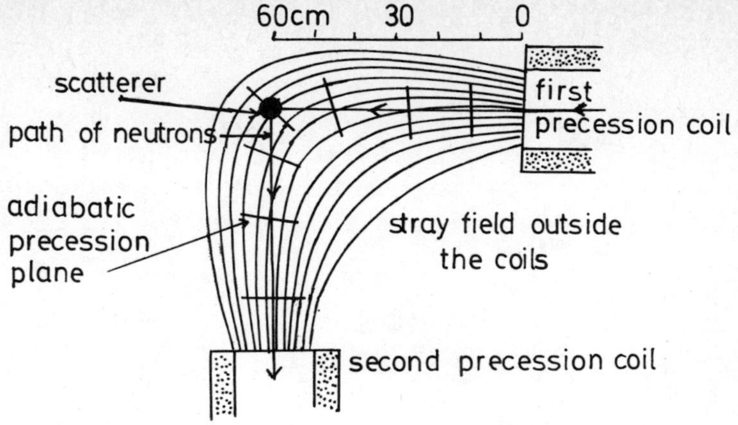

Fig.24 Path of the neutrons, change of the field direction and adiabatic following of the precession plane for 90^0-scattering

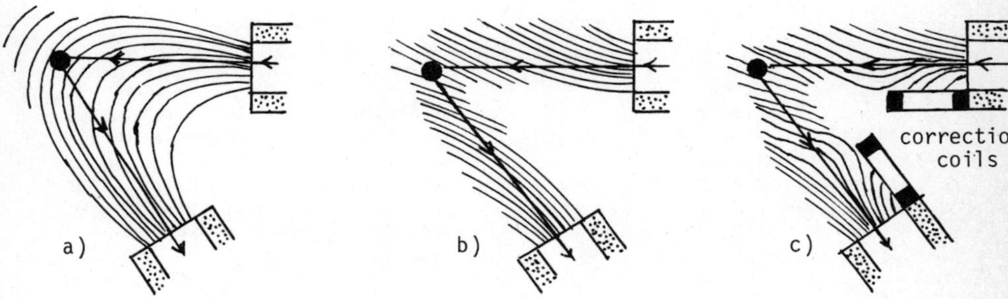

Fig.25 Scattering angle 135^0, field pattern on the neutron path with a) current in both guide fields in the same direction, b) current in both guide fields in opposite direction, c) current in both guide fields in opposite direction but with correction coils applied (285 turns, 216 mm diameter, 3.6 A, guide field current 180 A).

If the scattering angle is greater, e.g. 135^0, the field is shown in fig.25a. Here the rotation of the precession plane is very large and can be diminished by a reversal of the field in the second coil. This results in a field distribution given in fig.25b. Another disturbing effect for large scattering angles is that a current change in one coil also gives a field change in the range of the other coil, especially in the area of the coil exit. This mutual interference can be considerably reduced by the addition of two correction coils as indicated in fig.25c. The effect of this correction on the spin echo is described by F.Mezei (see last paper).

In the next stage, the second precession coil, the difference in precession for different λ is canceled (see fig.26). This cancellation is achieved by arranging that each neutron spin undergoes exactly the same number of "unwinding" precessions as it experienced in the first guide field, i.e. the line integral of this second guide field must exactly match that of the first. A small coil of 27 turns (called the symmetry coil, requiring typically ∿1 A) wound on the outside of the precession coil is used to finally "tune" the field in the second arm to achieve this condition. Thus using the symmetry coil the endposition, E, can be rotated and becomes broader the further away it is from the exact 'echo' position (see fig.27).

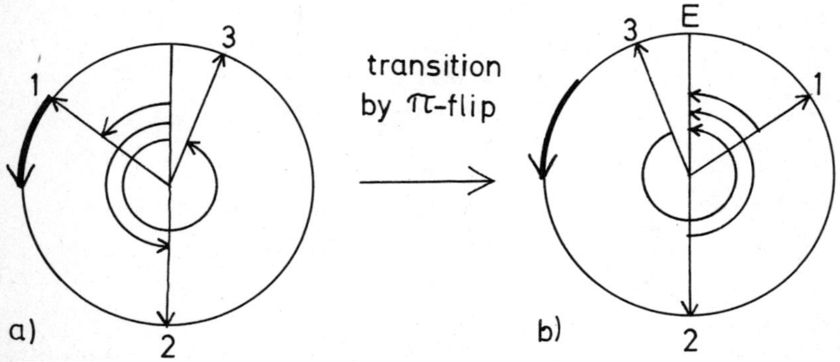

Fig.26 Difference in precession for different wavelengths λ in the first guide field (a) is cancelled in the second guide field (b). For example in a) spin 1 is mostly retarded, spin 3 mostly advanced and after flipping in b) spin 3 is mostly retarded and spin 1 is mostly advanced, and thus after traversing the same guide field as the first all spins run through the same angles they did in the first guide field. The differences are just cancelled, and all spins have the same end position E.

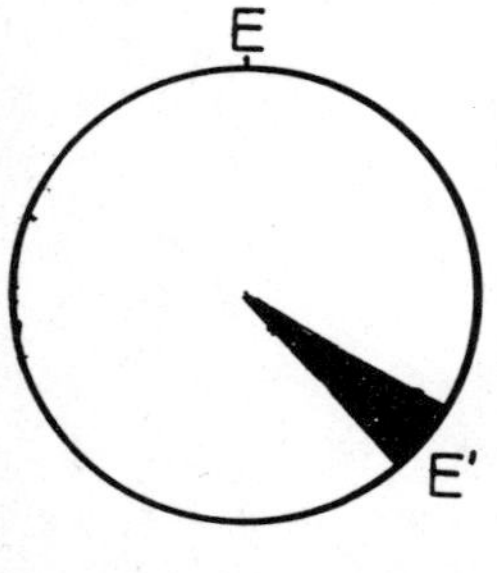

Fig.27 Broadening and change of direction of the end position E to E' of the spin directions at the exit of the second guide field by a small field added to the second guide field = <u>effect of a current change in the symmetry coil</u>

The next stage is now a <u>second π/2 - coil</u>. This π/2-coil brings the spin back to longitudinal direction if the phase is correct i.e. if the end position E of the spin directions in fig.26 and fig.27 points in the direction of the π/2-coil axis as shown in fig.28a. In this case one has the maximum count rate. If the point E is moved to E' by the symmetry coil, as in fig.27, but to the direction antiparallel to the π/2-coil axis, then this gives the lowest count rate, because than the neutron spin after traversing the π/2-coil points in the direction antiparallel to the guide field(fig.28b) and will be rejected by the analyser at the next stage. In the other cases the count rate is somewhere in between, giving a cosine-curve behaviour for a monochromatic beam and a cosine with a decrease in amplitude by the broadening of the spot E' for a polychromatic beam. This is what we know as SPIN-ECHO, and is shown in fig.29. The final step is achieved at the analyser which transmits only the components of spin parallel to the guide field.

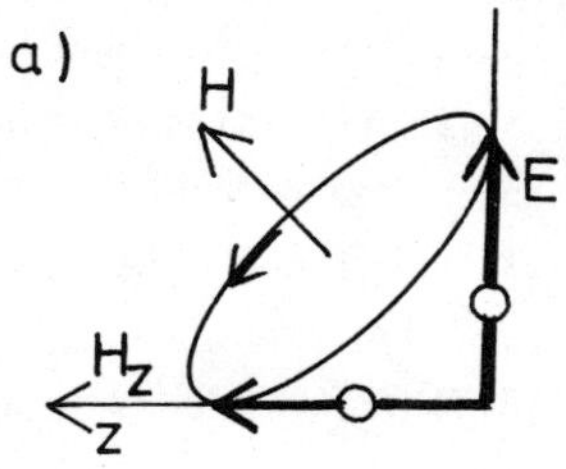

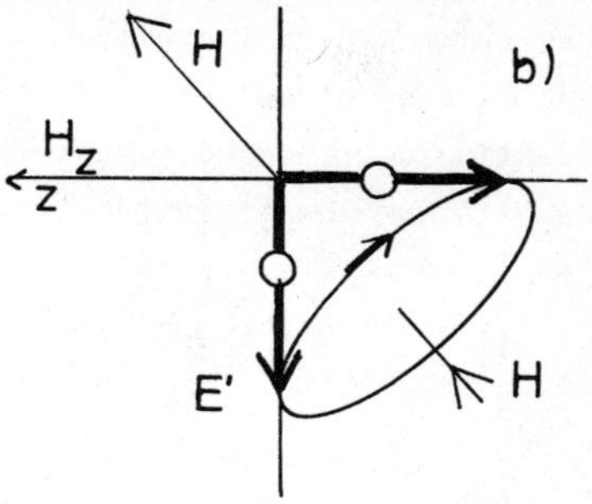

Fig.28 a) The second π/2-coil brings the spins of direction E of fig.26 back to the longitudinal direction z, parallel to the guide field H_z. The analyser transmits this spin direction giving maximum countrate

Fig.28 b) If the spin E is shifted by the symmetry coil as in fig.27 so that E' points in the antiparallel direction, the π/2-coil brings it to the direction antiparallel to the guide field and the analyser rejects it giving minimum count rate.

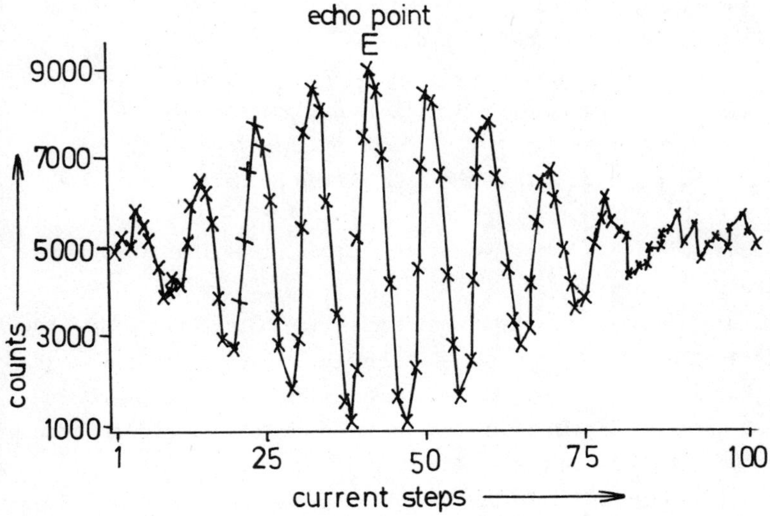

Fig.29 Spin-echo obtained by changing the field in the second guide field or in a small additional coil of 27 turns wound on the outside of the precession coil (the so called symmetry coil) in steps of 0.05 A

Until now we made an approximation by considering only the components of spin in the xy-plane. As we now understand the principles of spin-echo, it is also possible to see a little more quantitatively the behaviour of the flipper, of the second guide field, and of the second $\pi/2$-coil in the real (not approximated) case. To do this we can continue the considerations of fig.22 using the stereographic projection in the Wulff net. In fig.22 we saw that the guide field will distribute the spin directions of all the neutrons over the whole circumference in a zone that is given in fig.30 by the black line between AA' and BB'. Consider the line AB. Neutrons found to have spin directions corresponding to this line will in fact be those with the highest and also with the lowest limits of velocity of the incoming beam. This phenomenon can be explained by referring to fig.22a where it can be seen that the fastest neutrons are rotated through somewhat less than 90^0 and the slowest ones a similar amount greater than 90^0. Thus it can be seen in fig.22b that these two limits both correspond to the lower limit of the shaded band. Similarly the upper limit of the band (A'B') corresponds to neutrons having a velocity equal to the "tuned" value of the $\pi/2$-coil (normally the mean velocity of the beam spectrum).

Flipping by the π-coil can be considered to take place conterclockwise around an axis at the center of the diagram if the field H_π is chosen to come normally out of the paper. We must now consider what happens to neutrons at A. These will have both the highest and the lowest velocities possible; The fastest precess around the circumference of the diagram to a point A_f ($=180^0-18^0+$difference between A and A'), and the slowest correspondingly to a point A_s ($=180^0+18+$difference between A and A'). Similarly starting at B the fastest precess round to B_f and the slowest to B_s.Intermediate velocities will then fall somewhere between A_f and A_s (B_f and B_s).It is interesting to note that neutrons starting at A' will always precess round to B' and vice versa. Neutrons not starting at the circumference of the diagram will behave as described above and fall somwhere within the area marked by dotted lines $B_sA_fA_sB_f$.

The action of the second guide field is to smear this area out to the area whose limit is marked by the solid lines $B_sA_fA_sB_f$ (c.f. fig.20 and 22).In the process however all the spins are gradually being brought back into phase until at the end point E they are all lined up again (almost). This lining up is illustrated in the xy-plane in fig.26b, which is valid only for neutrons having a velocity equal to the tuned value of $\pi/2$ and π-coils In a beam having a spread of velocities (in our case ±10%) these will be realigned such that the faster ones arrive a little past the endpoint and the slower ones a little before it. This amounts to a mirror image of the distribution shown in fig.18a. It should be noted that errors are introduced because in the worst case we are about 20^0 from the ideal rotation of 180^0. This means that after the flipping the spin does not lie in the xy-plane as assumed in fig.23. Projection of this worst case back onto the xy-plane does,however, yield a maximum error of only $\pm 1.1^0$ which is of the order of the line thickness in fig.30 and can thus be ignored in this treatment.

The deviation of up to 20^0 away from the xy-plane can be clearly seen as the butterfly shape at the center E of fig.30 where the fastest neutrons have been sorted to have spin directions corresponding to the left-hand edge of the shape and the slowest the right-hand edge. It is obvious that neutrons corresponding to the center of the shape have a velocity equal to the tuned value. This butterfly shape does tend to give a wrong impression because in fact most neutrons lie near the center of the shape and are also grouped towards the xy-plane since not only do most neutrons have velocities near the average value for the beam but also only a fraction of those having extreme velocities are brought back into the worst-case positions.

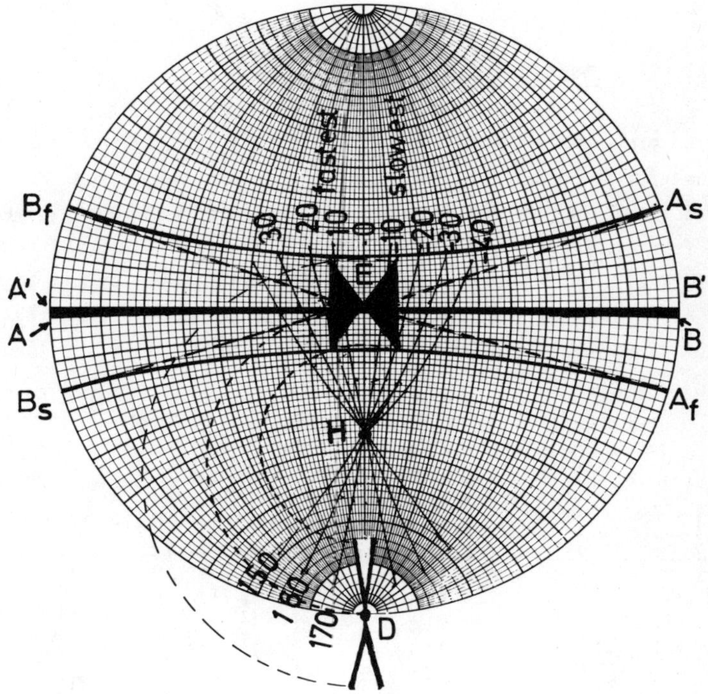

Fig.30 Stereographic projection of the effect of the flipper coil, the second guide field, and the second $\pi/2$-coil. The black area AA'B'B is the distribution of spin directions arriving at the flipper (fig.22b). The flipper produces a distribution bounded by the dotted lines $B_fEA_sA_fEB_s$. The second guide field orders the spins into the butterfly shape at E. (The area bounded by the solid lines $B_fA_sA_fB_s$ is an intermediate distribution found at some point in the second guide field). Finally the second $\pi/2$-coil causes a precession around H to produce the cross shape at D (see text).

The then following $\pi/2$-coil brings this distribution back into a cross shape as shown in fig.30 at D in stereographic projection. This is then the polarisation of the beam that enters the analyser. The effect of the $\pi/2$-coil on the distribution in E is indicated by the dashed lines. They are the stereographic projections of the precession cones about the field H. The lines that are orthogonal to it are lines of equal rotation about the direction H in steps of 10 degrees. One sees that above and below the ideal end position E the angular deviation is larger and smaller giving an effective broadening of the spot in this range of an x-form.

6. SOME PROPERTIES OF THE SPIN ECHO

To visualize the properies of spin echo it is advantageous to show its connection with a very powerful mathematical tool, the Fourier transformation. We know from the previous section, that an incident beam with a sharp wavelength gives as a spin echo a cosine function. The visualisation of this is given in fig.31. Normally the Fourier transform of a sharp wavelength is not a cosine function, it is a complex function with a real and an imaginary part, because the one wavelength λ_0 is not a symmetric function. The cosine corresponds to a symmetrically supplied spectrum as in fig.31a.

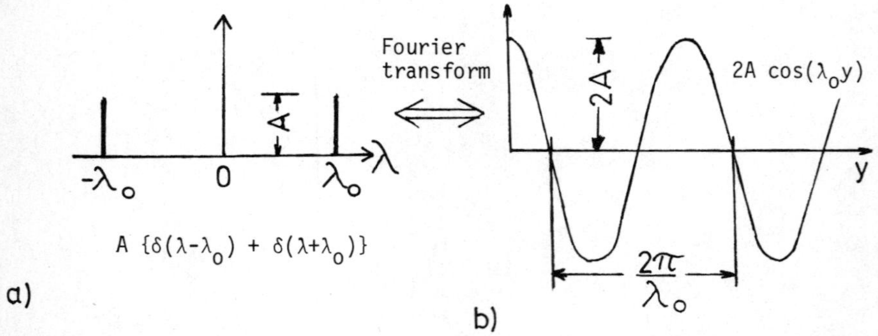

Fig.31 Correspondence between a symmetric monochromatic wavelength spectrum (a) and its Fourier transform - a cosine function (b), as in the spin echo signal

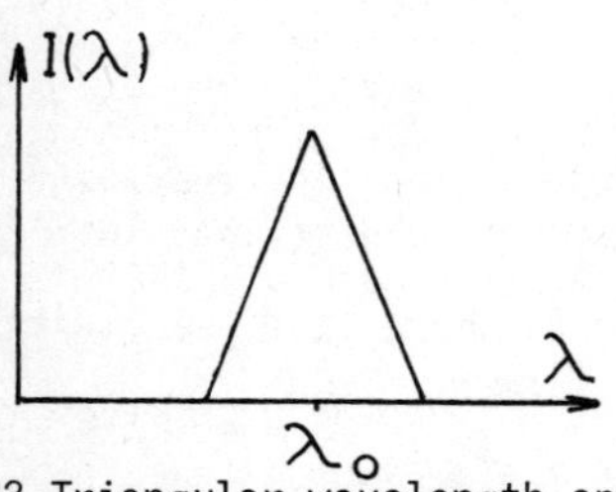

Fig.32 Triangular wavelength spectrum of the incident neutrons, that one would like to use (spectrum of a mechanical selector)

Let us now consider the result from a non-monochromatic beam as shown in fig.32. In fact one of the main advantages of spin echo is the fact that the full intensity of just such a beam can be used without loss of high resolution. It is possible to represent this spectrum as a convolution integral with the Dirac δ-function:

$$J(\lambda) = \int\{ \delta(\lambda-\lambda_0) + \delta(\lambda+\lambda_0) \} J(\lambda_0)\, d\lambda_0 = \delta(\lambda)\otimes J(\lambda)$$

This is shown diagramatically in fig.33. The convolution theorem of the Fourier transform allows us to readily find the Fourier transform of the convoluted functions i.e. the spin-echo (33f) of $J(\lambda)$ (33e) is simply the product of the cosine function (33b) for the wavelength λ_0 (33a) with the Fourier transform (33d) of the spectrum $J(\lambda)$ (33c). One can immediately see that the envelope of the spin echo in fig.33f is simply the Fourier transform (33d) of the incident spectrum. If the spectrum is not in itself symmetric the Fourier transform of fig.33c also has a sine transform, i.e. an imaginary part and the modulus of the Fourier transformed function has to be used to obtain the observed envelope.

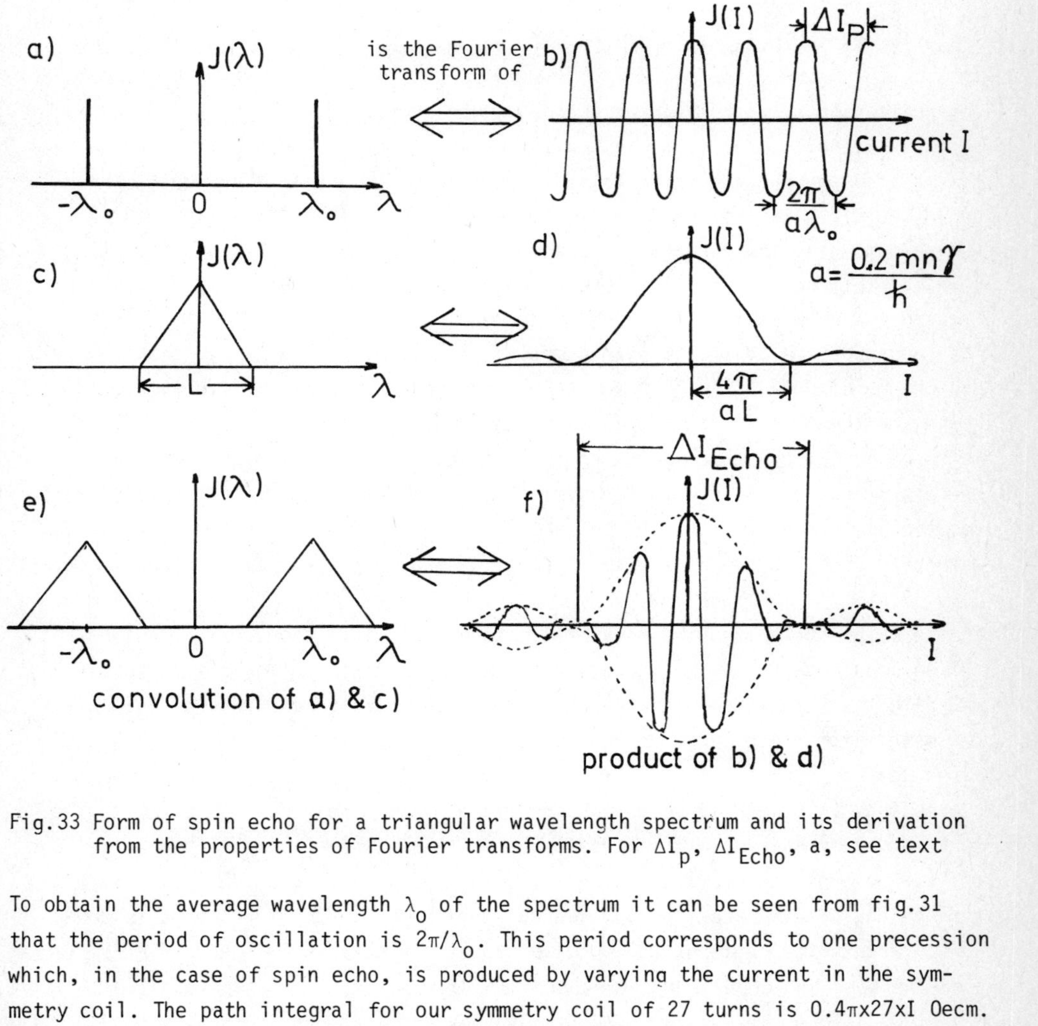

Fig.33 Form of spin echo for a triangular wavelength spectrum and its derivation from the properties of Fourier transforms. For ΔI_p, ΔI_{Echo}, a, see text

To obtain the average wavelength λ_0 of the spectrum it can be seen from fig.31 that the period of oscillation is $2\pi/\lambda_0$. This period corresponds to one precession which, in the case of spin echo, is produced by varying the current in the symmetry coil. The path integral for our symmetry coil of 27 turns is $0.4\pi\times 27\times I$ Oecm. Combining this with the results of fig.10 the current change ΔI_p for one full precession, or one full period in fig.31 (if the y-axis corresponds to the current I in the symmetry coil with n turns) is given as:

$$\Delta I_p = \frac{2\pi}{\lambda_o} \frac{\hbar}{2m\gamma n} = \frac{107.97}{n\lambda_o[\text{Å}]} \text{ Ampère} \quad \text{or for n=27 turns} \quad \Delta I_p = \frac{3.998}{\lambda_o[\text{Å}]} \text{ Ampère}$$

with λ_o = average wavelength of the spectrum, $\hbar$ = Planck constant divided by 2π, m = mass of the neutron, γ = gyromagnetic constant of the neutron = $1.832 x 10^4 s^{-1} Oe^{-1}$, n = number of turns of the symmetry coil. If ΔI_p for a full precession is known, this relation gives λ_o in Å. The width ΔI_{Echo} of the envelope of the echo in fig.33f is $8\pi/(aL)$ and thus gives the width L of the incoming spectrum as

$$L = 8\pi/(a\Delta I_{Echo}) \quad \text{with } a = 0.2m\gamma n/\hbar \quad \text{and } m,n,\gamma,\hbar \text{ the same as above.}$$

The relative width is $\Delta\lambda_o/\lambda_o = L/\lambda_o = 4\Delta I_p/\Delta I_{Echo}$.

In the scattering process the neutrons can lose energy or gain energy. For this inelastic scattering one is interested in the energy change and in the line shape of this scattering process. If a total energy change occurs at the sample, the second guide field has to have a different magnitude to that of the incoming guide field to give the echo point as shown in fig.26. This is because the flight time in each of the fields is different. Spin echo measures this time difference, and thereby a wavelength difference, as a current difference of the two guide fields (because time of flight is proportioal to λ, and $\Delta I/I = \Delta\lambda/\lambda$). Thus a wavelength change of only 2% (0.16 Å for λ=8Å) would need a current difference of 2% (i.e. 4A in the main 400 turn coil, carrying 200A, or 4x400/27=60A in the 27 turn symmetry coil). It can easily be seen for upscattering (energy gain) of a few meV, the necessary current change in the symmetry coil would be too large, necessitating an additional current source for the second guide field coil itself. The relative difference of the two currents gives a direct measure of the inelastic energy change.

This also helps us to find the possible instrumental resolution. Using neutrons of λ = 8Å one can obtain 6000 full precessions with a current in the guide field of 200 A. With the spin echo apparatus IN11 it is possible to measure a 5 deg. shift of the main maximum of the echo. Thus a relative wavelength change of 5:(6000x360) = $2 x 10^{-6}$ can be resolved. This is illustrated in fig.34, which shows an inelastic spin echo that is displaced 16 periods from the elastic spin echo by changing the current in the symmetry coil. The wavelength is λ=8Å. The elastic spin echo is found around zero current change i.e. at $\Delta\bar{H} = 0$ with a period of $\Delta\bar{H}$=18.95 Oecm ($\bar{H}=\int H\,dz$) corresponding to λ=8Å (see fig.10). The current in the guide field coils is 200A (≡500 Oe) and their length of 200 cm gives a path integral $\bar{H}=10^5$ Oecm. The displacement of the echo by 16 periods corresponds to 271.3 Oecm. As $\Delta I/I=\Delta\bar{H}/\bar{H}=271.3/10^5=\Delta\lambda/\lambda$ and $\Delta\lambda=0.00271\lambda$ one obtains for λ=8Å (≙1.278 meV) a wavelength change of $\Delta\lambda$=0.0217Å, or a relative energy change of $\Delta E/E \simeq 2\Delta\lambda/\lambda = 5.4 x 10^{-3}$.

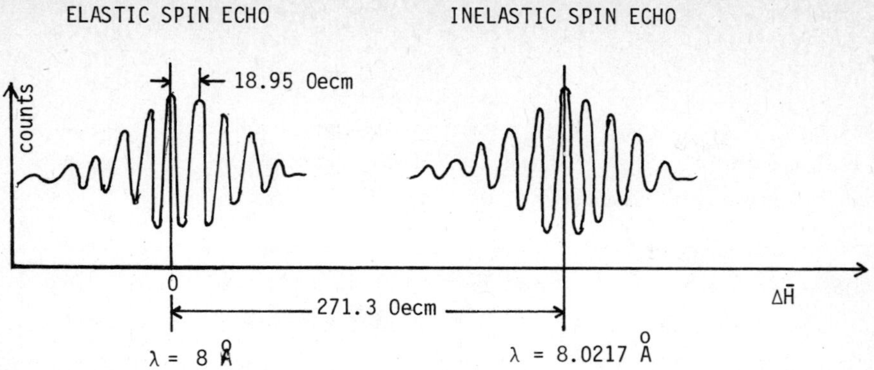

Fig. 34 Elastic ($\Delta\bar{H}$ = 0) and inelastic ($\Delta\bar{H}$ = 271.3 Oecm, $\bar{H}=\int Hdz$) neutron spin echo for $\bar{H}=10^5$ Oecm (200 A in the 2 m long coil of fig. 19 and 21), with the corresponding wavelength change at λ=8Å incident wave. The $\Delta\bar{H}$ = 271.3Oecm corresponds to 16 periods i.e. $\Delta\bar{H}$ = 16x18.95 Oecm in this figure. The relative wavelength change results from $\Delta\lambda/\lambda = \Delta\bar{H}/\bar{H} = 271.3/10^5$. The corresponding relative energy change is $2\Delta\lambda/\lambda = 5.4\cdot10^{-3}$ i.e. for λ = 8 Å ≙1.27 meV, ΔE = 6.91 μeV, with a very well separated echo.

In the inelastic scattering process the wavelength changes Δλ for neutrons of a polychromatic beam are not necessarily the same. A modification of the spin echo envelope can thus occur. Since the Fourier transform of the echo gives the spectrum of the scattered beam we can get rough information about upscattering and width of this spectrum, using the method shown in fig. 33. But this is not all: the scatterer can, for a given wavelength λ, cause λ to change by varying amounts Δλ with different probabilities S(Δλ) and $\int S(\Delta\lambda)d(\Delta\lambda) = 1$. It can be described by a spin echo line shape S(Δλ) as given in fig. 35. The principle of the measurement of this line shape S(Δλ) involves measuring the change of the central amplitude of the spin echo as a function of the guide field current. We again apply the Fourier transform representation of the spin echo. If there is a line shape dependence on

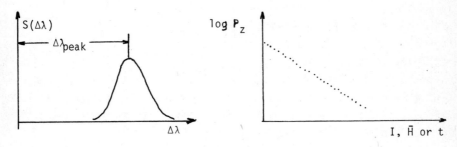

Fig. 35 a) Line shape S(Δλ) of an inelastic scatterer as a function of the wavelength change Δλ e.g. a Lorentzian (displaced).
b) Log-lin plot of the Fourier transform of the line shape in a)

$\Delta\lambda$, we have to convolute the spin echo signal also with $S(\Delta\lambda)$ resulting in

$$\delta(\lambda) \otimes J(\lambda) \otimes S(\Delta\lambda)$$

The whole spin echo in this case is the product of the cosine function, the Fourier transform of the incident spectrum and the Fourier transform of the line shape. As the convolution process is associative, we can first measure

$$\delta(\lambda) \otimes J(\lambda)$$

by changing the current in one of the coils to obtain the spin echo (or in the symmetry coil as described in fig.27 or 34). The amplitude of this spin echo, i.e. the difference between the maximum and minimum countrates at the center of the echo, depends now also on the Fourier transform of the line shape. Changing the current in both coils simultaneously, i.e. changing the resolution by using different numbers of precessions in the guide fields and measuring the change of the resulting amplitude of the spin echo as a function of the guide field current, we obtain a pointwise representation of the envelope of the Fourier transform of $S(\Delta\lambda)$ of the scattering process without any deconvolution, i.e. without assuming an apriori line shape $S(\Delta\lambda)$. For more details in a special case see [9].

In this paper I did not quote references because nearly all of it has come from oral discussions with F.Mezei, J.B.Hayter and T.Springer whom I want to thank. I want to thank especially P.Dagleish for tidying up the whole paper by clarifying many points and correcting my English. I refer the reader to the following publications for more information on the subject.

REFERENCES

[1] F.Klein, The mathematical Theory of the Top, Princeton 1897, cited in H.Goldstein, Klassische Mechanik, Akademische Verlagsgesellschaft,Frankfurt/M 1972

[2] J.L.Synge and B.A.Griffith, Principles of Mechanics 1959 p.387

[3] Ch.Schwink and O.Schärpf, Z.Physi B 21, 305 (1975)
O.Schärpf, J.Appl.Cryst. 11, 631 (1978)

[4] F.Mezei, Z.Physik 255, 146 (1975)

[5] J.B.Hayter, Z.Physik B 31, 117 (1978)

[6] F.Mezei, Neutron inelastic scattering, Conf.Proc.(IAEA, Vienna, 1978)vol.1,p125

[7] F.Mezei, Proc. 3rd Int.School on Neutron Scattering, Alushta, USSR (1978)

[8] F.Mezei, Nucl.Instr. and Methods, 164, 153 (1979)

[9] J.B.Hayter and J.Penfold, Z.Physik B 35, 199 (1979)

[10]R.Pynn, J.Phys.E 11, 1133 (1978)

THEORY OF NEUTRON SPIN-ECHO SPECTROMETRY

J.B. HAYTER

Institut Laue - Langevin,
156 X, 38042 Grenoble Cédex, France.

ABSTRACT

An introduction to the theory of neutron spin-echo spectrometry is presented for the non-specialist. The relationship between polarisation and the scattering law is first discussed. The practical realisation of a spin-echo configuration is then considered, and it is shown how to relate neutron counts to the calculated polarisation. Scanning techniques to measure either wavelength spectra or high resolution inelastic spectra are derived, and methods of data correction are demonstrated.

1. INTRODUCTION

This paper presents a general introduction to the theory of neutron spin-echo spectrometry /1/, particularly for the non-specialist who may be an intending user of spin-echo. We first present a physically transparent calculation of the response of an ideal quasielastic spectrometer, and show that there are basically two types of spin-echo experiment. The first, the asymmetric scan, allows measurement of a direct or scattered wavelength spectrum. The second, symmetric scan is used for high resolution inelastic spectroscopy.

The general response of a real spectrometer is next considered, and it is shown how to calculate the spin-echo polarisation from the sample scattering law for any given spectrometer configuration. We then discuss how a spin-echo measurement is performed, and relate the calculated polarisation to measured intensities. Finally, corrections to the data are considered and some examples are given of how spin-echo and polarisation analysis may be combined for this purpose.

Only salient points of the theory are presented, and the reader is referred to original papers for more detail where necessary. Specialist uses of spin-echo, such as suppression of thermal diffuse scattering /10/ or three-axis spectrometry /5,8/ have also been omitted from the discussion. The reference list /1-14/ provides an introductory bibliography of the literature of neutron spin-echo.

2. THE FUNDAMENTAL EQUATION OF NEUTRON SPIN-ECHO

Neutron spin-echo spectroscopy is based on a polarisation analysis spectrometer configuration in which the analysed polarisation is a function of the inelasticity of the sample scattering. We shall start by deriving this fundamental relationship between the scattering law and the spin-echo polarisation. The guide-field will be taken as the z-direction. (On the ILL spin-echo spectrometer IN11, this is also the beam direction.) Since the analyser projects the polarisation onto this field direction, we are only interested in the z-component of the final polarisation, P_z. We first consider quasielastic scattering in an ideal spectrometer, in which the spin-turn coils operate independently of wavelength. The real case and the extension to a general scattering law will then be considered in section 2.2.

2.1 Ideal Quasielastic Spectrometer Response

Provided the sample performs no spin-projection operations (we exclude polarisers as samples, for example), the spectrometer may be considered schematically as a net spin-turn operation, described by an operator $\hat{\underline{\underline{T}}}$ /6/ (see Fig. 1(a)). A neutron with polarisation (0,0,1) after the polariser P_1 will arrive at the analyser P_2 with polarisation $\hat{\underline{\underline{T}}} \cdot (0,0,1) = (T_{13}, T_{23}, T_{33})$. The probability of transmission by the analyser is thus T_{33}, so that the final polarisation average over the beam is $< P_z > = < T_{33} >$.

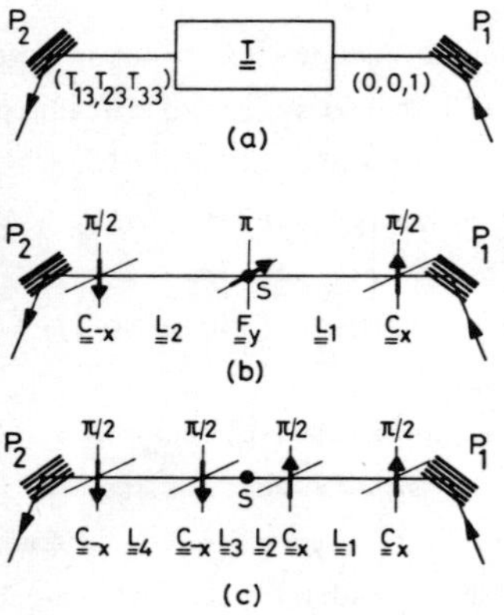

Fig. 1. (a) Schematic spin-turn analysis experiment. (b), (c) spin-echo configurations. S : sample.

The detailed spectrometer arrangement need not concern us at this point. We only note that it consists of a sequence of $\pi/2$ spin-turn coils (each described by an operator $\hat{\underline{\underline{C}}}$ /6/) and π spin-turn coils (operator $\hat{\underline{\underline{F}}}$), with Larmor precession regions (operator $\hat{\underline{\underline{L}}}$) between them ; examples are given in Figs. 1(b) and 1(c). (A $\pi/2$ coil starts or stops Larmor precession, while a π coil, which is a spin-flipper, reverses the precession direction. It is the formal $\pi/2 - \pi - \pi/2$ sequence of Fig. 1(b) which gives neutron spin-echo its name.) The spin-echo operator $\hat{\underline{\underline{T}}}$ is then just the product of all such operators, which are 3 x 3 rotation matrices, and the (ideal) configuration is chosen to make $\hat{\underline{\underline{T}}} = \underline{\underline{1}}$, independent of wavelength, for elastic scattering. This is spin-echo focussing /1/. The wavelength independence decouples resolution from beam monochromatisation, so that a broad incident beam spectrum may be used for high intensity. (On IN11, the incident beam may have a spectral width of up to 50 % in wavelength.)

This focussing condition means that the incident polarisation is transmitted unchanged by elastic scattering. We now seek to calculate how the polarisation is altered by inelastic scattering processes.

We assume for the purpose of demonstration a single precession region in each arm of the spectrometer (Fig. 1(b)). For an incoming wavelength λ, let there be $N(\lambda)$ precessions in the first arm (before scattering) and $N(\lambda+\delta\lambda) + \Delta N(\lambda+\delta\lambda)$ in the second arm (after scattering), where $\delta\lambda$ is the change of wavelength on scattering and $\Delta N(\lambda)$ is the difference between the numbers of precessions in the two arms at a given wavelength ; ΔN will be called the *asymmetry*. If the Larmor precession regions are parametrised by field integrals J and $J+\Delta J$, respectively,

$$N(\lambda) = 2|\gamma|\mu_N mJ\lambda/h^2 \tag{1}$$

where $\gamma = -1.9130$, μ_N is the nuclear magneton, h is Planck's constant, and m is the neutron mass. (For λ in Å and J in oe-cm, $N(\lambda) = 7.37 \times 10^{-3}\ J\lambda$.) A similar relation holds between ΔN and ΔJ.

It is convenient to relate these quantities to the mean incident wavelength λ_o, so that $N(\lambda) = N_o\lambda/\lambda_o$, where $N_o = N(\lambda_o)$, and similarly for ΔN.

The precession directions are opposite in the two arms, so that the analyser transmits the projection of the difference in the corresponding precession angles. In the present example we are concerned with small wavelength shifts and small asymmetries ; to first order in ΔN and $\delta\lambda$ we then obtain

$$T_{33} = \cos\left[2\pi\ (N_o\delta\lambda + \Delta N_o\lambda)/\lambda_o\right] \tag{2}$$

as the required projection.

The incident beam will in general be polychromatic with normalised spectrum $I(\lambda)$. The final average polarisation is obtained by integrating over all incident wavelengths and all possible wavelength changes :

$$< P_z > = \int_o^\infty I(\lambda)d\lambda \int_{-\lambda}^{\infty} W(\lambda+\delta\lambda)\ P(\lambda,\delta\lambda)\ \ T_{33}\ d(\delta\lambda) \tag{3}$$

where $P(\lambda,\delta\lambda)$ is the probability of $\lambda \rightarrow \lambda+\delta\lambda$ on scattering, and the window function $W(\lambda)$ is the probability that a neutron of wavelength λ will be transmitted by the spectrometer in the absence of spin-echo. We assume the spectrometer only transmits wavelengths in the neighbourhood of $I(\lambda)$, and take $W(\lambda) = 1$ for quasielastically scattered neutrons and $W(\lambda) = 0$ for all others. The limits on the second integral may then be formally extended to $(-\infty, \infty)$.

If the normalised quasielastic scattering law is $S(Q, \omega)$, $\delta\lambda$ is related to the energy transfer ω to first order by

$$\delta\lambda = (m\lambda^3/2\pi h)\ \omega \tag{4}$$

and the scattering probability is $P(\lambda,\ \delta\lambda)d(\delta\lambda) = S(Q,\ \omega)d\omega$. Equation (3) becomes

$$< P_z > = \int_0^\infty I(\lambda)d\lambda \int_{-\infty}^\infty S(Q,\omega)\ T_{33}\ d\omega \tag{5}$$

with T_{33} now given by (2) and (4). Here, Q is the momentum transfer ; $Q = (4\pi/\lambda)\sin\theta$ for quasielastic scattering through angle 2θ.

Equation (5) is the fundamental relationship between the spin-echo polarisation $< P_z >$ and the scattering law S(Q,ω) ; we shall see in the next section that it is in fact quite general provided T_{33} is correctly calculated.

For quasielastic scattering, we may take S(Q, ω) as an even function of ω, since the detailed balance factor is effectively unity. On expanding the cosine in (2), the sine integral appearing in (5) will then be zero, and we finally obtain from (4) and (5)

$$< P_z > = \int_0^\infty I(\lambda)\cos(2\pi\Delta N_o\lambda/\lambda_o)d\lambda \cdot \int_{-\infty}^\infty S(Q,\omega)\cos[\omega t(\lambda)]d\omega \tag{6}$$

where $t(\lambda) = N_o m\lambda^3/h\lambda_o = 1.863 \times 10^{-16}\ J\lambda^3$ sec for J in oe-cm and λ in Å. (On IN11 at $\lambda_o = 8$ Å, $t \leq 10$ nsec.)

Figure 2 shows a typical calculated spectrometer response to diffusive scattering as a function of N_o, the number of precessions in the first arm, and ΔN_o, the asymmetry between the arms. Note that the response has been converted to a relative intensity ; the relationship between $< P_z >$ and the measured intensity will be considered in section 3.

Reference to (6) and Fig. 2 shows that there are two limiting cases of particular interest :

(i) $N_o = 0$. Equation (6) reduces to

$$<P_z> = \int_0^\infty I(\lambda)S(Q)\cos(2\pi\Delta N_o\lambda/\lambda_o)d\lambda \tag{7a}$$

(remembering that Q is a function of λ), and a scan of the asymmetry ΔN_o measures the Fourier cosine transform of I(λ)S(Q), the effective scattered spectrum. (Since S(Q, ω) is normalised, (7a) is normalised to $S(Q_o)$, where $Q_o = Q(\lambda_o)$.) Fourier inversion then provides a useful means of obtaining the effective scattered spectrum in cases where the broad incident spectrum has been hardened by sample absorption, or modified by structure in S(Q). Such processes may significantly shift the mean

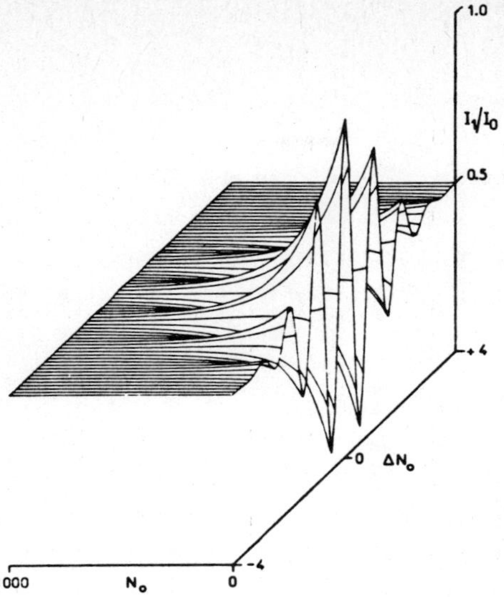

Fig. 2. Spectrometer response to a diffusive scatterer (calculated from (6) /11/).

scattered wavelength, and the asymmetric scan allows the effective Q to be determined at a given scattering angle.

In practice, N_o need only be chosen small enough for the quasielastic scattering not to have caused significant depolarisation. (To measure the incident spectrum, we may take $S(Q, \omega) = \delta(\omega)$ in (6), and any value of N_o may be used.) Comparison of incident and scattered spectra provides a fast estimate of S(Q) in, for example, small angle scattering.

(ii) $\Delta N_o = 0$. This is the symmetric, focussed configuration, used for high resolution inelastic scans. Equation (6) reduces to

$$< P_z > = \int_0^\infty I(\lambda)d\lambda \int_{-\infty}^{\infty} S(Q, \omega) \cos [\omega t(\lambda)] d\omega$$

$$= \int_0^\infty I(\lambda) F [Q, t(\lambda)] d\lambda \qquad (7b)$$

where F(Q, t) is the real part of the intermediate scattering law. The time-variable t is normally scanned by varying N_o.

The integration over λ in (7b) has little effect in many cases of interest. For diffusive scattering, for example, $F(Q, t) \sim \exp [-DQ^2 t(\lambda)]$; since $Q^2 \sim \lambda^{-2}$ and $t(\lambda) \sim \lambda^3$, the final λ dependence is weak. In favorable cases, F(Q, t) may not depend on λ at all. Certain types of small-angle scattering /9/, for example, involve $Q^3 t$ and all λ dependence vanishes.

2.2 General Real Spectrometer Response

In a real spectrometer, the action of the spin-turn coils depends on wavelength. This means $\hat{\underline{\underline{T}}}$ will no longer be the unit matrix in the elastic focussed configuration, but will contain off-diagonal elements which cause the integral transforms in (6) to be no longer simple cosine transforms. Further, even in the monochromatic case, (5) will no longer be valid with T_{33} given by (2) for a general (non-quasielastic) scattering law. The principle of the calculation, however, remains unchanged : we replace the T_{33} of (2) by a matrix element calculated explicitly for the particular configuration, and then integrate over all λ and $\delta\lambda$ values to obtain $< P_z >$ for a given $S(Q, \omega)$, N_o and ΔN_o /6/.

Taking the configuration of Fig. 1(b), for example,

$$\hat{\underline{\underline{T}}}(\lambda, \omega) = \hat{\underline{\underline{C}}}_{-x}(\lambda+\delta\lambda) \cdot \hat{\underline{\underline{L}}}_2 (N_o+\Delta N_o, \lambda+\delta\lambda) \cdot \cdot \hat{\underline{\underline{F}}}_y(\lambda) \cdot \hat{\underline{\underline{L}}}_1(N_o, \lambda) \cdot \hat{\underline{\underline{C}}}_x(\lambda) \tag{8}$$

Explicit forms for the operators are given in /6/. The notation provides the basis for a convenient shorthand description of any spin-echo configuration ; the operators are written from right to left in the order they are traversed by the beam. For this purpose, the specific operator notation may be dropped, writing $C_x = \hat{\underline{\underline{C}}}_x$, etc. Further, the direction of the first $\pi/2$ coil may be chosen without loss of generality as the reference **x**-direction, and the subscript may be omitted if the complementary configuration is denoted by a bar.

The configurations of Figs. 1(b) and 1(c), for example, are then unambiguously described by $\bar{C}L_2F_yL_1C$ and $\bar{C}L_4\bar{C}L_3L_2CL_1C$, respectively. In the case of the commonly used configurations of the type shown in Fig. 1(b), the presence of a Larmor precession region in each arm is implied ; the configuration may simply be written $\bar{C}F_yC$.

Since (8) will usually be evaluated numerically, $\delta\lambda$ may be calculated exactly for each ω and λ. The mean final polarisation is then

$$< P_z > = \int_0^\infty I(\lambda)d\lambda \int_{-\infty}^\infty S(Q, \omega)\, T_{33}\, (\lambda, \omega)d\omega \tag{9}$$

with $T_{33}(\lambda, \omega)$ calculated from (8) or its equivalent.

Equation (9) is the fundamental general equation, and may be evaluated as exactly as numerical integration permits. It is valid for general $S(Q, \omega)$, and the spectrometer transmission function $W(\lambda)$ may be included in the integration, if desired. It is to be noted, however, that (6) always provides a useful basis for a first look at spin-echo data.

In summary, for a given $S(Q, \omega)$ we may calculate the spectrometer response as a function of N_o, the number of precessions in the first arm, and ΔN_o, the asymmetry between the two arms. Scanning N_o with $\Delta N_o = 0$ measures a transform of the scattering law, while scanning ΔN_o at small N_o measures a transform of the effective wavelength spectrum. Recovery of the latter by a reverse integral transform in the general case is discussed in detail in /11/.

3. NEUTRON SPIN-ECHO POLARISATION MEASUREMENT

We now consider how a spin-echo configuration may be realised in practice. Intensity relations between $< P_z >$ and the neutron countrate will then be derived, specifically taking into account the possible presence of spin-flip scattering.

3.1 Spectrometer Alignment

Alignment of a neutron spin-echo spectrometer consists of two distinct phases : tuning the spin-turn coils to operate at the mean wavelength λ_o, then finding the field setting which gives exact precession symmetry between the two arms.

The first phase is quickly accomplished by tuning each spin-turn coil separately, with the other coils switched off. This tuning procedure aims to set the guide field at the coil and the field in the coil to produce an exact π or $\pi/2$ turn (as desired) at the mean wavelength. The method, which is straightforward, is described in detail in /6/ and will not be repeated here.

For a given spin-turn coil geometry and a given value of λ_o, the guide and coil fields are fixed absolutely. When the main spectrometer guide fields are altered (to scan N_o, for example, *via* (1)), the field at the spin-turn coil will generally be changed as well. To maintain a correct spin-turn, this field change must be cancelled by an offset correction coil ; the spin-turn coil alignment phase also consists of calibrating the associated correction coils as a function of the main field. (On the IN11 spectrometer, all coils are air-wound with no magnetic material, so that all corrections are linear and easily calibrated.)

All spin-turn coils are now switched on together to produce a spin-echo configuration. At this point, N_o is known (from (1) and the known main field geometry), but ΔN_o will in general be unknown, especially when the sample and π-turn coil are not coincident (see Fig. 1(b)). The symmetric position is now found by scanning ΔN_o, using an extra coil placed on the second arm.

This coil generates a field parallel or antiparallel to z, thereby increasing or decreasing the guide field in the second arm. Varying the field produced by this coil therefore varies the asymmetry, ΔN_o, independently of N_o. With the spectrometer set on the main beam, scanning this coil will therefore measure an integral transform (ideally, the Fourier cosine transform) of the incident wavelength spectrum /11/. An example is shown in Fig. 3(a). The centre of such a scan, corresponding to $\Delta N_o = 0$, is easily detected, and the periodicity calibrates the coil. Such scans are repeated at each main field value at which a symmetric scan of N_o will be made. (On IN11, asymmetry corrections are linear for the reasons already noted.)

Detailed analysis /6/ of spin-echo configurations shows that they form complementary pairs, related by a reversal of the current (i.e. field) in one $\pi/2$ coil.

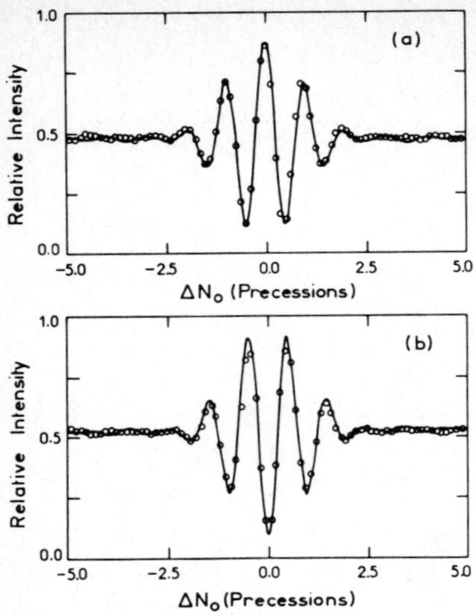

Fig. 3. Asymmetric spin-echo scans.
(a) Configuration CF_yC
(b) Complementary configuration CF_yC.

(Physically, this corresponds to a change of π in total precession angle, and hence reversal of the final polarisation. The choice of whether the final polarisation is positive or negative may be made *a priori* and is discussed in /6/.) Figures 3(a) and 3(b) show a corresponding pair of asymmetric scans, and it is seen that one is the complement of the other. We now show how $< P_z >$ may be evaluated from such complementary intensities, first for non-spin-flip and then for general scattering.

3.2 Coherent or Isotopic Incoherent Scattering

The general intensity relations of polarisation analysis, together with the measurement of polariser/analyser and π-turn coil efficiencies, are discussed in /6/ and we only quote relevant results here. We take the spectrometer to have known polariser and analyser efficiencies P_1 and P_2 (in fact we only need the product P_1P_2), and to have a π-turn coil (flipper) of known efficiency f. (Note that for a peaked incident beam spectrum of FWHM $\Delta\lambda$ centred on λ_o, $f = 0.5 + (1 - \cos \pi\delta)/\pi^2\delta^2$ gives a very good approximation for f, where $\delta = \Delta\lambda/\lambda_o$ /6/. On IN11, $f = 0.94$ and $P_1P_2 = 0.90$ with $\delta = 0.4$.) All intensities are assumed corrected for background, for example by subtraction of a blank run.

Let I^E and $\bar{I}^E$ be a complementary pair of spin-echo intensities (e.g. points corresponding to the same ΔN_o in Figs. 3(a) and 3(b)). These intensities correspond to the "spin-up" and "spin-down" intensities measured in a conventional polarisation analysis experiment, and we define an effective polarisation in the same way :

$$P_{eff} = (I^E - \bar{I}^E) \;/\; (I^E + \bar{I}^E) \tag{10}$$

The effective polarisation is the product of all polarisation effects in the beam, so that

$$< P_z > = (I^E - \bar{I}^E) / P_1P_2 (I^E + I^{-E}) \qquad (11)$$

As a complement to (11), a calculated $< P_z >$ may always be converted to a relative intensity using the general relation

$$I^E/I^\uparrow = (1 + P_1P_2 < P_z >) / (1 + P_1P_2) \qquad (11')$$

where $I^\uparrow$ is the intensity with all spin-turn coils switched off /6/.

The self-normalisation in such intensity ratios is inherent to all the forms of spin-echo analysis which will be presented here.

3.3 Spin-Incoherent Scattering

If some or all of the scattering is spin incoherent, a proportion α of the neutrons will be spin-flipped by the sample, diminishing or even reversing the spin-echo signal. In the case of an **isotopically pure sample, α = 2/3, although multiple** scattering may reduce this. In the general case of several different isotopes or chemical species, $0 \leq \alpha \leq 2/3$ /15/. This means that $< P_z >$ as defined by (11) would contain the effect of spin-flip as well as the sample dynamics, and two samples with identical dynamics but different isotopic composition would give different results. We prefer to decouple dynamic effects from spin-flip effects, and to calculate $< P_z >$ in such a way that spin-flip, if present, has been removed from the data. (This also has important consequences for data correction, to be discussed in the next section.)

Spin-flip may be integrated into (8) by including a flipper of efficiency α at the sample position. However, nuclear spin-flip reverses all three components of the polarisation /16/, and the corresponding operator is just $(1 - 2\alpha)\, \underline{\underline{1}}$. This commutes with everything, and hence may be placed immediately after the polariser for the purposes of calculation. The result is

$$< P_z > = 1 - 2f\, (I^\uparrow - I^E) / (I^\uparrow - I^\downarrow) \qquad (12)$$

where $I^\uparrow$ is the intensity with all spin-turn coils off, and $I^\downarrow$ the intensity with only the π-turn, of efficiency f, operating.

Equation (12) yields $< P_z >$ *free from any effects of spin-flip*. (The α-dependence is implicit in the normalisation to $(I^\uparrow - I^\downarrow)$, but α need not be known since it does not enter explicitly.)

Applying (12) to the complementary spin-echo intensities I^E and $\bar{I}^E$ (which correspond to $< P_z >$ and $- < P_z >$, respectively), we finally arrive at the important general result

$$< P_z > = f\,(I^E - \bar{I}^E)\ /\ (I^{\uparrow} - I^{\downarrow}) \tag{13}$$

Equation (13) requires more measured values than (11), but must be used wherever spin-flip effects are present. If desired, α may be measured from the flipping ratio

$$R = I^{\uparrow}/I^{\downarrow} = \left[1+P_1P_2(1-2\alpha)\right]/\left[1+P_1P_2(1-2\alpha)(1-2f)\right] \tag{14}$$

4. CORRECTIONS TO THE DATA

Real neutron spin-echo data requires correction for both magnetic inhomogeneities in the spectrometer and for parasitic scattering from the sample (for example, Bragg contamination). We now show that the former may be removed by a simple normalisation procedure, and give examples of how a combination of spin-echo and polarisation analysis permits treatment of the latter in certain cases.

4.1 Field Inhomogeneities

Inhomogeneities in the guide field mean that neutrons of the same wavelength may have different N_o values, depending on the paths they have taken through the spectrometer. This effect becomes important in high resolution symmetric scans, and in general means that the measured polarisation $< P_z >_M$ differs from the $< P_z >$ calculated using (6) or (9).

Consider a symmetric spectrometer setting, where $\Delta N_o = 0$ for neutrons taking a mean path through the spectrometer. Let δN_o be the change in N_o induced by inhomogeneities over a different path. Provided δN_o is nowhere large, the arguments which led to (6) may be used to derive

$$< P_z >_M = \int_0^{\infty} I(\lambda)d\lambda \int_{-\infty}^{\infty} S(Q,\omega)d\omega \cdot$$
$$\cdot \int_{-\infty}^{\infty} C(\delta N_o) \cos\left[\omega t(\lambda) + 2\pi\delta N_o\lambda/\lambda_o\right] d(\delta N_o) \tag{15}$$

where $C(\delta N_o)$ is the inhomogeneity distribution and $\int\cdot d(\delta N_o)$ represents integration over this distribution. Since the inhomogeneities are uncorrelated with either $I(\lambda)$ or $S(Q,\omega)$, we may consider this integral separately. On expanding the cosine, the sine integral goes to zero, since positive and negative path differences are equally likely and hence $C(\delta N_o)$ is an even function. The cosine integral is just the cosine transform of the inhomogeneity distribution, and if $C(\delta N_o)$ is fairly sharply peaked about zero, its transform will be a relatively insensitive function of λ. We therefore ignore its λ dependence over the spectral width of $I(\lambda)$, finally obtaining

$$< P_z >_M = H \cdot < P_z > \tag{16}$$

where $< P_z >$ is the value which would be obtained in the absence of inhomogeneities and

$$H = \int_{-\infty}^{\infty} C(\delta N_o) \cos (2\pi\delta N_o \lambda/\lambda_o) \, d(\delta N_o) \tag{17}$$

The inhomogeneity transform H may be measured by using a purely elastic scatterer, for which $< P_z >^s = 1$, where the superscript s refers to the standard elastic scatterer.

We may thus obtain the true $< P_z >$ from the measured values by a simple normalisation to the standard :

$$< P_z > = < P_z >_M / < P_z >_M^s \tag{18}$$

where the subscript M refers to a measured value. Note that the measured values must both be evaluated in the same way ; that is, using either (11) or (13) for both standard and sample. In particular, if only one of the two is a spin-flip scatterer, (13) must be used for both.

Reference to (11) or (13) and (18) shows that the machine parameters P_1P_2 **or** f cancel in this procedure, and hence need not be measured. Equations (13) and (18) provide our general master formula for spin-echo data analysis.

4.2 Bragg Contamination of Inelastic Scattering

The broad wavelength spectrum which may be used in spin-echo means it is often impossible to avoid Bragg contamination in a scattered spectrum. In cases of strong contamination the only practical recourse is to use a filter or monochromator. When the contamination is weak, however, it may be desirable to gain the extra intensity which comes from the broad spectrum, and to correct the data for the Bragg contribution.

Reference to (7a) shows that, if $I(\lambda)$ is broad, $< P_z >$ is quickly damped in an asymmetric scan (see Fig. 3). In contrast, the Fourier components of any δ-functions present in $I(\lambda)$ will continue to oscillate as ΔN_o increases. An asymmetric scan will therefore show Bragg contamination in the form of continuing oscillations at values of ΔN_o where $< P_z >$ would normally be damped to zero. If there is only a single Bragg peak involved, the corresponding cosine component may be measured and subtracted from $< P_z >$.

If several Bragg peaks are involved, the above procedure is less reliable, and polarisation analysis (*always available on a spin-echo spectrometer !*) provides an alternative procedure.

As an example, consider the case of Bragg contamination in an otherwise purely spin-incoherently scattered spectrum. For the incoherent part $\alpha = 2/3$, and we will use a non-spin-flip standard for calibration. Define γ as the ratio of Bragg to total scattering, and note that (12) is valid provided we take

$$\begin{aligned} I^{\uparrow} &= I^{\uparrow}_{inc} = I^{\uparrow}_{M} - I^{\uparrow}_{B} \\ I^{\downarrow} &= I^{\downarrow}_{inc} = I^{\downarrow}_{M} - I^{\downarrow}_{B} \\ I^{E} &= I^{E}_{inc} = I^{E}_{M} - I^{E}_{B} \end{aligned} \tag{19}$$

where M denotes a measured value ; B and inc denote Bragg and incoherent contributions, respectively. Assuming no multiple scattering, the corrected intensities are then given by

$$I^{\uparrow}_{inc} = (1-\gamma)(3-P_1P_2)\ I^{\uparrow}_{M}\ /\ (3+P') \tag{20}$$

$$I^{\downarrow}_{inc} = (1-\gamma)\ [3-P_1P_2(1-2f)]\,I^{\downarrow}_{M}/[3+P'(1-2f)] \tag{21}$$

and

$$I^{E}_{B} = [3\gamma(1+P_1P_2)/(3+P')]\cdot(I^{E}_{s}\ I^{\uparrow}_{M}/I^{\uparrow}_{s}) \tag{22}$$

where $P' = P_1P_2(4\gamma-1)$ and s denotes the (non-spin-flip) standard scatterer. The value of γ is obtained from the measured flipping ratio (P_1P_2 and f being known quantities):

$$R_M = I^{\uparrow}_{M}/I^{\downarrow}_{M} = (3+P')/[3+P'(1-2f)] \tag{23}$$

4.3 Mixed Coherent and Incoherent Inelastic Scattering

As a final example, we consider the case of inelastic coherent scattering in the presence of spin-flip incoherence. This will often occur, for example, in small-angle inelastic scattering from colloidal or polymeric systems.

We first assume that the incoherent scattering is purely spin-flip ($\alpha = 2/3$) and obtain (cf. (11'))

$$I^{E}_{coh}\ /\ I^{\uparrow}_{coh} = (1+P_1P_2< P_z >_{coh})/(1+P_1P_2) \tag{24}$$

$$I^{E}_{inc}\ /\ I^{\uparrow}_{inc} = (3-P_1P_2<P_z>_{inc})/(3-P_1P_2) \tag{25}$$

Now assume that the coherent and incoherent scattering have the same dynamics. Since $< P_z >$ is defined as a function of the dynamics alone (independent of spin-flip), $< P_z >_{coh} = < P_z >_{inc} = < P_z >$. Then

$$I^{E}_{M}\ /\ I^{\uparrow}_{M} = (3+P'< P_z >)/(3+P') \tag{26}$$

with $P' = P_1P_2(4\gamma-1)$, where γ is the ratio of coherent to total scattering. (Note that (23) still applies.) It may be shown from (26) that (12) applies in this case also.

The result may be generalised by extending the definition of γ to be the ratio of non-spin-flip to total scattering. Since spin-incoherent scattering gives 2/3 spin-flip, we have by definition $\alpha = 2(1-\gamma)/3$, where α is the proportion of spin-flip from the sample. Hence $(4\gamma-1) = 3(1-2\alpha)$, and (26) again reduces to (12) or (13).

We therefore conclude that the master formula given by (13) and (18) is general for mixed scattering, provided all scattering processes involved have the same dynamics.

ACKNOWLEDGEMENT

Over the years during which the theory of neutron spin-echo has evolved, I have benefitted from discussions with colleagues too numerous to mention, but I wish to acknowledge particularly A. Heidemann, W.M. Lomer, S.W. Lovesey, F. Mezei, R.L. Mössbauer, J. Penfold, R. Pynn and T. Springer.

REFERENCES

1. Mezei, F. : Z. Physik 255, 146 (1972)
2. Hayter, J.B. : In : Proceedings of the Neutron Diffraction Conference, Petten : Reactor Centrum Nederland Report RCN-234 (1975)
3. Hayter, J.B. : In : Proceedings of the Conference on Neutron Scattering (ed. R. Moon) Gatlinburg : ORNL Report CONF-760601-P2 (1976)
4. Mezei, F. : Physica 86-88B, 1049 (1977)
5. Mezei, F. : In : Proceedings of the International Symposium on Neutron Scattering, Vienna : IAEA (1977)
6. Hayter, J.B. : Z. Physik, B 31, 117 (1978)
7. Hayter, J.B. : In : Neutron Diffraction (ed. H. Dachs) Berlin : Springer (1978)
8. Pynn, R. : J. Phys. E 11, 1133 (1978)
9. Richter, D., Hayter, J.B., Mezei, F. and Ewen, B. : Phys. Rev. Letts. 41, 1484 (1978)
10. Hayter, J.B., Lehmann, M.S., Mezei, F. and Zeyen, C.M.E. : Acta Cryst. A 35, 333 (1979)
11. Hayter, J.B. and Penfold, J. : Z. Physik B 35, 199 (1979)
12. Hayter, J.B. : J. Magn. Mag. Materials (accepted for publication, 1979)
13. Mezei, F. and Murani, A.P. : J. Magn. Mag. Materials (accepted for publication, 1979)
14. Mezei, F. : Nucl. Instr. Meth. 164, 153 (1979)
15. Copley, J.R.D. and Lovesey, S.W. : In : Proceedings of the 3rd International Conference on Liquid Metals, Bristol (1976)
16. Marshall, W. and Lovesey, S.W. : Theory of Thermal Neutron Scattering, Oxford : Clarendon Press (1971)

THE IN11 NEUTRON SPIN ECHO SPECTROMETER

P.A. DAGLEISH, J.B. HAYTER and F. MEZEI

Institut Laue-Langevin,
156X, 38042 Grenoble Cédex,
France

INTRODUCTION

The first full-fledged Neutron Spin Echo spectrometer was built at the Institut Laue-Langevin on the H142 cold neutron guide and it is called IN11. The basic design was made back in 1973 and IN11 was meant to serve as much as an experimental facility for development and testing, as a regular user instrument. Consequently, the only design feature which was pushed to an optimum is flexibility of both the hardware and the software, whereas for all the other features like resolution, neutron intensity etc. safe and fairly inexpensive middle-of-the-road solutions have been adopted. With a large number of magnetic field coils (about 20 used at any one time in order to perform up to 11 different functions) and a mechanical construction which contains no iron within 1 m distance of the beam in order to avoid magnetic disturbances, IN11 is rather particular for a neutron scattering machine. The unusual task of building it was mastered with excellence by the technical services of the ILL, in particular by the teams of Guy Gobert for the mechanics and Yves Lefebvre for the electronics. The instrument turned out to be less expensive than other inelastic scattering machines, mainly because it contains fewer moving parts and nothing was pushed to the technological limits. It was technically completed by early 1977 and after a learning and testing period it became available for users on a full-time schedule by mid-1978.

The general lay-out and operation of IN11 has been described in the paper by Otto Schärpf above. Here we will only give a summary of its main characteristics and a short description of the computer control system. The first of the authors (PAD) is mostly responsible for the computer operating system, the second (JBH) for the programming and the velocity selector, and the third (FM) for the mechanical design. The overall responsibility for the 1.4 M FF project was shared by JBH and FM.

1. SPECIFICATIONS OF IN11

The schematic lay-out of IN11 is shown in Fig. 1. It can be seen that the neutron momentum transfer and resolution is determined by the "background spectrometer", consisting of the velocity selector and the 3.5 m long free flight paths both before and after the sample. This corresponds to a single detector small angle scattering instrument. The smallest usable scattering angle is 2° and the incoming neutron wavelength λ can be varied between 4 and 9 Å. The velocity selector usually produces a $\frac{\Delta v}{v}=18$ % full-width-half-maximum (FWHM) monochromatic beam with a roughly triangular wavelength distribution, and the monochromatization can be changed somewhat, between 10 and 25 %

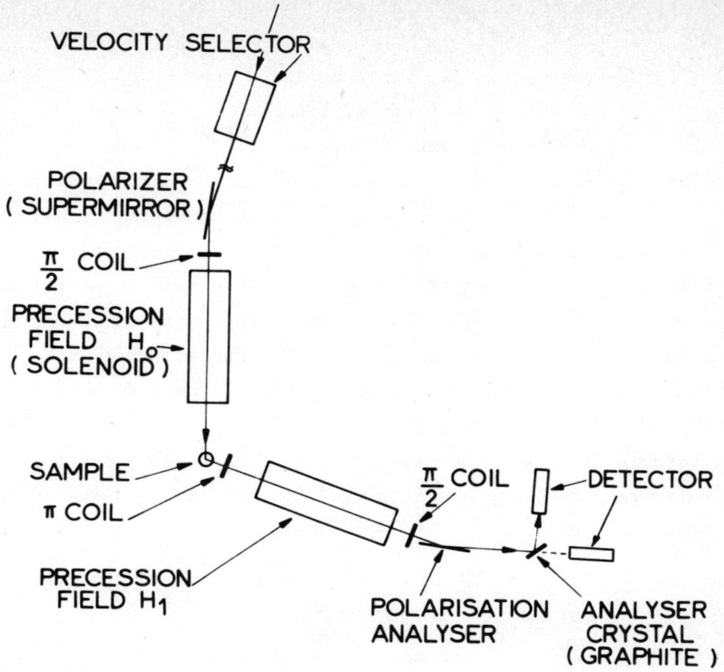

Figure 1.

The schematic lay-out of IN11 Neutron Spin Echo spectrometer. The 6 m distance between the velocity selector and the polarizer is bridged by a circular neutron guide tube. The distance between the polarizer and the sample, and between the sample and the analyser is 3.5 m. The precession field solenoids are 2 m long. In normal operation the optional graphite analyser is removed and the "straight" detector is used.

FWHM, by slightly rotating the axis of the selector drum with respect to the beam direction. If better monochromatization is required, $\frac{\Delta v}{v}$=3 % can be obtained by putting a graphite crystal analyser after the supermirror polarization analyser (cf. Fig. 1).

The H142 cold neutron guide provides a 30x30 mm cross section incoming beam at the velocity selector which is reduced to a Ø 30 mm area by a Braunschweig type 6 m long circular neutron guide tube between the selector and the polarizer. The beam is diaphragmed to Ø 35 mm at the sample and the entrance window of the analyser corresponds to Ø 30 mm cross section. Thus, in view of the 3.5 m free flight paths before and after the scattering, both the incoming and the outgoing beam collimation is 0.5° FWHM, both vertically and horizontally. This collimation is essential for experiments at small scattering angles, but it means a quite small detector solid angle for large angle diffuse scattering studies. The scattering angle ϑ is geometrically limited to a maximum of 140°.

Both of the Larmor precession field solenoids (H_o and H_1 in Fig. 1) consist of three 66 cm long sections with 21 cm internal diameter, which dimensions were found quite reasonable from the point of view of manufacturing and power requirements. Although the solenoids were designed for producing 1.5 kØe maximal field at 48 kW power dissipation, in practice, however, they are most conveniently used between 5 and 500 Øe with activating currents between 2 and 200 Amp, respectively. Table I indicates the corresponding limits of the t time Fourier parameter of the NSE scan at various neutron wavelengths (for convenience 1/t is presented in energy units). The table also contains the available momentum transfer ranges, while the momentum resolution is given by the equation in the bottom line. If necessary, all of the limiting values given here, with the exception of the maximum momentum transfer, can be exceeded by a factor of 2-3 with the help of some specific modifications of the usual spectrometer configuration we are considering (e.g. additional diaphragms, field correction coils etc.).

Table I.

IN11: available energy resolution and momentum transfer

Neutron wavelength λ (Å)	NSE scan limits for 1/t (μeV)		Momentum transfer ($Å^{-1}$)	
	Min.	Max.	Min.	Max
4	0.550	55.0	0.055	2.95
5	0.282	28.2	0.044	2.36
7	0.103	10.3	0.031	1.69
9	0.048	4.8	0.024	1.31
Momentum resolution (FWHM)	$\Delta K = \frac{2\pi}{\lambda}[4(\frac{\Delta v}{v})^2 \sin^2\frac{\vartheta}{2} + 0.0002]^{1/2}$			

Note that the useful 1/t energy range of the NSE scans, as given in the table, applies for line shape studies. On the other hand, if a particular line shape can be assumed, considerably smaller effects can be easily detected. For example the linewidth of a Lorentzian can be measured even if it is 10-20 times less than the lower limits given in second column of the table, provided that 10^4 neutron counts can be collected in the scan. Furthermore, the figures apply for the investigation of both quasi-elastic scattering effects and optical type elementary excitations. (For the latter case provision was made for running the spectrometer at any reasonable $H_o/H_1 \neq 1$ ratio, too.)

The combined polarization efficiency of the supermirror polarizer and analyser system varies between 90 and 97 %, depending on the details of the spectrometer configuration used. For the supermirror the optimal neutron beam reflection angle (cf. Fig. 1) depends on the neutron wavelength, therefore this angle has to be adjusted to the utilized wavelength for both the analyser and the polarizer. One such setting covers a ±20 % wavelength band, thus the 4-9 Å range can be taken care of by three pre-established and easily reproducible settings.

The highest polarized flux, which is obtained between 4 and 5 Å wavelength, amounts to $5 \cdot 10^6$ neutrons/cm^2sec at the sample for 18 % FWHM monochromatization. It gives 1 count/sec detector counting rate for a 5 % isotropic scatterer sample. At $\lambda = 9$ Å the flux is about 8 times less. Background is of order 1 count/min.

2. IN11 CONTROL SYSTEM

The spectrometer is controlled through a single-crate CAMAC system driven by a PDP11-20 (see Fig. 2). The system architecture is based on the DEC RT11-F/B operating system, which provides general utilities such as editing, file-handling and program compilation /assembly/ linking. The system occupies about 5K of the 28K words of memory available, and supports foreground (F/G) and background (B/G) programs in a real-time environment.

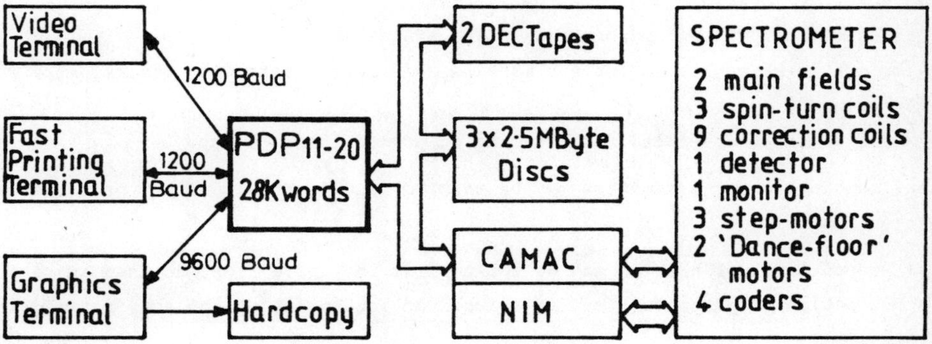

Figure 2.

IN11 Spin Echo spectrometer hardware configuration

The fundamental machine and user interface program, MAUI, runs in the F/G with absolute priority. Asynchronous or synchronous liaison with the B/G program is possible, allowing the latter access, for example, to the current dataset during data collection. MAUI comprises about 32K words of code overlaid into a 13K word region. The main routines in MAUI are written as Fortran modules, with all Macro control routines being Fortran callable.

The B/G area is of order 10K words, in which 15-20K words of overlaid code are typically run. Utility programs, BASIC, data-analysis, graphic and other general programs run in this area. Altogether, typically 50-60K words of code are thus run in the 28K of core available; the system is highly efficient, in that the user is unaware of the overlaying and receives immediate response to all input.

The user controls the machine by direct command. MAUI provides totally free-format command interpretation, so that typing errors are minimized; over 140 plain-language error messages allow easy correction of those errors which occur. Commands are simple English language mnemonics which may always be typed in full for clarity, but which are structured so as to minimize the amount of typing if desired. Comments may be interspersed at any point, and commands may be typed ahead. At present there are 50 basic commands, which are structured at 5 levels:

(i) simple utility commands
- <u>any</u> machine parameter may be set directly (The user never touches the electronics!);
- a single or multiple count may be initiated;
- files may be printed and/or checked for syntax;
- the status of any aspect of the machine and current settings, limits, etc. may be printed;
- messages may be scheduled for printing at preset times of day.

(ii) Any combination of parameters may be scanned simultaneously with a preset count at each setting.

(iii) A second level of nesting may be included in the above loop, so that any other combination of parameters may be scanned at each point of the scan defined in (ii).

(iv) More sophisticated scans may be defined in a JOB file, executed by a RUN command. There may be any number of (named) JOB files. The RUN command may be used to execute only a sub-section of the JOB file, if desired, and the counting times may be modified by a global scale-factor at run-time. Machine settings may be defined by polynomials (as a function of main-field and/or angle), or may be contained in preset tables on disc. Output data may be directed to disc, teletype or both. Comments contained in the JOB file are printed as they occur at run-time.

(v) Any combination of the above commands may be saved in an indirect command file; There may be any number of these (named) files. Execution of such a file replaces input from the user terminal by indirect command file input, so that any conceivable sequence of operations of which the machine is capable may be stored in such a file, and executed by a single command. If RUN commands are included, data-file naming may be made automatic, so that the same indirect command file may be executed repetitively for different samples requiring the same measurement sequence, and data will be uniquely named on disc for each RUN.

The command interpreter in MAUI performs all the obvious limit checking, ensures that necessary JOB files are present and that the data-file name does not yet exist, etc., before allowing a command to be executed. These checks are also performed when JOB or indirect command files are defined. The system is inviolable from the keyboard, as far as is known. (That is, it has so far proved 100 % secure over 3 years' experience.) This factor, plus the simplicity and flexibility of the command structure, makes it particularly easy to learn for the new user.

The B/G is available at all times for data analysis, graphics, and general BASIC or FORTRAN programming. For experiments following a predetermined style of data collection, programs exist to examine data during collection and to analyse it fully and graph it once collected. (Full cursor-controlled graphics are provided.) Fast Fourier analysis is also available as a routine tool. Data may be transferred by DEC-tape to the PDP-10 mainframe computer for further multiple-parameter analysis, if desired.

A particular feature of the data-analysis programs is that only a general sequence of data collection is required for their use, so that the user is free to determine how long and how often each point in the sequence is counted. The data are stored in ASCII files which are readable by the user; analysis programs read data lines as character strings, and decide how to decode each line after examining its textual content, maintaining file readibility between user and machine.

In a typical experiment, the machine is 98.5 % efficient; that is, over a 24--hour period only about 20 minutes are lost in setting, printing and accessing disc, the remaining time being spent counting. From the user's viewpoint, the B/G terminal is available effectively full-time for data-analysis; a user typically has hard-copy graphic output of fully analysed data within 5 minutes of completing a run. The flexibility of the modular software further allows easy inclusion of new ideas, and the system is continuously evolving as new techniques of NSE emerge.

ACKNOWLEDGEMENT

The authors are indebted to many colleagues from the ILL technical services for their generous cooperation and help in both the construction and the maintenance of IN11. It would be difficult to mention them all by name, still, we cannot conclude this paper without paying a tribute to MM. Brancaleone, Chevalier, Faudou, Gobert, Just, Lefebvre, Loppe, Messoumian, Pastor, Thurel and Tshoffen. Finally, special thanks are due to MM. Mössbauer, Jacrot and Lomer, Directors of the ILL at the time, for their decision to launch this novel project.

CHAPTER II:

Neutron Spin Echo Experiments

DYNAMICS OF DILUTE POLYMER SOLUTIONS

L.K. Nicholson and J.S. Higgins

Department of Chemical Engineering,
Imperial College of Science & Technology

and

J.B. Hayter

Institut Laue-Langevin

Neutrons scattered by nucleii undergoing slow motion e.g. the internal motion within polymer chains, lose or gain very small amounts of energy. It is therefore the quasi-elastic region of the neutron scattering spectrum which is of interest and in particular the time correlation function (or intermediate scattering law S(Q,t)) which is ideally required to define the motion.

The neutron spin echo spectrometer (IN11) at the ILL facilitates the measurement of very small energy changes (down to 10 neV)on scattering from a sample, by changing and keeping track of neutron beam polarization non-parallel to the magnetic guide-field (1). The resultant neutron beam polarization, when normalized against a standard (totally elastic) scatterer is directly proportional to the cosine Fourier Transform of the scattering law $S(Q,\omega)$, which is to say the time correlation function is measured directly.

Dilute solutions of deuterated polystyrene (PSD) and deuterated polytetrahydrofuran (PTDF) in carbon disulphide, and of their hydrogenous counterparts (PSH and PTHF respectively) in deuterated benzene were investigated in the range $0.027\ \text{Å}^{-1} \leq q \leq 0.16\ \text{Å}^{-1}$, at 30°C. The molecular weights, radii of gyration (R_g) and concentrations of the samples are listed in Table I.

TABLE I

Polymer	M.W.	R_g	Solvent	Conc %
PSH	50×10^3	∿ 70	C_6D_6	∿ 3.0
PSD	50×10^3	∿ 70	CS_2	0.8 & 2.4
PTHF	60×10^3	∿140	C_6D_6	∿ 3.0
PTDF	60×10^3	140	CS_2	0.8 & 2.4

The types of motion expected for a polymer molecule in dilute solution as the scattered wave vector increases are summarised in Fig. 1. Three main regions are defined by the points $q = 1/R_g$ and $q = 1/a$ (where 'a' is a step-length within the polymer).

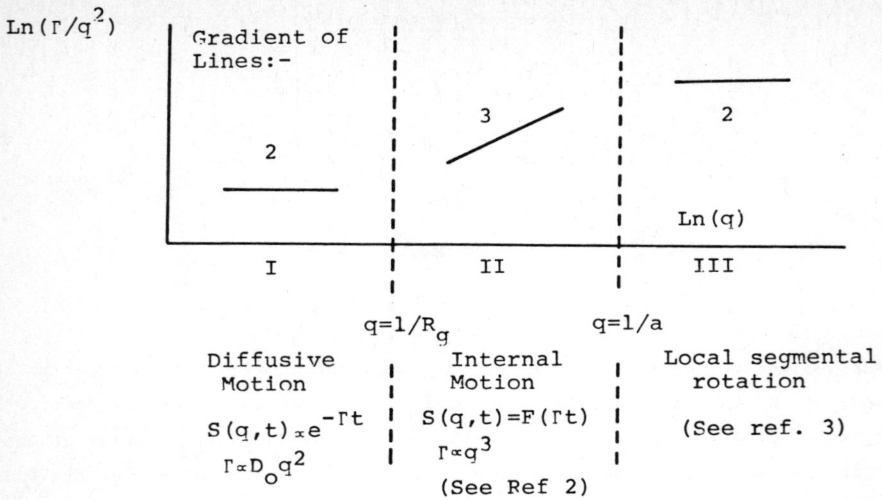

Figure 1 : Schematic diagram to indicate the types of motion expected for a polymer molecule in dilute solution as the scattered wave vector increases.

At low q diffusive motion of the whole molecule is observed but as q increases, distances within the molecule are explored and internal motion dominates. Finally, at high enough q, rotation of individual bonds will become apparent. The low q region is well understood and internal motion, based on a Rouse bead-spring model has been discussed by Dubois-Violette and de Gennes (2) (although not for the good-solvent conditions of our experiments). More recently the transitions from I-II and II-III, and the high q region III have been extensively studied and correlation functions are now available (3).

Our experimental q-range corresponds to the upper end of region II. As a first good approximation therefore the correlation functions obtained were fitted to S(q,t) as calculated by Dubois-Violette and de Gennes (2) for the case of coherent scattering from infinitely long chains in the dilute limit. S(q,t), which is a universal function of the (effectively) normalized time coordinate, (Γt), has the form :

$$S(q,t) = F(\Gamma t) = (\Gamma t)^{2/3} \int_0^\infty \exp(-(\Gamma t)^{2/3} u(1+h(u)))du \qquad \text{(i)}$$

$$\text{where } h(u) = \frac{4}{\Pi} \int_0^\infty \frac{\cos y^2}{y^3} \quad (1 - \exp(-u^{-3/2} y^3))dy$$

The correlation time, which is the only variable to be fitted, is given by : -

$$\Gamma = \frac{\sqrt{3}}{6^{2/3}\Pi} \quad \frac{kT}{\eta} \quad q^3 \qquad \text{(ii)}$$

where η = solvent viscosity.

An example of the fitted raw data is shown in Fig. 2.

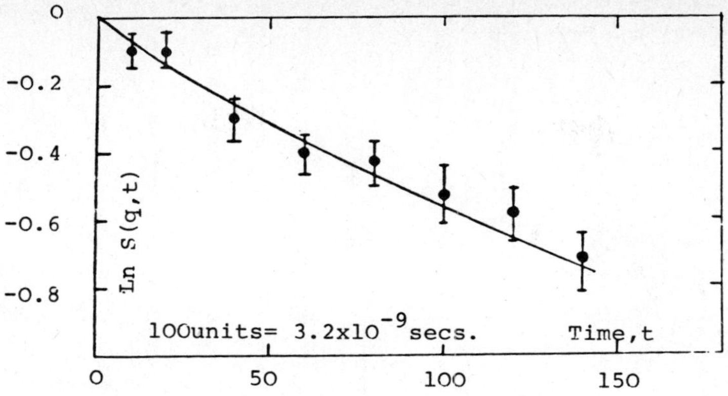

Figure 2 : The figure shows an example of some results obtained from a 3% solution of PTHF in C_6D_6, at q = .082 $Å^{-1}$, the solid line represents the fitted theoretical curve, evaluated from the expression for S(q,t) due to Dubois-Violette and de Gennes (2) for $\Gamma = 1.5 \times 10^8 s^{-1}$.

As may be seen from equation (ii) a q^3 dependence of the correlation time, Γ, is anticipated. Figs (3) and (4) show plots of $\ln(\sqrt{2}\Gamma)$ versus ln(q), and it is seen that in all cases the data exhibits q^3 behaviour at the higher q values measured. The differences in the absolute values of Γ for a given polymer in different solvents reflect the differences in solvent viscosity (η). Within the limits of experimental error, values of η are found to be reasonably good. (See table 2).

Although the higher q data follow the expected behaviour, the values at low q deviate from q^3. It was anticipated that the effect of overall molecular size would not take over as the dominant motion, in the form of q^2 dependent diffusion, until much lower values of q. In the dilute limit the diffusion coefficient, D_o, is related to a hydrodynamic radius, R_H, by the Stokes Einstein equation : -

$$D_o = k_B T / 6\Pi\eta R_H$$

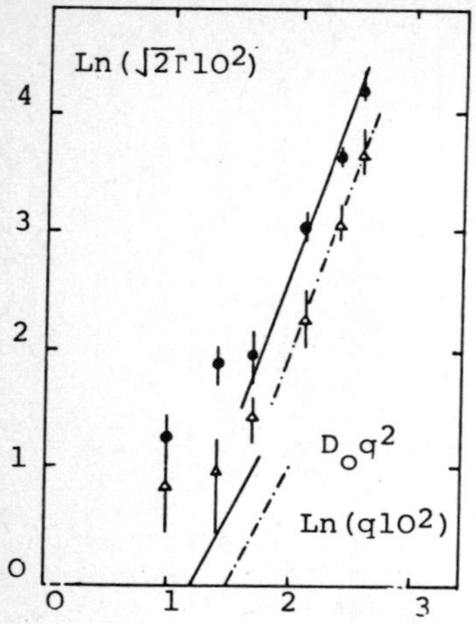

Figure 3

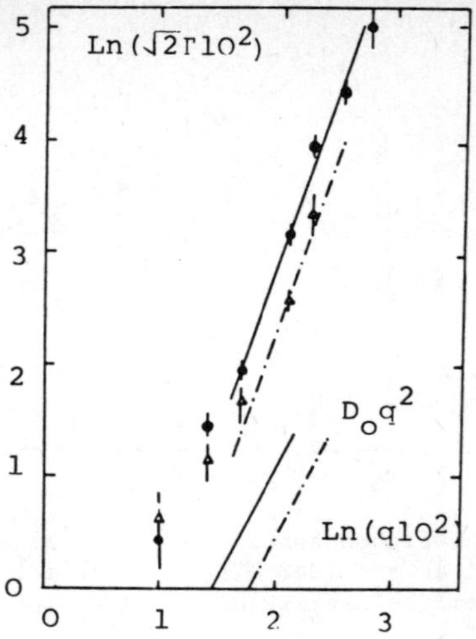

Figure 4

The figures show ln - ln plots of Γ (in μeV) versus q (in $Å^{-1}$) for 3% solutions of : in Fig. 3 • PSD in CS_2 and Δ PSH in C_6D_6 ; and Fig. 4 • PTDF in CS_2 and Δ PTHF in C_6D_6. The lines through the higher data points are of gradient 3, indicating q^3 behaviour of the correlation time Γ. The lines marked D_oq^2 show the expected level for the observation of diffusive motion (see text).

Table 2

Polymer	Solvent	η theor. (cp)	η expl. (cp)
PSD	CS_2	.36	.42
PSH	C_6D_6	.65	.79
PTDF	CS_2	.36	.34
PTHF	C_6D_6	.65	.57

R_H/R_g has been calculated as a function of temperature (4) and has been shown to agree well with results from light scattering experiments (5) for which $qR_g << 1$. However, all values of the ratio R_H/R_g fall within the limits $1 > (R_H/R_g) > 0.5$. As the values of R_g for our samples have been determined independently, it is possible to fix an upper limit for D_o using $R_H = R_g/2$. In figures 3 and 4 it is seen that the experimental points lie well above the anticipated level for diffusive motion. In recent years there have been various theoretical predictions of the concentration dependence of the diffusion coefficient (6,7). Altenberger and Deutch (8) for example have derived the following expression for the concentration dependence of the diffusion coefficient using a hard sphere approximation : -

$$D = D_o(1+2.0\emptyset)$$

where $\emptyset$ is the volume fraction of macromolecules in solution.

This expression indicates that the low q deviations observed in the data might be explained by a concentration dependent interparticle effect of this type, increasing the rate of diffusive motion.

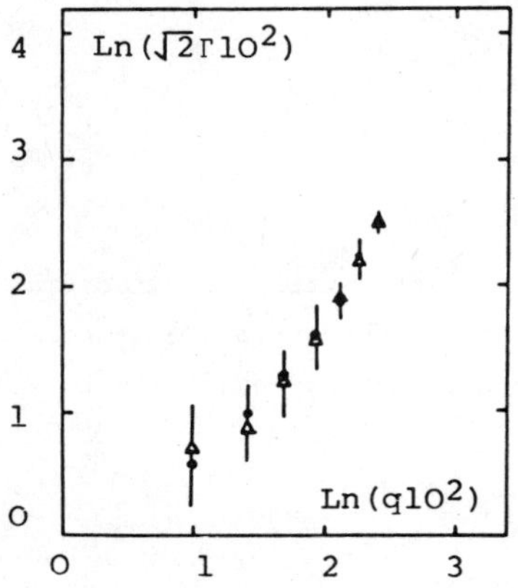

Figure 5

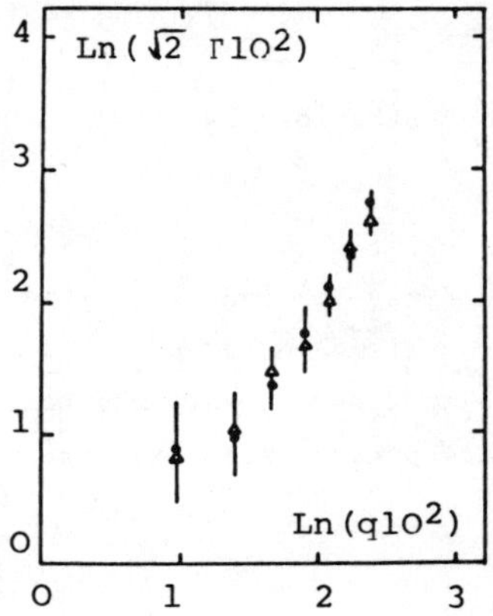

Figure 6

Results obtained for Fig. 5, Δ 2.4% and • 0.8% solutions of PSD in CS_2 and Fig. 6, Δ 2.4% and • 0.8% solution of PTDF in CS_2.

Further experiments were therefore carried out, repeating measurements made at the original concentrations and reducing this concentration by a factor of three. The results depicted in Figs 5 and 6 show no concentration dependence whatsoever, indicating that there is no apparent interparticle interaction. We are forced to conclude that the low q data cannot be accounted for by any physically reasonable motion within the samples.

Further investigations into the significance of this part of the data are to be continued.

REFERENCES

1. These proceedings constitute a general reference to the neutron spin echo technique
2. Dubois-Violette E. and de Gennes P.G. : Physics Vol. 3, No. 4, 181 (1967)
3. Akcasu A.Z., Benouna M. and Han C.C. : to be published in Polymer
4. Akcasu A.Z. and Han C.C. : Macromolecules 12, 276 (1979)
5. Adam, M. and Delsanti M. : Journal de Phys. 37, 1045 (1976)
6. Cummins, H.Z. p. 306 and Pusey, P.N. p. 408 in : - 'Photon Correlation and Light Beating Spectroscopy', Proceedings of the NATO Advanced Study Institute, Ed. H.Z. Cummins and E.R. Pike, Plenum, New York, 1974
7. Pusey, P.N. : J. Phys. A, 11, 119 (1978)
8. Altenberger, A.R. and Deutch, J.M. : Journal Chem. Phys., 59, 894 (1973)

Appendix E. to this volume reports on a similar experiment by Richter et.al.(editor)

DYNAMICS OF MICELLE SOLUTIONS

J.B. HAYTER

Institut Laue-Langevin,
156 X, 38042 Grenoble Cédex, France

and

J. PENFOLD

Neutron Beam Research Unit,
Rutherford Laboratory, Chilton, Oxon., England

ABSTRACT

Neutron and X-ray small-angle scattering have provided extensive structural studies of micellar systems in recent years. Neutron spin-echo now provides a tool to complement these studies with dynamic time-domain measurements. We discuss the use of this technique and present some first results.

1. INTRODUCTION

For many years light scattering, and in particular photon correlation spectroscopy, has been used to study the dynamics and structure of colloidal suspensions, where both the particle sizes and the spatial correlation lengths are of the order of the wavelength of light /1-3/. More recently, attention has been focussed on macroionic solutions in which the dominant species are micelles ; that is, clusters of charged surface-active molecules (see, for example, /4/).

The characteristic dimensions in these systems are of order 50 - 100 Å, and small-angle X-ray and neutron scattering are appropriate to their study. The contrast variation available with neutron scattering is particularly valuable, since the experimental conditions may be chosen so that only the micelles show any contrast against the average solution background.

Until recently, these small-angle scattering studies have been purely structural. The measurements are often severely hindered by interference between the single particle structure factor $S_o(Q)$ and the structure factor $S(Q)$ corresponding to intermicellar correlations.

Dynamic effects between particles, on the other hand, depend on interparticle correlations and hence only involve $S(Q)$; they are therefore of considerable assistance in understanding static structure, apart from their intrinsic dynamical interest. Neutron spin-echo /5/ now provides, for the first time, sufficient energy resolution to undertake such dynamical studies in neutron small-angle scattering. We present in this paper an introduction to this use of neutron spin-echo, and show the first experimental results.

2. THEORETICAL CONSIDERATIONS

The long-range nature of Coulomb forces gives rise to long-range spatial correlations between ions in electrolyte solutions, even at low volume fraction. If one of the ionic species is large and highly charged compared with all the others, these *macroions* will dominate the solution properties, the remaining ions merely forming a neutralising background which determines the screening of the potential and sets the zero of energy for the system.

Although the time-averaged distribution of ions (and therefore charge) around a given (spherical) macroion in an isotropic solution will be symmetric, local concentration fluctuations which break this symmetry mean that the macroion experiences strong fluctuating forces. These produce instantaneous accelerations which contribute to the macroion diffusion. Since the fluctuating forces result from perturbations of local structure, we expect their average effect to be related to the radial distribution function g(r) around the central ion. In consequence, the macroionic diffusion coefficient D will depend on the time-averaged structure S(Q) of the solution.

Micelles self-assemble from ionic solution when the concentration of surface-active molecules exceeds a critical micellar concentration (cmc), which is typically of order 10^{-2} M. Intrinsically, therefore, micellar solutions are always concentrated, in the sense that the average interparticle separation is of the order of the range of the potential, and there is always interaction between particles. The usual practice of assessing the single-particle scattering by extrapolation to infinite dilution is therefore unreliable, since the micelles dis-assemble below the cmc. This complicates the structural study of micelle solutions by small-angle scattering, because the scattered intensity depends on the single particle (infinite dilution) structure factor $S_o(Q)$ as well as on the inter-micellar structure factor S(Q) ; for spherical micelles the scattered intensity has the form /6/

$$d\sigma/d\Omega = K.\ S_o(Q).\ S(Q) \tag{1}$$

where K is a contrast factor. Even when the micelle structure and hence $S_o(Q)$ is known, $S_o(Q)$ becomes extremely small in these systems at Q values beyond the first peak in S(Q) ; the latter is therefore difficult to extract reliably from the data.

The dynamics of a macroion system may be characterised by two relaxation times /7/ ; τ_B, corresponding to the rapidly fluctuating Brownian motion, and a much longer interaction time τ_I, which corresponds roughly to the time a particle takes to move a distance of order the interparticle separation. If measurements are made in a time-domain régime where $\tau_B \ll t \ll \tau_I$, Pusey /7,8/ has shown that the initial decay of the normalised coherent intermediate scattering function F(Q,t) is given by

$$\begin{aligned} dF(Q,t)/dt &= -D_oQ^2/S(Q) \\ &= -D_{eff}Q^2 \end{aligned} \tag{2}$$

where the effective macroion diffusion constant is

$$D_{eff} = D_o/S(Q) \tag{3}$$

and D_o is the (infinite dilution) Stokes-Einstein diffusion coefficient, typically $< 10^{-6}\ cm^2 s^{-1}$ for micellar systems. (The analogy between (2) and the familiar de Gennes narrowing is discussed in /7,9/.) When hydrodynamic effects become important, (3) becomes more complicated due to the solvent "backwash" from one particle moving another. A first estimate /9/ gives

$$D_{eff} = D_o\,[1 + H(Q)]/S(Q) \tag{4}$$

where $H(Q) = 0.75\,[S(Q) - 1]$.

For micellar systems, where interparticle interaction may be strong even at small volume fraction, (3) is expected to be valid near the cmc, where hydrodynamic effects should be small. Measurement of $D_{eff}(Q)$ then gives direct access to the interparticle interactions, and the range of the interaction potential may be varied independently of macroion concentration by changing the ionic strength of the solution /10/ (adding a neutral salt, for example). In combination with static measurements, this allows, in particular, determination of the charge on the macroions, since a reasonable analytic model of S(Q) is now available for DLVO potentials /11/. Measurements at higher concentrations then provide an estimate of the strength of hydrodynamic effects.

Neutron spin-echo permits direct time-domain measurement of the normalised intermediate scattering function F(Q,t). (For an introduction, see /12/.) For micellar systems, τ_B is typically $\sim 10^{-12}$s and $\tau_I \sim 10^{-7}$s, so that the spin-echo measurement time-scale $10^{-10} < t < 10^{-8}$s satisfies the requirements of (2). Further, effects such as excitation of spherical vibrational modes /13/ or exchange due to interparticle encounters /14/ are calculated to be at frequencies which are too high or too low to interfere with the measurement.

Under these conditions, the polarisation measured in a symmetric neutron spin-echo scan is given to a good approximation by

$$\langle P_z \rangle = \exp\,(- D_{eff}\, Q^2\, t) \tag{5}$$

Since all of the usual contrast variation techniques are also available, neutron spin-echo is thus ideally suited to these measurements.

3. EXPERIMENTAL

The well-characterised system /4/ sodium dodecyl sulphate (SDS)/water was chosen for a first experiment, since SDS micelles are known to be spherical over a wide concentration range /15/ and the single particle structure factor $S_o(Q)$ is fairly accurately calculable. (It is interesting to note, however, that although this system is amongst the most studied in the literature, there is still no agreement on details such as the aggregation number in the micelle or the micellar charge.) Solutions were prepared in D_2O (to maximise contrast) at concentrations of 0.04, 0.1, 0.2, 0.4, 0.6, 0.8 and 0.9M. (For SDS at 298 K, cmc = 8×10^{-3}M.) A second series of solutions was prepared at the same concentrations in 0.1 M NaCl/D_2O. The Debye screening wavevectors for the two series were κ = 0.03 and κ = 0.1 $Å^{-1}$, respectively.

Measurements were performed at two temperatures (303 K and 313 K) on the ILL spin-echo spectrometer IN11, using the $\bar{C}F_yC$ configuration /12/.

Three types of measurement were made :

(i) A 2θ-scan in the range $2° \leq 2\theta \leq 14°$ ($0.03 < Q < 0.2$ $Å^{-1}$) to measure $d\sigma/d\Omega$ (2).

(ii) An *asymmetric* spin-echo scan /16/ at each of eight 2θ values in the above range. (2θ = 2, 3, 4, 6, 8, 10, 12 and 14°.) This allowed determination of the effective scattered wavelength spectrum, and hence the corresponding Q value, at each angle.

(iii) A *symmetric* spin-echo scan at each 2θ value to measure the effective diffusion constant at the corresponding Q. From (5), D_{eff} is obtained from the slope of a log-plot of $< P_z >$ *versus* t ; the plot will be linear provided there is no interference from effects with different relaxation times.

The incident beam had mean wavelength λ_o = 8.2 Å and $\Delta\lambda/\lambda_o$ = 0.4. Total measurement times (for the eight Q values) were of order 12 - 18 hours per sample. Inhomogeneity corrections /12/ were made by normalising the data to the polarisation measured on a standard elastic scatterer : deuterated polystyrene (polyethylene) in a hydrogenous polystyrene (polyethylene) matrix.

4. RESULTS AND DISCUSSION

Figure 1 shows typical 2θ scans for low and high concentrations, each at two ionic strengths. The effect of changing the screening of the potential is clearly seen ; in particular, the compressibility (given by the zero-Q limit) increases as the screening increases, in agreement with theoretical prediction /11/. The loss of intensity beyond the peak is due to the single particle structure factor $S_o(Q)$, as discussed in section 2.

It is this structure in $d\sigma/d\Omega$, due to interparticle interference, which makes the mean scattered wavelength a function of scattering angle, and necessitates the asymmetric spin-echo scans to determine the corresponding Q values when the incident spectrum is broad.

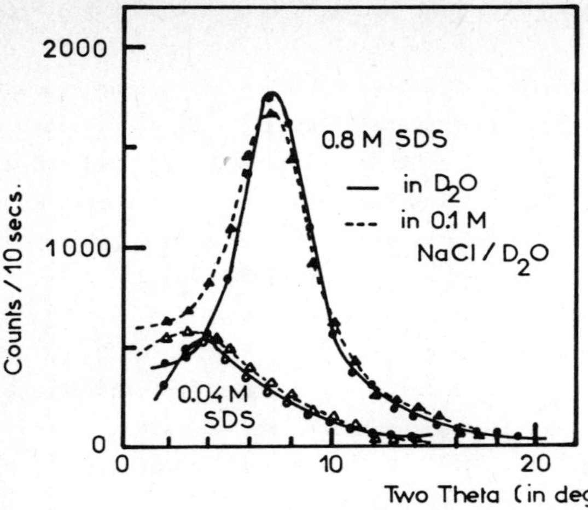

Fig. 1. Scattered intensity from SDS at two concentrations. —— in D_2O. ---- in 0.1 M $NaCl/D_2O$.

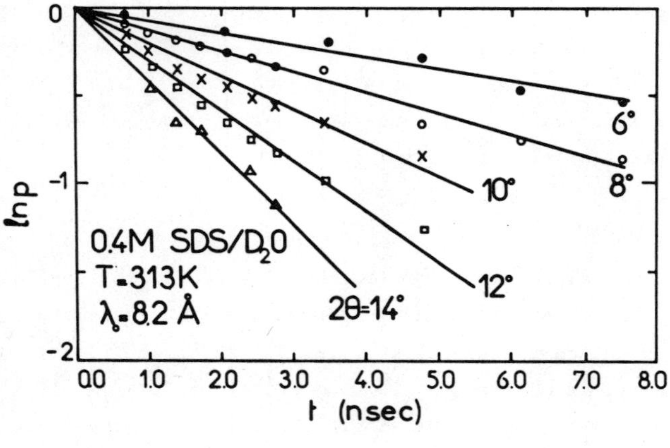

Fig. 2. Logarithm of the spin-echo polarisation as a function of t at several scattering angles 2θ. The lines correspond to pure exponential decay.

Symmetric scan neutron spin-echo data for an intermediate concentration is presented in Fig. 2. The polarisation was calculated allowing for incoherent spin-flip processes, as described in /12,17/. (The incoherent contribution was weak, and the 2/3 spin-flip means that its contribution to the spin-echo signal is in any case highly damped.) Errors have not been shown for reasons of clarity. The absolute error in the polarisation is about 0.04 ; at this level it is only statistically limited, and could be improved with longer counting times. The fact that all curves extrapolate to unit polarisation at the origin provides a global check on the consistency of the data.

The linearity of Fig. 2 (at this statistical level) confirms the validity of (5). This provides a convenient (and more conventional) form of data presentation. Equation (5) is Fourier transformed, yielding a Lorentzian scattering law, and the Lorentzian half-width γ is presented as a function of Q^2.

Figure 3 shows the inelastic data corresponding to the higher concentration of Fig. 1, presented in this form. The effective diffusion coefficient is $D_{eff} = \gamma/Q^2$; Lorentzian widths for typical values of D_o are also presented. $S_{max}(Q)$ is the momentum transfer at which the macroionic solution structure factor is calculated to be a maximum /11/ ; note that this will not in general coincide with the maximum in $d\sigma/d\Omega$. The correlation between D_{eff} and the solution structure is evident.

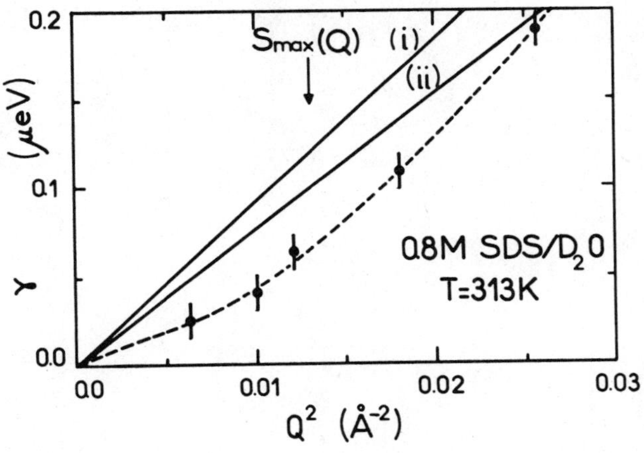

Fig. 3. Lorentzian half-width γ (μeV) as a function of momentum-transfer squared ; (i) and (ii) correspond to Stokes-Einstein diffusion for particles of hydrodynamic radii 25 Å and 30 Å, respectively.

These are only preliminary results from a first experiment, and it would be premature to discuss them in detail at this stage. We believe, however, that they demonstrate the utility of neutron spin-echo time-domain measurements in the study of micellar systems, where the corresponding quasielastic broadenings may be only a few tens of nano-eV. A more detailed study of the SDS system is in progress.

ACKNOWLEDGEMENTS

We have benefitted from discussions with J.M. Loveluck, S.W. Lovesey, R.H. Ottewill, P. Schofield and J.W. White. The standard elastic samples were kindly provided by B. Ewen, J.S. Higgins, K. Nicholson and D. Richter.

REFERENCES

1. Berne, B.J. and Pecora, R. : Dynamic Light Scattering, New York : Wiley (1976)
2. Pusey, P.N. : In : Photon Correlation and Light Beating Spectroscopy (eds. H.Z. Cummins and E.R. Pike) New York : Plenum (1974)
3. Brown, J.C., Pusey, P.N., Goodwin, J.W. and Ottewill, R.H. : J. Phys. A 8, 664 (1975)
4. Tanford, C. : J. Phys. Chem. 78, 2469 (1974)
5. Mezei, F. : Z. Physik 255, 146 (1972)
6. Kostorz, G. : In : Neutron Scattering in Materials Science (ed. G. Kostorz) New York : Academic (1978)
7. Pusey, P.N. : J. Phys. A 8, 1433 (1975)
8. Pusey, P.N. : J. Phys. A 11, 119 (1978)
9. Ackerson, B.J. : J. Chem. Phys. 64, 242 (1976)
10. Verwey, E.J.W. and Overbeek, J.Th.G. : Theory of the Stability of Lyophobic Colloids, Amsterdam : Elsevier (1948)
11. Hayter, J.B. and Penfold, J. : To be published.
12. Hayter, J.B. : Theory of Neutron Spin-Echo : These proceedings.
13. Schofield, P. : In : Proceedings of the Int. Conf. on Neutron Scattering, Vienna : IAEA (1977)
14. von Smoluchowski, M. : Z. Phys. Chem. 92, 129 (1917)
15. Reiss-Husson, F. and Luzatti, V. : J. Phys. Chem. 68, 3504 (1964)
16. Hayter, J.B. and Penfold, J. : Z. Physik B 35, 199 (1979)
17. Hayter, J.B. : Z. Physik B 31, 117 (1978)

TENTATIVE USE OF NSE IN BIOLOGICAL STUDIES

Y. ALPERT

Institut de Biologie Physico-Chimique
13, rue Pierre et Marie Curie, 75005 Paris, France

INTRODUCTION

There is increasing interest in internal dynamics of biological molecules and supermolecular assemblies which are now assumed to participate in functional mechanisms. Internal dynamics of large molecules comprises many types of motion of different time scale, duration, space localization. Several methods are used to study the various types of phenomena, among which spectroscopic methods prevail. Clotting of milk, allosteric transitions between two extreme stable conformations as illustrated by oxyhemoglobin to deoxyhemoglobin conformational change, give examples of transient phenomena which in these two cases are internal molecular rearrangements. The non-transient, equilibrium motions extend over a wide range of frequencies, say 0 to 10^{14} sec^{-1}. Localized vibrational modes of individual bonds occur at 10^{13} - 10^{14} sec^{-1} and are a matter for infrared and Raman spectroscopy. The usual NMR range (10^{-7} to 10^{-12} sec -1) typically allows observation of aliphatic side-chain motion. Of greater biological importance are those motions which imply larger fractions of a molecule. They may be vibrations of the peptide backbone as a whole, radial vibrations of a globular protein figured out as a pulsating sphere, periodical oscillations of a subunit relative to the others, fluctuations of the angle between two helical segments of a protein, periodical deformations of a closed structure which facilitate ligand penetration towards its binding site, or other collective motions. Hvidt & Nielsen(1) studying hydrogen exchange on native proteins in solution, suggested that a "breathing" of the molecules occurs during which atoms of the interior are transiently exposed to the solvent. Frauenfelder et al(2) and Sternberg et al(3) report on structural fluctuations in the crystalline state; these observations arise from static crystallographic studies hence they have no bearing on time scale determination. There is no report on experimental evidence of similar conformational fluctuations in the liquid state. However a theoretical study by Mc Cammon et al(4) provides the magnitude, correlations and decay of fluctuations about the average structure of a folded globular protein, bovine pancreatic inhibitor. These authors suggest that internal motions in the folded protein have a fluid-like nature at ordinary temperatures, and they conclude that the fluctuations in atomic positions are on a time scale $t > 1 \times 10^{-12}$ sec . This is indeed the only numerical value that has been associated to biomolecular internal dynamics (if we omit studies of transient conformational changes). The lack of both theoretical predictions and experimental measurements of the frequencies of collective motions of large molecules leads the author to simply assume that they have low-frequency modes which should be found between 0 and 10^9 sec^{-1}. This assumption is consistent with the failure of previous attempts to observe these movements and with Mc Cammon and coworkers' estimate.

Because it has a high energy resolution at small momentum transfer, the Neutron Spin Echo (NSE) spectrometer[(5)] IN11 recommends itself to a tentative approach to the study of collective motions in biology. We have so far considered three systems representative of different kinds of biological objects: a protein, a virus, a cell. After a short presentation of the samples and of the experimental conditions, we will describe and discuss the results of the experiments.

1. SURVEY OF THE EXPERIMENTAL CONDITIONS

A. Sample description

The three systems differ in size, shape, structure, complexity, and biological function.

Hemoglobin (Hb) the wellknown oxygen carrier is a globular, tetrameric protein of about 60 Å in diameter, which exists in two stable conformations depending on whether it bears oxygen or not. The relative positions of the subunits change on going from the oxygenated to the deoxygenated state as does the number of contacts and salt-bridges between them.

Bromegrass Mosaic Virus (BMV) is a small, spherical, RNA virus which also exists in two structural states: it has a compact form (diameter = 260 Å) at low pH and a swollen form (diameter = 300 Å) at higher pH. In compact BMV the protein capsid is closed around the RNA core whereas holes within the capsid become accessible to RNA in the swollen virus. A similar shape transition occurs in many small RNA viruses but their functional relevance is unknown. In the search for an understanding of the virus mechanism: how do viruses liberate their RNA and become infectious, finding a difference of rigidity of the two states would be a valuable information.

Red cells (from human blood) are much bigger, flat-disk shaped objects with a diameter of 7 - 8 μ with a thickness of 2.4 μ at the periphery and 1 μ at the center. The cells contain mainly hemoglobin at a concentration of 35 % (w/v). Experiments were undertaken with the hope of "viewing" Hb dynamical behaviour inside the cells in its physiological surroundings and at its physiological concentration.

B. Sample handling

All the samples have been prepared in or dialyzed against buffers of D_2O. Unless otherwise specified, standard procedures have been followed and will not be described here.

Hemoglobin concentration was normally adjusted to 100-120 mg/ml but measurements have also been performed at 160 mg/ml and 40 mg/ml. Deoxyhemoglobin was prepared by adding sodium dithionite to oxyhemoglobin. Hb samples were in 0.1 M phosphate buffer at pD = 7.2 .

BMV (kindly given by B. Jacrot) was 40 to 90 mg/ml. Compact BMV was prepared in 5 mM acetate buffer at pD = 5.4 , swollen BMV in 0.1 M phosphate + 5 mM EDTA at pD = 7.5 .

Red cells (free of heparine) carefully washed several times with 0.9 % NaCl were simply transferred to the experimental cell.

In these experiments we used ordinary quartz cells 5 or 6 mm thick, and the sample temperature was usually 22°C.

C. Experimental procedure

For full details on the principles and experimental specifications of NSE the reader is referred to the preceeding chapters in this volume and to the original papers reproduced in its Appendix. We will just recall the very practical way of using this technique: how to collect and evaluate the data.

Neutron scattering reflects molecular motions through the variation of the positions of the scattering particles. Analysis of the scattering spectra $S(Q,\omega)$, i.e. determination of the position, shape and width of the lines, usually allows identification of the nature of these motions. Q is the magnitude of the scattering vector $Q = \frac{4\Pi}{\lambda} \sin\theta/2$, and θ is the scattering angle.

In the NSE experiment the measured quantity is the final polarization of the scattered neutron beam $P(Q,H)$ as a function of the applied magnetic field H . $P(Q,H)$ is proportional to the real part of the intermediate scattering function $S(Q,t)$. Hence it is the Fourier transform of the scattering law which is directly obtained and $P(Q,H)$ behaves as a relaxation process versus time (the magnetic field is converted to a time variable through a conversion factor which is determined by instrumental specifications).

A large, complex molecule in solution which suffers internal collective motions also performs diffusive displacements through the solvent and these diffusion processes also interact with the neutron beam. The final energy spectrum or rather here the polarization variation will contain the contributions of all types of motions.

2. RESULTS AND DISCUSSION

All the data are normalized to polystyrene taken as a standard purely elastic scatterer. Blanks were recorded with buffer in the cell and accounted for in the analysis though their effect was almost negligible. The scattered intensity was measured at fixed Q as a function of the guide field intensity H.

The polarization is plotted in a logarithmic scale. In this way a linearity of the plot (ln P vs. H) would indicate the lorentzian nature of the scattering law with an exponential Fourier transform $\exp(-\Gamma t)$ of width Γ expressed in energy units.

A. Hemoglobin

Fig. 1 shows the final polarization P observed with oxy- or deoxyhemoglobin at various scattering angles. Both states of the protein give comparable measurements at scattering angles θ between 2° and 10° (12° and 15° in some runs) giving Q values between 0.026 and 0.132 $Å^{-1}$ at an incident wavelength of 8.3 Å . The data fit reasonably well with an exponential decay function at all angles. The lorentzian widths Γ are obtained from the slopes of the ln P lines.

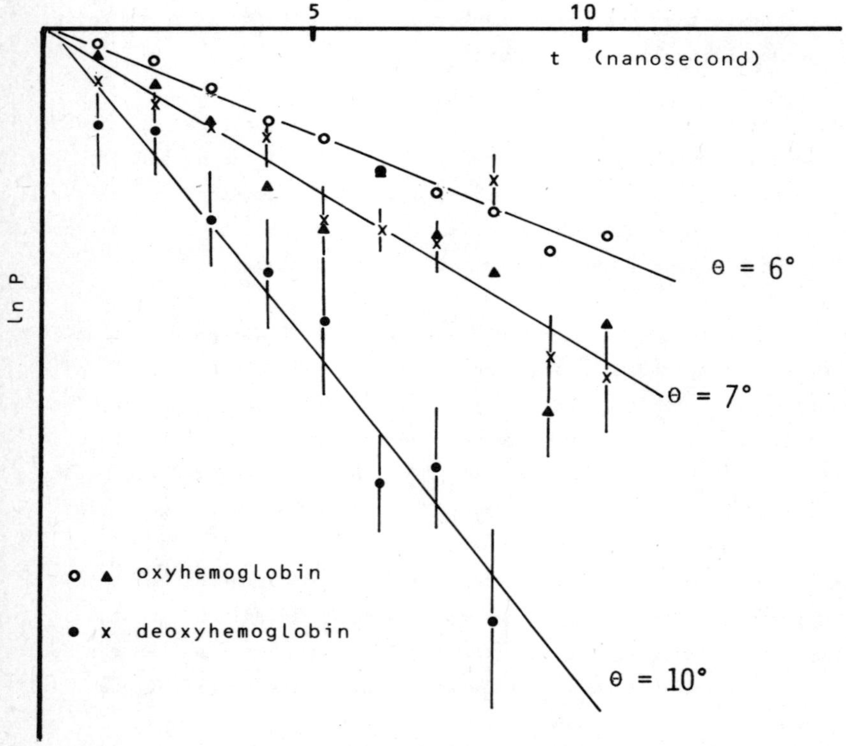

Fig. 1 Final polarization P of the scattered neutron beam for hemoglobin at various angles. For clarity, only some of the errors bars are shown. The time values have been calculated from magnetic field intensity. DeoxyHb cannot be distinguished from oxyHb.

The linewidth Γ is proportional to the squared scattering vector Q^2 at angles larger than 6° at concentration c = 160 mg/ml , and larger than 4° at concentration c = 120 mg/ml. At smaller angles the ratio Γ/Q^2 increases considerably (Fig. 2). Hemoglobin solutions show inelastic scattering of the neutron beam with exponential decay and linewidth proportional to Q^2 : this behaviour is characteristic of pure diffusive translation of the scattering particles with a macroscopic diffusion constant $D = \Gamma/Q^2$.

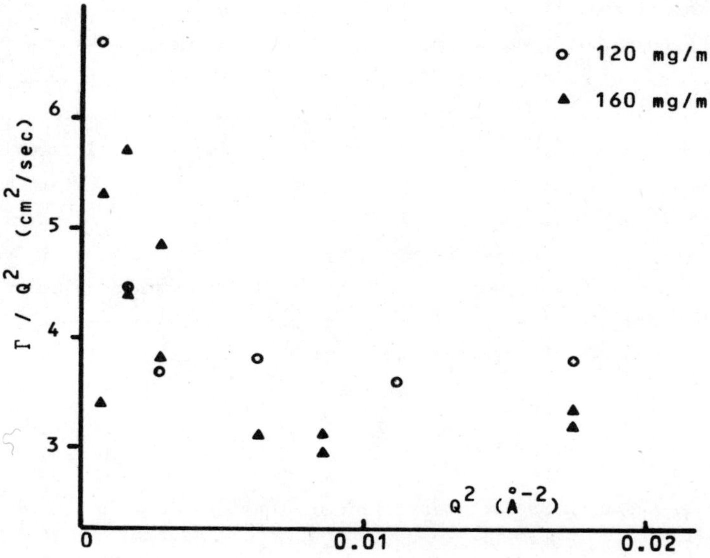

Fig.2 Effective diffusion constant $D = \Gamma/Q^2$ as a function of Q^2

Table I

Θ°	4	5	6	7	8	9	10	
Γ (neV)	6.7		15.7		26.5		43.2	120 mg/ml
D (cm^2/sec)	3.7		3.8		3.6		3.8	
Γ (neV)	9.1		12.8	17.0			37.4	160 mg/ml
D (cm^2/sec)	4.9		3.1	3.0			3.2	

(Some of the numerical values are presented in this Table, all of them are plotted in Fig.2)

The diffusion constant of hemoglobin is known from various methods of which we consider here quasielastic light scattering measurements[6,7] which have been extended up to very high concentrations. The results, including those obtained from macroscopic methods, are in good agreement in the infinite dilution limit, giving $D_o \simeq 7 \times 10^{-7}$ cm^2/sec at 20°C in water. Jones et al[6] find a linear variation of D with concentration, valid up to 350 mg/ml. According to Veldkamp and Votano[7] a much faster decay of D is observed at high concentrations. However these two groups indicate nearly identical values in our concentration range, which we can use as a comparison with our own data. After correction for the viscosity of D_2O instead of H_2O the predicted diffusion constant is 5.2×10^{-7} cm^2/sec at c=120 mg/ml and 5.0×10^{-7} at c=160 mg/ml. From our measurements, averaging the results from 4° at 120 mg/ml and from 6° at 160 mg/ml we find (see Table I) smaller numbers : $\Gamma/Q^2 = D = 3.7 \times 10^{-7}$ and 3.0×10^{-7} cm^2/sec at 120 and 160 mg/ml, respectively.

The drop in diffusion constant in D_2O is probably indicative of intermolecular interactions or aggregations which would be more important than in ordinary water at the same concentrations. This analysis might be easily checked by performing new quasielastic light scattering experiments with hemoglobin solutions in D_2O in the same concentration range. The increase of D at small Q is more difficult to understand. It might actually be an artifact, since the relative accuracy of the data is getting worse towards the smallest scattering angles used.

B. Bromegrass Mosaic Virus

The results obtained with BMV are less clear as they differ according to the series of measurements.

Due to the size of BMV (260 to 300 Å), diffusive translation of this molecule is slow enough and cannot produce an observable inelastic scattering of the neutrons at angles smaller than 6°, and did not affect our measurements which were all performed at scattering angles between 2 and 6° where the intensity is more favourable (the first scattered intensity minima were found at 2.25° and 2° as expected for spherical particles of 260 Å and 300 Å in diameter, respectively). Compact BMV produced almost pure elastic scattering in all cases. Swollen BMV gave inelastic, Q-dependent scattering in one set of measurements, and almost purely elastic scattering in the other series (the shift in the positions of the intensity extrema proved that the diameter had augmented and the virus was really swollen). We do not know presently which result should be more reliable as in one case, the instrumental resolution was not at its best, whereas in the other case a phaseshift problem occurred during the measurement of the swollen virus.

C. Red Cells

Due to its large size, a cell as a whole is seen by the neutron beam as a macroscopic, motionless object and we were tempted to look at hemoglobin diffusion inside the cells in order to get informations on the molecular behaviour and molecular informations. Though very preliminary, the results are worth of enough interest to be mentioned here.

At scattering angles between 3° and 9°, inelastic scattering was observed and the experimental spectra show exponential decay. The linewidth has a constant value $\Gamma = (4.6 \pm 0.2)$ neV from 5° to 7°, a very small value (1.6 neV, very inaccurate) at smaller angles, and from 8° upwards Γ increases proportional to Q^3 . The region of constant linewidth corresponds to a flat part of the angular distribution of scattering. These data reflect an organized structure of the protein molecules within the erythrocytes. Two remarks could be made here:
(i) to the author's knowledge this brings the first experimental evidence for this intracellular protein organization in normal blood (in sickle cells it is known that a number of Hb molecules aggregate to form fibers)
(ii) looking at the linewidth value of 4.6 neV we notice that the NSE spectrometer energy resolution was definitely required for the experiment to be successful.

CONCLUSION

Of these three attempts to use NSE in biological studies, we conclude that the NSE spectrometer provides a high enough sensitivity to successfully observe inelastic scattering or quasielastic scattering from biological solutions. The existence and frequency measurement of collective motions did not appear in these experiments. The most prominent result concerns the observation of an organized liquid arrangement of molecules inside the red cells of normal human blood. This suffices to prove that NSE spectrometry is a promising tool in the field of biology.

I wish to thank F. Mezei for his helpful collaboration and fruitful discussions all throughout this work. I acknowledge J. Hayter and O. Schaerpf who participated at various stages of the experiments, and B. Jacrot who really initiated this work.

REFERENCES

(1) A. Hvidt and S.O. Nielsen, Adv. Prot. Chem. 21, 287 (1966)
(2) H. Frauenfelder, G.A. Petsko and D. Tsernoglou, Nature 280, 558 (1979)
(3) M.J.E. Sternberg, D.E.P. Grace and D.C. Phillips, J. Mol. Biol. 130, 231 (1979)
(4) J.A. McCammon, B.R. Gelin and M. Karplus, Nature 267, 585 (1977)
(5) F. Mezei, in Neutron Inelastic Scattering 1977 (IAEA, Vienna, 1978) p. 125
(6) C.R. Jones, C.S. Johnson, Jr., and J.T. Penniston, Biopol. 17, 1581 (1978)
(7) W.B. Veldkamp and J.R. Votano, J. Phys. Chem. 80, 2794 (1976)

A PROPOSAL FOR DETERMINING h/m_n WITH HIGH ACCURACY

W. Weirauch, E. Krüger and W. Nistler
Physikalisch-Technische Bundesanstalt,
Bundesallee 100, D-3300 Braunschweig,
Federal Republic of Germany

Abstract

A precision determination of h/m_n (h Planck constant, m_n neutron mass) would improve our knowledge of the fundamental constants. It can be carried out by measuring the wavelength and the velocity of reactor neutrons. Several authors have proposed methods for this experiment. Here, a method is described in which a beam of polarized neutrons is modulated by periodically changing the direction of the polarization vector. First experimental tests of the modulation principle are reported.

1. Introduction

In the following, an experiment proposed for determining the ratio h/m_n (h Planck constant, m_n neutron mass) is described. A relative uncertainty of about $1x10^{-6}$ is expected. h/m_n has not so far been precisely measured.

The recommended values of the fundamental constants which are published every few years are the results of least-squares adjustments. The measured values of constants and of combinations of these are the input parameters. The last adjustment was carried out by Cohen and Taylor [1)] in 1973. h/m_n can be calculated from the values of h and m_n given by these authors. Proceeding from the corresponding variance-covariance matrix, a relative uncertainty of $1.7x10^{-6}$ can be calculated for the result. This is about the same as expected for the measurement. Nevertheless, the experiment would be of great importance for altering the status of h/m_n from that of a deduced quantity to that of an additional input parameter of the least-squares adjustment. The reliability of the adjustment could thus be clearly improved.

The wavelength λ and the velocity v of a neutron are related by the de Broglie equation:

$$h/m_n = \lambda v \ . \quad (1)$$

Hence, h/m_n can be determined by measuring both the wavelength and the velocity of neutrons.

Several methods have been proposed for this experiment, in particular by Stedman [2], Mezei [3] and Weirauch [3-5]. In all of them, λ shall be determined by Bragg reflection of the neutrons in a most perfect single crystal, and v shall be measured by a time-of-flight method. However, various techniques have been proposed for carrying out the velocity measurement with the desired high accuracy.

The method finally proposed by Weirauch [5] is briefly described in this paper. Furthermore, first measurements carried out for testing the feasibility of the experiment are here reported.

2. The proposed experiment

2.1. The principle of the measurement

For measuring the velocity of the neutrons, a beam of polarized neutrons is modulated by periodically changing the direction of the polarization vector. After a path of several meters the neutrons are backscattered by a silicon single-crystal. They thus pass the modulator for a second time, and the beam is again modulated. The modulator defines the beginning as well as the end of a flight path. The flight time follows from the phase difference between the two modulations.

The wavelength of the neutrons is determined by the Bragg reflection in the silicon crystal. Silicon should be used because almost perfect crystals can be grown from this material and its lattice parameter is known with the required small uncertainty [6-8].

2.2. The beam of polarized neutrons

A polarized and monochromatic neutron beam is produced, for example, by the reflection in a polarizing crystal (see fig. 1). The polarization vector $\vec{p}$ is assumed to be parallel to a magnetic guide field (magnetic induction $\vec{B}_o$).

The neutrons traverse a $\pi/2$ coil C1, the axis of which is perpendicular to $\vec{B}_o$ and to the axis of the neutron beam. When they leave the coil, $\vec{p}$ is parallel to the coil induction $\vec{B}_{C1}$. Coils of this type have first been used by Mezei [9].

$\vec{B}_{C1}$, the beam axis and $\vec{B}_o$ define the x, y and z coordinates given in fig. 1.

The neutrons traverse the modulating assembly (No.3) and, after several meters, a second $\pi/2$ coil. Both these arrangements should be disregarded for the moment.

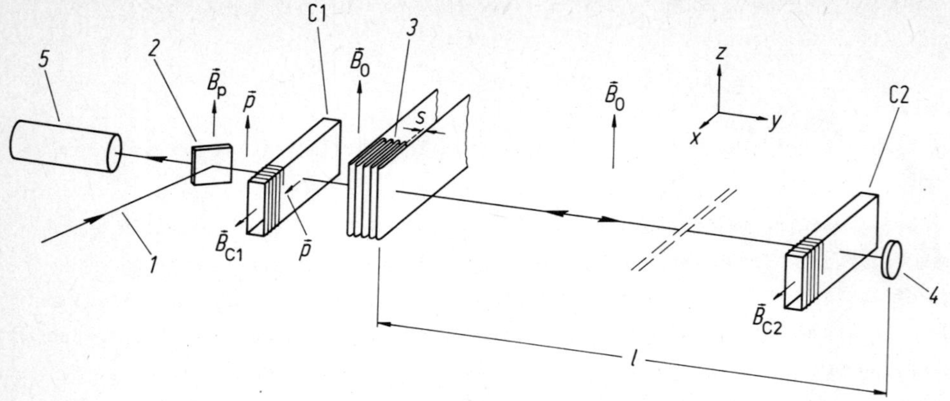

Fig. 1. Arrangement for measuring h/m_n. (1) neutron beam entering the apparatus, (2) polarizing crystal, (3) meander coil, (4) silicon single-crystal, (5) detector. (C1,C2) $\pi/2$ coils. $\vec{B}_o$ magnetic induction of the guide field, $\vec{B}_p$ magnetic induction magnetizing the polarizing crystal, $\vec{p}$ polarization vector in case the meander coil is not operating.

The neutrons are then reflected in the silicon crystal mentioned above and return to coil C1.

Along the entire path, from leaving coil C1 to reaching it again, $\vec{p}$ precesses in the guide field. B_o and the distance between coil C1 and the silicon crystal have to be such that the entire precession angle is an integer multiple of 2π. Hence, $\vec{p}$ is again parallel to the x axis when the neutrons enter coil C1 for the second time.

In coil C1, $\vec{p}$ is again turned into the direction of the z axis. The neutrons are then incident on the polarizing crystal. For simplicity, this is assumed to operate ideally, i.e. its polarization efficiency and its reflectivity for neutrons with the optimal spin direction are to equal 1. Thus all the neutrons are reflected and none reach the detector.

2.3. Modulating the neutron beam

The assembly used for modulating the neutron beam consists of an aluminium sheet, bent to form a meander (see fig. 1, No.3). It is therefore referred to as meander coil in the following.

The meander coil forms a close sequence of single coils, each having one turn. If an electric current is passed through them, the magnetic inductions in neighbouring coils are antiparallel. If alternating current is used, there is a phase difference of π between the magnetic inductions in neighbouring coils. An arrangement of this type, supplied with direct current, was used by Drabkin et al. [10] for the construction of a velocity filter for polarized neutrons.

To modulate the direction of $\vec{p}$, the meander coil is supplied with high-frequency alternating current. The coil field is parallel to the guide field. The thickness s of the single coils is such that the neutrons pass each of them in half a period of the alternating current. As a consequence, the neutrons find the same variation of the magnetic induction in all the single coils.

$\vec{p}$ precesses in the meander coil due to two parallel magnetic fields, the guide field and the coil field. The first part of the precession has already been taken into account. Hence, only the precession caused by the coil field is considered in the following.

The magnetic inductions in the single coils are assumed to vary sinusoidally with time. As a consequence, the Larmor frequency of $\vec{p}$ also varies sinusoidally with time, while the neutrons traverse a single coil. The angle by which $\vec{p}$ precesses during the passage depends on the time at which the neutrons enter the coil.

The precession angle is the same in all the single coils. In the whole meander coil it adds up to the angle

$$\Phi(t) = \hat{\Phi} \cos 2\pi\nu t , \qquad (2)$$

$$\hat{\Phi} = 4\mu_n \hat{B}_m N/h\nu \qquad (3)$$

(ν frequency of the electric current, t time, μ_n magnetic moment of the neutron, $\hat{B}_m$ amplitude of the magnetic induction in the meander coil, N number of single coils of which the meander coil consists).

In principle, the modulation could be produced by only one single coil. This case has been discussed in a previous paper [11] but it cannot be applied here, as follows from eq. (3): in the proposed experiment $\hat{\Phi} \geqslant 0.6\pi$ and $\nu \approx 1$ MHz is required. Because $\hat{B}_m$ is limited for technical reasons to a few mT, these values can only be achieved with $N \gg 1$.

2.4. Measuring the velocity of the neutrons

When the neutrons leave the meander coil for the first time, the direction of $\vec{p}$ is modulated as given by eq. (2). $\vec{p}$ is at any time in the xy plane (see fig. 1).

Returning from the silicon crystal, the neutrons traverse the meander coil for a second time; the beam is again modulated. The total modulation depends on the phase difference between the two modulations and thus on the flight time of the neutrons.

After the second modulation, the neutrons traverse coil C1 in which $\vec{p}$ is turned into the yz plane. The modulation is then such that $\vec{p}$ changes its direction within this plane. As a consequence, a time-dependent neutron current traverses the polarizing crystal and reaches the detector. Here, the mean neutron current $\bar{I}$ is measured.

$\bar{I}/I_0$ has been calculated as a function of the flight time, using the parameter $\hat{\Phi}= 0.6\ \pi$. I_0 is the neutron current after the reflection in the silicon crystal. The flight time is proportional to the distance l between the meander coil and the silicon crystal. Hence, $\bar{I}/I_0$ can also be given as a function of this quantity: fig. 2.

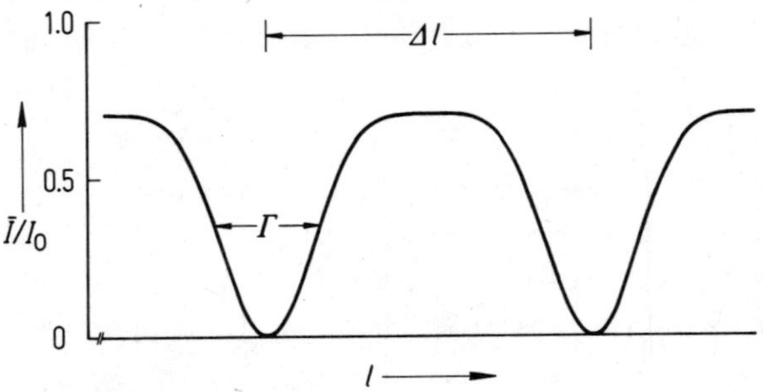

Fig. 2. Calculated ratio $\bar{I}/I_0$ as a function of the distance l. Parameter: $\hat{\Phi} = 0.6\ \pi$.

Going from a minimum of $\bar{I}(l)/I_0$ to the next one, the flight path is enlarged by $2\Delta l$ and the phase difference by 2π. From the latter, it follows that the flight time is enlarged by one period of the modulation. The velocity of the neutrons is therefore:

$$v = 2\Delta l \nu. \tag{4}$$

To find Δl with sufficient precision, an integer multiple of it has to be measured. For this purpose, two minima of $\bar{I}(l)/I_0$ should be determined, the l values of which differ by several meters. For altering l, the meander coil has to be moved.

2.5. Data suited for the experiment

The wavelength $\lambda = 0.25$ nm is well-suited for the experiment. It corresponds to the (331) reflection in the silicon crystal. $\nu = 1.0$ MHz is a suitable modulation frequency. The thickness of the single coils then has to be s = 0.8 mm.

Δl equals s. The full width of the dips at half-minimum of $\bar{I}(l)/I_0$ is $\Gamma = 0.3$ mm (see fig.2). The l values of the two minima used in the experiment will probably differ by about 10 m. They then have to be determined with an uncertainty of about 3% of Γ in order to measure h/m_n with a relative uncertainty of 1×10^{-6}.

2.6. Eliminating long-term instabilities of the guide field by spin echo

In section 2.2 it is required that $\vec{p}$ precesses by an integer multiple of 2π on the way from coil C1 to the silicon crystal and back again. If this condition is not met, the maximum and minimum values of $I(1)/I_0$ differ in general by less than shown in fig. 2. The precession angle and thus the guide field have therefore to be constant during the entire measuring time, which will run into many weeks. This difficulty can be avoided by means of spin echo. But because of the backscattering in the silicon crystal, the experimental arrangement has to be different from the one used by Mezei [9] in his original spin-echo experiment.

For producing the spin echo, the additional $\pi/2$ coil C2 is required. It is adjusted in the same way as coil C1.

On the way from coil C1 to coil C2, $\vec{p}$ precesses by the angle

$$\phi_1 = 2n\pi + \alpha \qquad (-\pi < \alpha \leqslant \pi) \tag{5}$$

due to the guide field. n is an integer. α gives the angle between $\vec{p}$ and the x axis. The direction of $\vec{p}$ at the surface of coil C2 is shown in fig. 3, a.

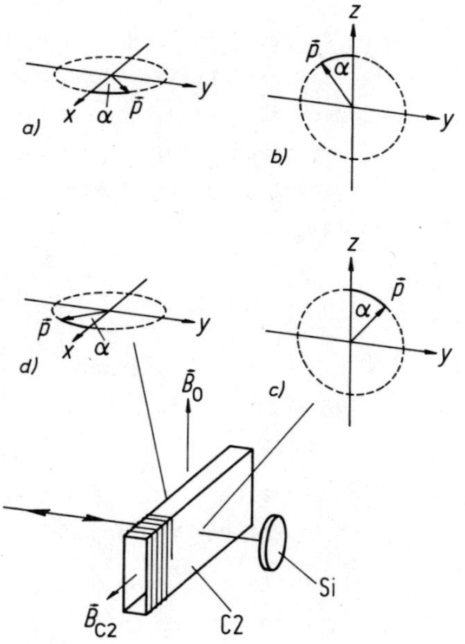

Fig. 3. Polarization vector $\vec{p}$ at the surfaces of the $\pi/2$ coil C2, (a,b) before the reflection in the silicon crystal Si, (c,d) after the reflection.

In coil C2, $\vec{p}$ is turned into the yz plane. Its component p_y is reversed. α becomes the angle between $\vec{p}$ and the z axis (see fig. 3,b).

Coil C2 is positioned at such a distance from the silicon crystal that $\vec{p}$ precesses by a small odd multiple of π while the neutrons traverse this distance twice. The precession results in the changing of the sign of α (see fig. 3,c).

While the neutrons traverse coil C2 for the second time, $\vec{p}$ is turned back into the xy plane. p_y is once more reversed (see fig. 3, d).

In fig. 3,d α has the opposite sign, as compared with fig. 3, a. Hence, ϕ_1 has been replaced by

$$\phi_2 = 2n\pi - \alpha. \tag{6}$$

On the way from coil C2 back to coil C1, $\vec{p}$ precesses again by ϕ_1. Between leaving and entering coil C1, it therefore precesses altogether by

$$\phi = \phi_2 + \phi_1 = 4n\pi. \tag{7}$$

Hence, $\vec{p}$ has the same direction at the beginning and at the end of the path, independent of α .

B_o must only be constant during the time the neutrons need for traversing twice the distance between coil C1 and the silicon crystal. Long-term stability is needed only where the $\pi/2$ coils are positioned and in the space between coil C2 and the silicon crystal. However, the requirements are low: the neutrons are in these regions for such short times that a variation of the Larmor frequency causes only a small change in the precession angle.

3. Experimental test of the modulation method

A meander coil consisting of N = 99 single coils was used for testing the modulation method. It was made of aluminium foil 60 µm thick. The single coils were of a thickness of s = 1.1 mm.

The measurements were carried out at a beam hole of the reactor in Braunschweig. Neutrons with the wavelength λ = 0.25 nm entered the apparatus shown in fig. 4.

By the reflection in the Heusler crystal H1, a beam of polarized neutrons was produced. $\vec{p}$ was turned into the xy plane in the $\pi/2$ coil C1. After traversing the meander coil, the neutrons entered a second $\pi/2$ coil (C2) in which $\vec{p}$ was turned into the yz plane.

The Heusler crystal H2 was for analysing the polarization of the modulated beam. The reflected neutrons were incident on a BF_3 counter which measured the mean neutron current $\bar{I}^{*}$.

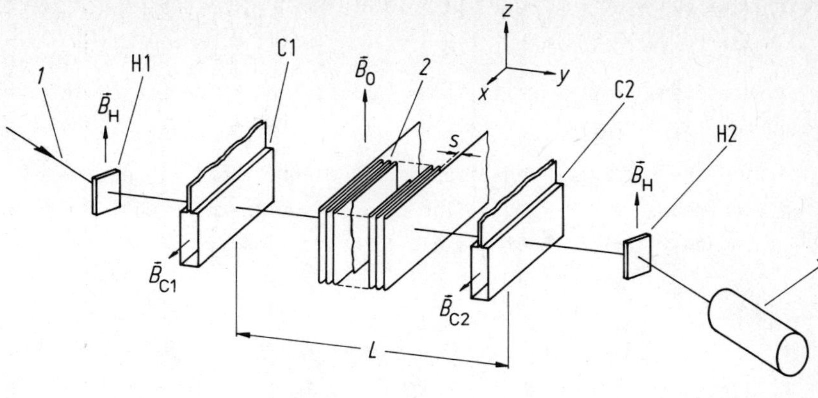

Fig. 4. Arrangement for testing the modulation method. (1) monochromatic neutron beam entering the apparatus, (2) meander coil, (3) BF_3 counter. (H1,H2) Heusler crystals, (C1,C2) $\pi/2$ coils. $\vec{B}_0$ magnetic induction of the guide field, $\vec{B}_H$ magnetic induction magnetizing a Heusler crystal.

The distance L between coils C1 and C2 was adjusted so that $\vec{p}$ precessed on this way by an integer multiple of 2π due to the guide field . $\bar{I}^*$, measured without modulation, is then maximum, and the modulation reduces it.

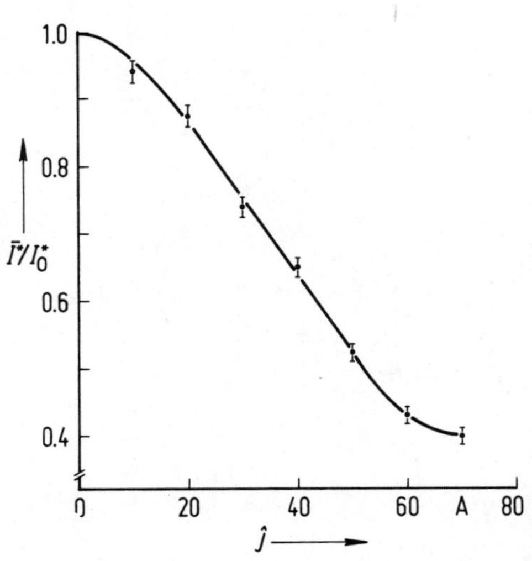

Fig. 5. Ratio $\bar{I}^*/I_0^*$ measured as a function of the amplitude $\hat{J}$ of the electric current in the meander coil. Modulation frequency : ν = 743 kHz.

$\bar{I}^*$ was measured as a function of the amplitude $\hat{J}$ of the electric current in the meander coil. The frequency was ν = 743 kHz. The result is shown in fig. 5, giving

$\bar{I}^*/I_o^*$ instead of $\bar{I}^*$. I_o^* is the neutron current measured without modulation. From the lowest value of $\bar{I}^*/I_o^*$ it can be concluded that a modulation amplitude of $\hat{\Phi} = 0.9\pi$ was achieved. This value is larger by 50% than the one used in calculating the curve in fig. 2.

For testing the resonance behaviour of the meander coil, $\bar{I}^*$ was measured as a function of the frequency ν : fig. 6. The curve shown in the figure has been calculated by fitting only the depth of the dip.

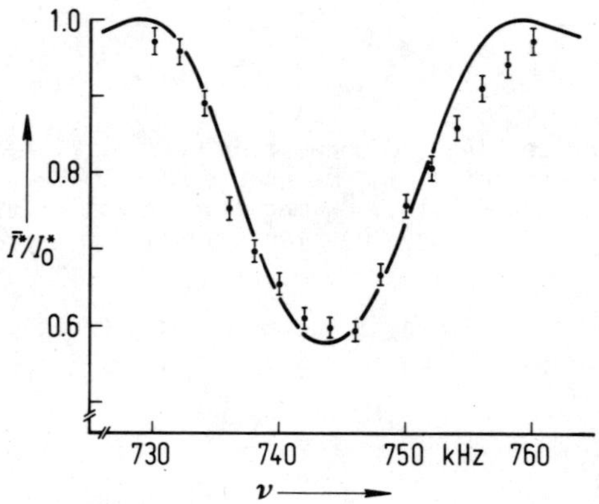

Fig. 6. Ratio $\bar{I}^*/I_o^*$ measured as a function of the modulation frequency ν. $\hat{J}$ = 40 A.

4. Conclusion

The method given here seems to be well-suited for determining h/m_n with high accuracy. It will therefore be used in carrying out the experiment at the high-flux reactor of the Institut Laue-Langevin in Grenoble.

Acknowledgements.

We should like to thank Messrs. W. Berne, E. Gelbke und H.-J. Trauscheid for their help in preparing and carrying out the measurements.

References

1. Cohen, E.R.; Taylor, B.N.: J. Phys. Chem. Ref. Data 2 (1973) 663-734.
2. Stedman, R.: J. Sci. Instr. (J. Phys. E) 1, ser. 2 (1968) 1168-1170.
3. Mezei, F.; Weirauch, W.: Research proposal No. 01-003 R (1975), Institut Laue-Langevin, Grenoble.
4. Weirauch, W.: Nucl. Instr. and Meth. 131 (1975) 111-117.
5. Weirauch, W. in: Fundamental Physics with Reactor Neutrons and Neutrinos, Inst. Phys.Conf. Ser.No. 42 (ed. T. von Egidy; The Institute of Physics, Bristol and London, 1978) 47-52.
6. Deslattes, R.D.; Henins, A.: Phys. Rev. Lett. 31 (1973) 972-975.
7. Deslattes, R.D.; Henins, A.; Bowman, H.A.; Schoonover, R.M.; Carroll, C.L.; Barnes, I.L.; Machlan, L.A.; Moore, L.J.; Shields, W.R.: Phys. Rev. Lett. 33 (1974) 463-466.
8. Deslattes, R.D.; Henins, A.; Schoonover, R.M.; Carroll, C.L.; Bowman, H.A.: Phys. Rev. Lett. 36 (1976) 898-900.
9. Mezei, F.: Z. Physik 255 (1972) 146-160.
10. Drabkin, G.M.; Trunov, V.A.; Runov, V.V.: Zh. Eksp. Teor. Fiz. 54 (1968) 362-366 (Soviet Physics JETP 27 (1968) 194-196).
11. Weise, K.; Mehl, A.; Weirauch, W.: Nucl. Instr. and Meth. 140 (1977) 269-274.

NEUTRON SPIN ECHO STUDY OF SPIN GLASS DYNAMICS

A.P. MURANI and F. MEZEI

Institut Laue-Langevin, 156X, 38042 Grenoble Cedex

Measurements of the time dependent spin correlation function S(κ,t) for a Cu-5 at. % Mn spin glass alloy have been performed over the time range $10^{-12} < t < 10^{-9}$ sec using, for the first time, a combination of the neutron spin echo and polarisation analysis techniques. The results contribute to our knowledge of spin glass dynamics, showing most clearly the evolving spectrum of relaxation times spreading over a wide range and extending into very long times with decreasing temperature.

Among the experimental techniques applied to study the spin dynamics of spin glass systems /1/, neutron scattering is the most powerful providing direct measurements of the imaginary part of the susceptibility χ"(κ,ω) or equivalently the scattering law S(κ,ω) The recent measurements on metallic 3d spin glass systems such as CuMn performed with the time-of-flight technique show that at high temperatures and in the dilute limit the Mn spins undergo rapid relaxations due to Korringa coupling with the conduction electrons. For alloys containing a few percent of solute atoms, the measurements of S(κ,ω) performed at various temperatures from well above to below the spin glass freezing temperature T_{sg} (i.e. the temperature of the ac susceptibility peak) show clear evidence of continuous evolution of correlations among the spins accompanied by the development of a whole spectrum of slow relaxation times with decreasing temperature, extension of the measurements into the μeV region, using the high resolution back scattering spectrometer (IN10) have enabled determination of the spectral weights in various regions of the relaxation spectrum /2/ although even the very high energy resolution of that spectrometer still limits information to correlation times shorter than $\sim 10^{-9}$ sec.

In the following we report measurements of spin dynamics of a Cu - 5 at. % Mn alloy using for the first time a combination of the neutron spin echo and polarisation analysis techniques which together permit direct measurement of the intermediate scattering law S(κ,t) over nearly 3 decades of time in the range $10^{-12} < t < 10^{-9}$ sec in a single experimental set up /3/. Furthermore the use of polarised neutrons enables measurements of purely magnetic scattering, eliminating all possible uncertainties associated with the previous unpolarised neutron experiments. Although the results obtained for the intermediate scattering law are completely consistent with the previous measurements they provide, nevertheless, some essential refinement to our knowledge of spin glass dynamics enabling, for example, direct comparison with theoretical model calculations of the time dependent spin correlation function for these systems.

The principle of the neutron spin echo technique was described several years ago /4/ and has been rediscussed in the first article. The method involves the measurement of neutron polarisation after scattering in the original direction of polarisation, say along the x-axis. This quantity gives directly the real part of the scattering law $S(\kappa,t)$ since,

$$\langle P_x \rangle = P \int_{-\infty}^{\infty} S(\kappa,\omega) \cos \omega t \, d\omega \Big/ \int_{-\infty}^{\infty} S(\kappa,\omega) d\omega = P \cdot \mathrm{Re}\, S(\kappa,t) \Big/ S(\kappa,0) \qquad (1)$$

where P is the incident polarisation, $S(\kappa,\omega)$ is the scattering law. The time t is given by $t = \pi n/E_o$, with E_o the incident energy and n the number of precessions carried out by the incident neutrons in the guide field.

For isotropic spin systems such as spin glasses $S(\kappa,t)$ is purely real so that a direct determination of the time correlation function can be performed by the spin echo technique. Furthermore, because of the isotropic distribution of the spins both in the paramagnetic and the spin glass states, the spin flip scattering from the sample can be used directly to achieve the spin echo without the use of the spin flipper coil at the sample necessary for measurements on non-magnetic systems. This has the advantage that only magnetic scattering is measured in the spin echo signal, eliminating possible uncertainties associated with separation of magnetic and non-magnetic scattering contributions. We discuss below how this can be achieved.

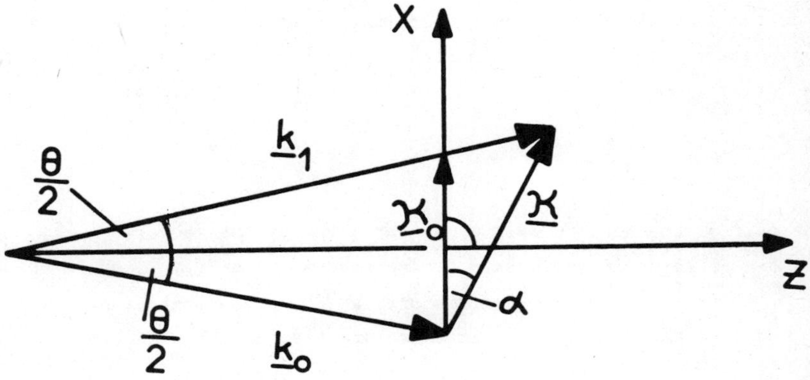

Fig. 1 The scattering diagram in the x-z plane. Neutrons are incident along $\underline{k}_o$ and are scattered through an angle θ along $\underline{k}_1$. In the diagram $\underline{\kappa}_o$ and $\underline{\kappa}$ represent the elastic and inelastic momentum transfer vectors, with an angle α between them.

The scattering diagram is shown in Fig. 1. The neutrons incident with wave vector $\underline{k}_o$ are scattered through an angle θ, with outgoing momentum vector $\underline{k}$, and wave vector transfer $\underline{\kappa}$. The axes are chosen such that the direction x is along the elastic scattering vector q_o and the z-axis bisects the angle θ ; y -axis is perpendicular to the plane of the diagram.

Thus, a neutron polarised initially along the x direction traverses a magnetic field in which it carries out n precessions in the x-y plane and arrives at the sample making an angle ϕ with the x-axis. The scattering by the sample which involves both spin flip and non spin flip processes causes a change of the spin direction as may be seen by examining the equation for the scattered beam polarisation $\underline{P}'$ given by /5 ,6 /

$$\underline{P}' = - \hat{\underline{\kappa}} \, (\hat{\underline{\kappa}} \cdot \underline{P}) \qquad (2)$$

where $\underline{P}$ is the initial polarisation. For elastic scattering this relation can also be expressed very simply by means of two thumb rules namely :

a) only the atomic spin components perpendicular to the scattering vector are effective in scattering neutrons, and

b) the atomic spin components perpendicular to the neutron spin cause spin flip scattering whereas those parallel to the neutron spin give non-spin-flip scattering.

Thus, since the scattering vector is in the plane perpendicular to the y-axis the y component of the neutron spin becomes completely depolarised due to equal amounts of spin flip and non-spin-flip scattering.

$$\text{i.e.} \quad P'_y = 0 = \frac{1}{2} P_y - \frac{1}{2} P_y \qquad (3)$$

also since the energy changes involved are rather small, the scattering is nearly elastic and the scattering vector is parallel to the x-axis, we have therefore spin flip scattering from both the y and z components of the atomic spins hence,

$$P'_x = - P_x = - \frac{1}{2} P_x - \frac{1}{2} P_x \qquad (4)$$

Thus, one half of the neutron spin component given by the vector $\underline{P}_1{}' = (\frac{1}{2} P_y , - \frac{1}{2} P_x)$ undergoes $\phi \rightarrow -\phi$ spin flip necessary for spin echo , see Fig. 2, while the other half given by $\underline{P}'_2 = (- \frac{1}{2} P_y, - \frac{1}{2} P_x)$ suffers only a 180° phase shift in the precession and does not give the echo. We obtain therefore a spin echo signal with a maximum amplitude of $\frac{1}{2}$, without using the spin flipper at the sample.

The isotropic magnetic character of the spin glass enables a very useful extension of the time range of measurement of $S(\kappa,t)$ to fairly short times by means of simple polarisation analysis of the scattered beam. This can be seen in the following way. For an inelastic scattering event the scattering vector $\underline{\kappa}$ makes an angle α with the elastic scattering vector $\underline{\kappa_o}$ so that the final polarisation of the beam measured in the initial polarisation directions x, y, z (carried out sequentially) will be given according to equation 2 by

$$P'_x = -P\cos^2\alpha \qquad \text{for } \underset{\sim}{P} \,//\, x$$

$$P'_z = -P\sin^2\alpha \qquad \text{for } \underset{\sim}{P} \,//\, z \qquad (5)$$

$$\text{and } P'_y = 0 \qquad \text{for } \underset{\sim}{P} \,//\, y$$

Evidently, the angle α is directly related to the inelasticity of the scattering, hence the measurements of the average polarisations $\langle P'_x \rangle$ and $\langle P'_z \rangle$ can provide a measure of the inelasticity as shown by Maleev /6/ and Drabkin et al. /7/. In fact, as shown

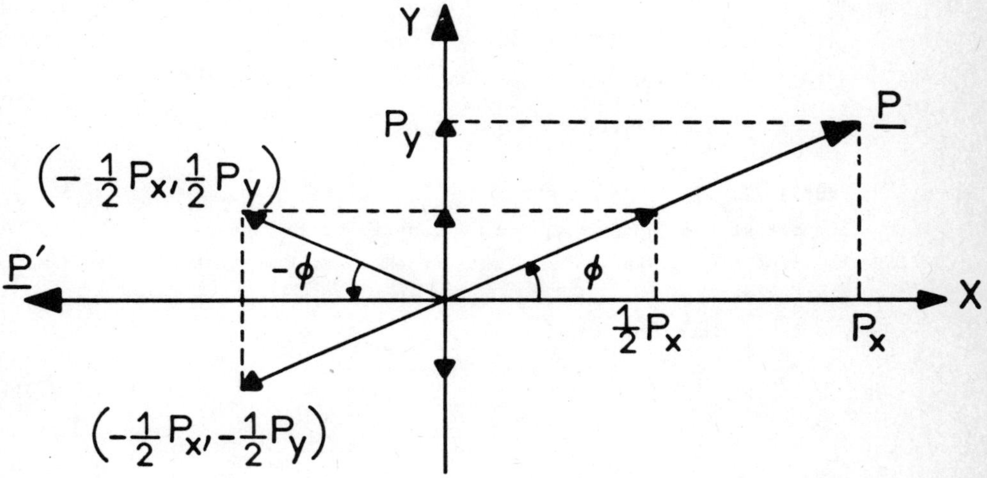

Fig. 2 The neutron spin components in the x-y plane. Neutrons arrive at the sample with their polarisation along $\underline{P}$ making an angle ϕ with the x-axis. After scattering the polarisation $\underline{P}'$ is along the x-axis, which can be resolved into the components $(-\frac{1}{2}P_x, \frac{1}{2}P_y)$ and $(-\frac{1}{2}P_x, -\frac{1}{2}P_y)$.

below, the difference $\langle P'_x \rangle - \langle P'_z \rangle$ provides to a good approximation a measurement of the intermediate scattering law Re $S(\kappa,t)$ at time $t = \text{ctg}(\theta/2)/2E_o$. This may be seen by the following. From Eq. 5 we have

$$\begin{aligned}\langle P'_x \rangle &= -P\langle\cos^2\alpha\rangle \\ &= -P\int_{-\infty}^{\infty} S(\kappa,\omega)\cos^2\alpha\, d\omega \Big/ \int_{-\infty}^{\infty} S(\kappa,\omega)\, d\omega \qquad (6)\\ &= -\frac{P}{2}\int_{-\infty}^{\infty} S(\kappa,\omega)(1+\cos^2\alpha)\, d\omega \Big/ \int_{-\infty}^{\infty} S(\kappa,\omega)d\omega\end{aligned}$$

similarly,

$$\langle P'_z \rangle = -\frac{P}{2}\int_{-\infty}^{\infty} S(\kappa,\omega)\ (1 - \cos 2\alpha)\ d\omega \Big/ \int_{-\infty}^{\infty} S(\kappa,\omega)\ d\omega \tag{7}$$

hence,

$$\langle P'_z \rangle - \langle P'_x \rangle = P \int_{-\infty}^{\infty} S(\kappa,\omega) \cos 2\alpha\ d\omega \Big/ \int_{-\infty}^{\infty} S(\kappa,\omega)\ d\omega \tag{8}$$

For small energy transfers we can linearize the dependence of α on ω by putting

$$\alpha \simeq \frac{\omega}{4E_o} \text{ ctg } (\theta/2) \tag{9}$$

Thus

$$\begin{aligned} \langle P'_z \rangle - \langle P'_x \rangle &= P \int_{-\infty}^{\infty} S(\kappa,\omega) \cos \left(\frac{\omega}{2E_o} \text{ ctg } (\theta/2)\right) d\omega \Big/ \int_{-\infty}^{\infty} S(\kappa,\omega)\ d\omega \\ &= P \text{ Re } S(\kappa,t) \Big/ S(\kappa,0) \end{aligned} \tag{10}$$

with $t = \text{ctg } (\theta/2)\ /2E_o$.

It turns out that the energy transfers ω contributing principally to $S(\kappa,t)$ in Eq. 10 for the value of t given by Eq. 11, are rather small so that the approximation in Eq. 9 is fully justified. We note also that the total scattering intensity is obtained very simply, since from Eq. 5 we have

$$\begin{aligned} P'_x + P'_z &= -P\ (\cos^2 \alpha + \sin^2 \alpha) \\ &= -P \end{aligned} \tag{11}$$

The measurements were made on the IN11 spin echo spectrometer using neutrons of incident wavelength 5.9 Å with a wavelength spread of 35 % fwhm. Thus, the measured $S(\kappa,t)$ is, in fact, averaged over the κ-range defined by the momentum resolution of the spectrometer. This κ-averaging, however, has no influence on the results since we find in the present case that $S(\kappa,t)/S(\kappa,0)$ is independent of κ within the accuracy of measurement. Furthermore, it can also be shown that for relaxation type spectra Eq. 2 still remains a good approximation with E_o corresponding to the average incoming neutron wavelength. The question of momentum resolution of the spectrometer aside, the κ for the measured correlation function is given by the elastic scattering vector since only small energy transfers are involved in its measurement in the time range $t > 10^{-11}$ sec.

In Fig. 3 we show the results for the intermediate scattering law $S(\kappa,t)$ measured at scattering angle $\theta = 5°$ ($\langle\kappa\rangle = .092$ Å^{-1}). In the abscissa we have used the logarithmic time scale in order to represent the data over a time range varying over several orders of magnitude. The data have been normalized to the total polarisation efficiency of the spectrometer, and yield directly the quantity $S(\kappa,t)/S(\kappa,0)$ where

$t = 0$ corresponds in effect to times smaller than 10^{-14} sec as seen from earlier neutron scattering experiments. We should mention that the data point for $t = 3\times10^{-12}$ sec has been obtained from simple polarisation analysis along the x and z directions, as discussed earlier, and the rest of the data points are obtained using the spin echo technique. In the diagram we have included a curve representing a simple relaxation process, given by the exponential $e^{-\Gamma t}$ for $\Gamma = 0.5$ meV. It should be noted that on such a diagram a simple exponential function always maintains the same shape, being displaced laterally for different values of Γ.

Within the limits of the experimental error bars the shape of $S(\kappa,t)$ is found to be independent of κ (over the momentum interval of 0.047 - 0.37 $Å^{-1}$ covered in the present measurements). It is emphasized that this observation holds for temperatures below 45 K where non zero values of the correlation function can be measured in the time range $3 \times 10^{-12} < t < 2 \times 10^{-9}$ sec covered in the present experiment. From our earlier measurements with the time-of-flight technique we have noted that the spectral width of $S(q,\omega)$ (measured above $|\omega| > 200$ μeV) becomes more and more q-independent with decreasing temperature. This is because the influence on the spin dynamics due to the rapid Korringa relaxation mechanism ($\tau_K \lesssim 10^{-11}$ sec for $T > 5$ K), which is necessarily present in metallic spin glass systems with 3d magnetic solute atoms, becomes relatively weaker compared with relaxation processes arising through solute-solute couplings as these become strongly established at low temperatures. Thus whenever the solute-solute couplings dominate, we observe a spectral distribution whose variation with time or frequency is almost q-independent. In other words, the present observations from the spin echo measurements of the approximate q-independent shape in time of the intermediate scattering law over the time-range which is far from the rapid Korringa regime is completely consistent with the earlier observations. We therefore express $S(\kappa,t)$ or $S(\kappa,\omega)$, valid for $t > 10^{-11}$ sec or $|\omega| < 200$ μeV as :

$$\begin{aligned} S(\kappa,t) &= S_1(\kappa)\, S(t) \\ \text{or } S(\kappa,\omega) &= S_1(\kappa)\, S(\omega) \end{aligned} \qquad (12)$$

$$\text{with } S(\omega) = \left(\frac{\omega}{1 - \exp(-\omega/kT)}\right) F(\omega) \simeq kT\; F(\omega) \quad \text{for } \omega < kT$$

where $S_1(\kappa)$ represents the spatial correlations between the spins and $S(t)$ the time dependent correlation function which in this case must coincide with the self correlation function $\langle S_i(0)\, S_i(t) \rangle$. Monte Carlo computer simulation calculations of Ising spin glasses /8/ assuming a Gaussian distribution of +ve and -ve near neighbour couplings yield self correlation functions which at first sight appear to be qualitatively similar to the present results. These, however, have the form

$$S(t) = \text{const} - \ln(t) \qquad (13)$$

which would be represented by straight lines in Fig. 3. This appears qualitatively to be the case of $\tau \geq 10^{-10}$ sec whereas the marked curvature in S(t) for $t \leq 10^{-11}$ sec is apparently a special pecularity of the Cu-Mn type of metallic spin glasses which reflects the influence of the rapid Korringa relaxation mechanism discussed above.

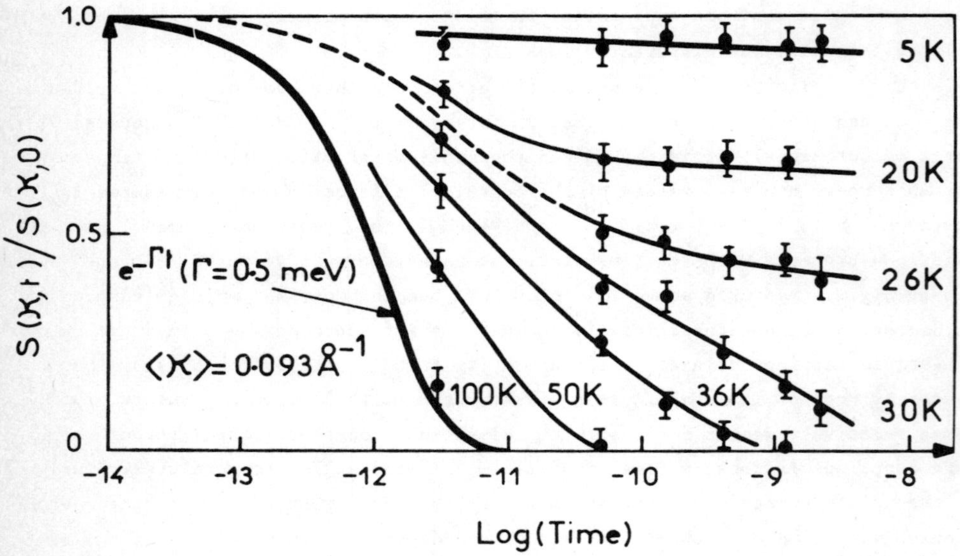

Fig. 3 The measured time dependent spin correlation function for Cu - 5 at % Mn sample at various temperatures. The thick solid line corresponds to the simple exponential decay $e^{-\Gamma t}$ with Γ = 0.5 meV. The thin lines are guides to the eye ; the dashed line gives a plausible behaviour of the correlation function in the short time regime.

The power of the spin echo technique lies in the measurement of relaxation type spectra and it is clear both from earlier measurements and the present results that the spin system can be best described in terms of a whole spectrum of relaxation times. We express therefore $S(\omega)$ in terms of a spectrum of Lorentzian functions, i.e.

$$S(\omega) = \frac{kT}{\pi} \int_0^\infty f(\Gamma) \frac{\Gamma}{\Gamma^2+\omega^2} d\Gamma \tag{14}$$

then,

$$\begin{aligned} S(t) &= \frac{kT}{\pi} \int_{-\infty}^{\infty}\int_0^\infty f(\Gamma) \frac{\Gamma}{\Gamma^2+\omega^2} \cos \omega t \, d\omega d\Gamma \\ &= kT \int_0^\infty f(\Gamma) \, e^{-\Gamma t} \, d\Gamma \\ &\simeq kT \int_0^{\Gamma=\frac{1}{t}} f(\Gamma) \, d\Gamma \end{aligned} \tag{15}$$

This is because if $f(\Gamma)$ is a slow function of Γ the contribution to the integral for $\Gamma > \frac{1}{t}$ is small. Now we define

$$kT\int_0^{\Gamma=\frac{1}{t}} f(\Gamma)\, d\Gamma = \int_t^{\infty} N(\tau)\, d\tau \qquad (16)$$

where $N(\tau)$ is the density of relaxation times τ.

$$\therefore\ S(\tau_1) - S(\tau_2) \simeq \left(\int_{\tau_1}^{\infty} N(\tau)\, d\tau - \int_{\tau_2}^{\infty} N(\tau)\, d\tau \right)$$

$$= \int_{\tau_1}^{\tau_2} N(\tau)\, d\tau$$

or simply,

$$\frac{d}{d\tau} S(\tau) = - N(\tau) \qquad (17)$$

Thus, under the above assumption of a slowly varying $f(\Gamma)$ or $N(\tau)$, the latter is given approximately by the derivative of $S(\tau)$.

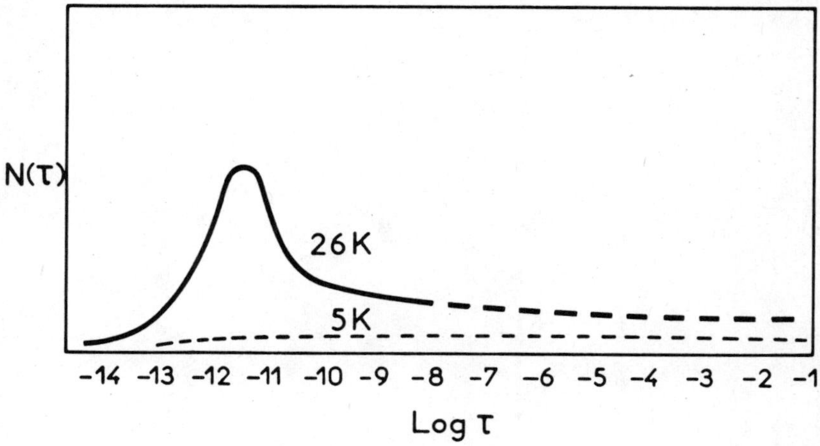

Fig. 4 Possible forms of the density of relaxation times $N(\tau)$ at T = 26 K and T = 5 K derived from the measured correlation functions shown in Fig. 3.

In Fig. 3 we have drawn a dashed curve joining smoothly with the curve through the measured data points for T = 26 K to indicate a possible variation of $S(\kappa,t)$ with t at short times. Using this, we derive a possible form of $N(\tau)$ with the help of equation 16 which is shown in Fig.4, where a form of $N(\tau)$ at 5K suggested by the present data is also included. As suggested previously /2/ it is clear that there is a progressive transfer of spectral weight from short times to longer times with decreasing temperature, and that even at the freezing temperature represented by the ac susceptibility maximum, a large fraction of the spectral weight resides at relatively high frequencies, due to the influence of the rapid Korringa relaxation mechanism on the spin dynamics of this alloy system.

In conclusion, we believe the present measurement of the intermediate scattering law $S(\kappa,t)$ over a time interval exceptionally large for a single experiment provides the most clear insight obtained up to now into the dynamics of spin glasses, and that the novel techniques we have used seem to be particularly well adapted for its measurement.

REFERENCES

/1/ A.P. Murani, J. de Phys. C6, 1517 (1978).
/2/ A.P. Murani, Solid State Comm. 33, 433 (1980).
/3/ F. Mezei and A.P. Murani, Journal of Magnetism and Magnetic Materials 14, 211 (1980) contains the first account of this study.
/4/ F. Mezei, Z. Phys. 255, 146 (1972)
/5/ O. Halpern and M.R. Johnson, Phys. Rev. 55, 898 (1939).
/6/ S.V. Maleev, Sov. Phys. JETP Lett. 2, 338 (1966).
/7/ G.M. Drabkin, A.I. Okorokov, E.I. Zabidarov and Ya. H. Kasman, Sov. Phys. JETP 29, 261 (1969).
/8/ K. Binder and K. Schröder, Phys. Rev. B14, 2142 (1976).

NEUTRON SPIN ECHO INVESTIGATION OF ELEMENTARY EXCITATIONS IN SUPERFLUID ^{4}He

F. MEZEI
Institut Laue-Langevin, 156X, 38042 Grenoble Cedex, France
and
Central Research Institute for Physics,
1525, Budapest 114 POB. 49, Hungary

INTRODUCTION

The present work represents the first experimental evidence for the application of Neutron Spin Echo (NSE) in high resolution study of both optical-like (non dispersive) and dispersive elementary excitations. The results obtained proved to be relevant contributions concerning the temperature dependence of the energy and linewidth of the roton excitation between 0.96 and 1.4 K; the temperature dependence of the linewidth of the 1.1 $Å^{-1}$ and 1.72 $Å^{-1}$ phonons and the suggested onset of three-phonon decay between 2.1 $Å^{-1}$ and 2.4 $Å^{-1}$. The energy transfer resolution achieved in this work was 10-40 times superiour to those in previous similar neutron scattering experiments.

In this paper most of the attention will be paid to the experimental aspects. In the first section the details of the NSE experiment are described, with particular emphasis on the first demonstration of the general scheme of NSE focussing, which involves the tuning of both the ratio of the precession fields H_o/H_1 and their geometrical assymetry ("tilt angle"). The second section gives the experimental results without, however, a detailed discussion of their significance for the understanding of superfluid ^{4}He, which will be published elsewhere.

1. THE NSE EXPERIMENT

The experimental set-up is shown in Fig. 1 and the dispersion relation $\omega_d(\kappa)$ of superfluid ^{4}He is illustrated in Fig. 2. The IN11 spectrometer at the Institut Laue-Langevin, described elsewhere in this volume, has been used with a graphite analyser crystal in front of the detector. This provided a good momentum resolution (±0.02 $Å^{-1}$) necessary in order to single out a particular part of the dispersion relation. The host spectrometer configuration, corresponding to something like a triple-axis-spectrometer with unusually long arms, includes the velocity selector (±10 %), the graphite energy analyser set to a fix outgoing wavelength 4.75 Å ±1.5 % and the collimations (0.5° both horizontally and vertically) determined by the Ø 30 mm beam cross section and the 3.5 m distances between the polarizer and the sample,and the sample and the polarization analyser. The resulting transmission function, i.e. the resolution ellipsoid of the host spectrometer, which singles out the $(\vec{\kappa},\omega)$ space volume probed by NSE, is illustrated by the dashed area in Fig. 2.

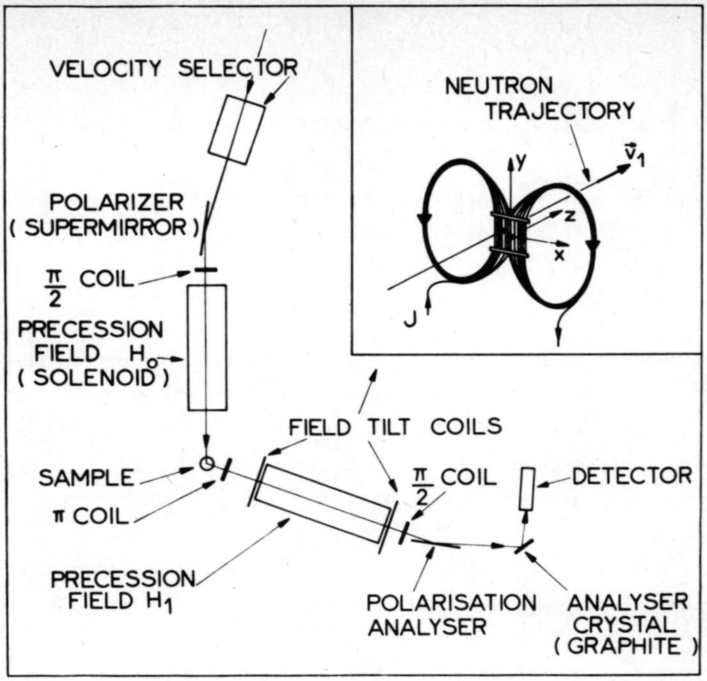

Fig. 1. Scheme of the IN11 NSE spectrometer at the ILL as used in the study of elementary excitations in superfluid ^4He. The "field tilt coils" shown in the insert are discussed later in the text.

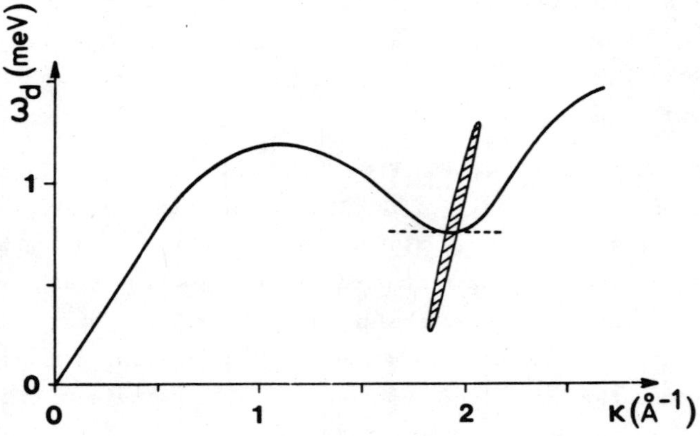

Fig. 2. Dispersion relation of the single-phonon excitations in superfluid ^4He and the transmission function characterizing the IN11 spectrometer configuration in Fig. 1 (shaded area).

Note that the good κ resolution was particularly important in the case of the roton minimum (κ = 1.925 $Å^{-1}$), as shown in the figure, since in the NSE focussing the slope of the dispersion relation $d\omega_d/d\kappa$ is properly taken care of (cf. Section 3 in the first paper in this volume, referred to as I. in what follows), but the curvature $d^2\omega_d/d\kappa^2$, and higher derivatives are neglected. The curvature has a maximum of 7.6 eV $Å^2$ at the roton minimum, which results in an effective 0.4 μeV full-width-half-maximum line broadening, as given by the Monte Carlo simulation I have performed to check all the experimental details. From the point of view of the data reduction procedure, however, this broadening just makes part of the instrumental resolution effects contained in the standard P^o_{NSE} signal (cf. Eq. (16) of I.), which could be directly determined using the roton line at the lowest available temperature, 0.96K, as standard for calibration. (Actually, the extrapolated roton linewidth at this temperature, 0.5 μeV, happens to be comparable to the resolution, and it has been corrected for in the final data evaluation). This calibration procedure assumes that the wave number of the roton minimum does not change appreciably within the 0.96 to 1.4K temperature range covered in the experiment. This was established in earlier neutron scattering work /1/ with a precision better than ±0.02 $Å^{-1}$, the actual figure required in this case in order to keep possible spurious effects below the statistical error. This calibration was checked by measuring the ordinary NSE signal for the elastic scattering from a standard quartz sample with the same diameter (42 mm) as the He sample can. Within error (∿ 0.4 μeV) the two procedures were in agreement.

The data have been corrected to the background which includes both fast neutrons and eventual traces of elastic scattering from the sample holder. This background was found to give zero NSE signal, and it only amounts to 10 counts/15 min. In addition, there was a sample scattering background too, coming from the multi-phonon scattering of the He specimen itself which occurs at energies just above /2/ the dispersion curve in Fig. 2. This effect together with the multiple scattering represents a very broad energy distribution, thus they give rise to complete dephasing of the precessions, i.e. they can not contribute to the observed precessing polarization signal P^{eff}_{NSE}. Multi-phonon scattering accounts for as much as 20 % of the 200 counts/15 min total counting rate for the roton, which is due to the big relative volume of the shaded transmission ellipsoid above the roton line in Fig. 2, and it could be quantitatively identified in the above calibration procedure as the unpolarized background in the NSE scans. The ratio of single to multi-phonon scattering cross sections was earlier found to be practically temperature independent below 1.4 K/3/, which has therefore been assumed in evaluating the data. Note that the present counting rates are very low compared with those obtained on conventional triple-axis machines (for example /4/∿6000 counts/15 min) which is partly due to the too long distances between polarizer and sample, etc. on IN11 not specifically designed for this type of work. Nevertheless, the available neutron intensity was sufficient to determine a phonon linewidth in an about 24 hours NSE scan with a resolution by far inaccessible to conventional methods, of course.

The parameters /2,5/ of the elementary excitations I have investigated in superfluid ^4He at saturated vapour pressure (SVP) are given for $T \sim 1K$ in Table I. below, together with the NSE parameters, i.e. the ratio of the precession field integrals $\ell_o H_o / \ell_1 H_1$, (which will be referred to as H_o/H_1 assuming $\ell_o = \ell_1$ in what follows) and the tilt angles ϑ_o and ϑ_1 which characterize the lateral assymetries of the precession field configurations, as discussed in Section 3 of I. The NSE parameters have been calculated for neutron energy loss scattering using Eqs. (27) and (28) of I.

Table I. NSE parameters for various phonons in ^4He for energy loss scattering and $\bar{E}_1$ = 3.63 meV final energy.

κ ($Å^{-1}$)	$\omega_d(\kappa)$ (meV)	$d\omega_d/d\kappa$ (meVÅ)	H_o/H_1	ϑ_o (°)	ϑ_1 (°)
1.10	1.20	0	1.535	0	0
1.72	0.85	-0.96	1.680	-6.1	-9.1
1.925	0.742	0	1.322	0	0
2.10	0.84	1.22	0.994	8.4	7.6
2.20	0.99	1.62	0.927	11.0	9.1
2.30	1.14	1.40	1.023	8.5	7.6
2.40	1.27	1.10	1.148	5.8	5.8

Note that the differences in H_o/H_1 come mostly from the differences of the slopes, cf. the second and fourth line in the table.

The calculated values have been experimentally verified in a few cases. The sample tuning curves shown in the upper part of Fig. 3 were taken by scanning H_1 with H_o fixed at a value giving rise to $\bar{\varphi}_o$ = 2000 rad mean incoming Larmor precession angle. The curves in the figure are envelopes of the NSE groups, which only depend on the monochromatization of the outgoing beam. They correspond actually to the NSE group shown in Fig. 2b of I., measured independently for the same monochromatization. The continuous lines in Fig. 3 show the expected shape of the curves and the arrows the calculated values given in Table I. Note that for the zero slope roton no field tilt is required $\vartheta_o = \vartheta_1 = 0$) which is an essential simplification.

In connection with Fig. 6 of I. I have discussed the originally suggested method for making the Larmor precession angle dependent on the direction of the neutron velocity, which is quantitatively described by the field tilt angle ϑ. For the solenoidal precession fields on IN11, however, a better adapted solution is offered by the "figure 8" coils developed in the early days of polarized neutron work at Kjeller /6/ and shown as "field tilt coils" in Fig. 1. When a neutron crosses the flat monolayer of wires at the center of the "8" it goes through the interior of a certain number n of winding loops (see the insert). For moderate activating currents J only the component

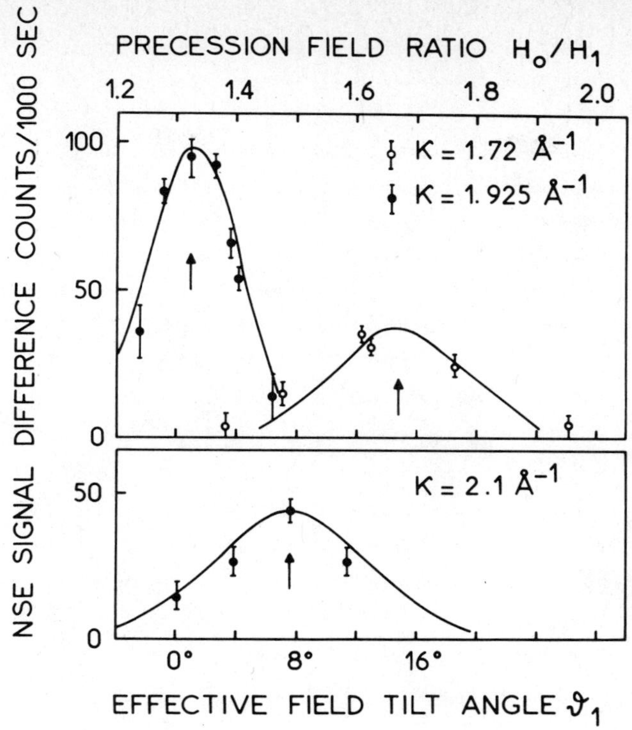

Fig. 3. Examples for tuning the NSE spectrometer (see text). The amplitude of the NSE oscillation signal, measured for excitations with different κ values as indicated, is plotted against the ratio of the precession fields H_o / H_1 (upper part of the figure, cf. Fig. 2 in I.) and against the effective tilt angle ϑ_1 (lower part). The arrows indicate the ideal values (cf. Table I.) and the lines correspond to the expected shape of the tuning curves.

of the "figure 8" coil field parallel to H_1 (i.e. the mean beam direction) has to be considered, and the precession field integral $H_1\ell_1$ will be modified for the particular neutron by $\pm 4\pi nJ/c$, where the sign corresponds to the sense of the windings traversed : clockwise to the right of the center (where n = 0) and counter-clockwise to the left. With two of these coils activated in the opposite sense at both ends of the precession field coil H_1 (Fig. 1) we find

$$\varphi_1 = \gamma_L \ell_1 H_1/v_1 + 4\gamma_L \pi NJ(x_1-x_2)/cv_1 \qquad (1)$$

where N is the number of windings per unit length in the wire layer, x_1 and x_2 are the x coordinates of the neutron trajectory at the first and second "figure 8" coils, respectively. Thus we obtain the directional dependence of φ_1 required by Eq.(24) of I.

with $\vec{n}_1 = (n_x, n_y, n_z) = (4\pi JN/cH_1, 0, 1)$. The corresponding field tilt angle is given as $\varphi_1 = \arctan(n_x/n_z)$. Note that $z||H_1$ is the mean beam direction. (In this first experiment no tilt coils were used for H_o, the beam cross section was rather reduced instead when necessary).

The lower part in Fig. 3 shows the first evidence for the general NSE focussing involving field tilt for a phonon group with finite slope. The NSE signal measured at different currents activating the field tilt coils nicely shows the maximum expected at the ideal value of the effective tilt angle (cf. Table I), as discussed in connection with Eq. (12) of I. Both the H_o/H_1 and the ϑ_o, ϑ_1 tuning behaviour shown in Fig. 3 illustrates the interesting novel feature of NSE that it is selective for the slope of the dispersion relation, which could be of particular utility in the study of crossing or hybridizing branches.

2. RESULTS

Sample NSE spectra are shown in Fig. 4 for the roton excitation, corrected to instrumental resolution, to the curvature of the dispersion relation and to multiphonon scattering, as described above. The straight lines represent in the semilog

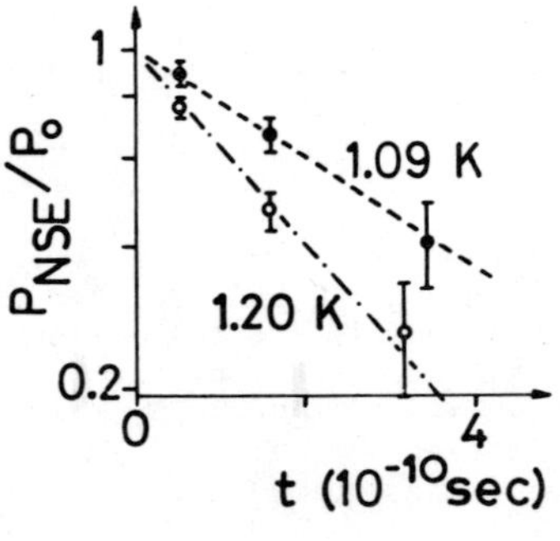

Fig. 4. Sample measured NSE spectra for the roton excitation. The straight lines correspond to Lorentzian fits with $\gamma = 20$ and 34 mK for T = = 1.09 and 1.20 K, respectively

plot the $\exp(-\gamma t)$ function corresponding to the Lorentzian line-shape with half width γ (cf. Eq. (15) in I.) where t has been determined by Eq. (29) of I. It is seen that the data are consistent with this line-shape, and exclude very different ones, e.g. the Gaussian which would give parabolas in the figure. (Note that the Lorentzian and damped harmonic oscillator line-shapes considered in Ref. 4 are indistinguishable for the small line-widths we are concerned with here). The Lorentzian shape has been therefore assumed throughout, and the present results on the temperature dependence of the roton-line width $\gamma(T)$ are presented in Fig. 5 together with the findings of previous vorkers. It is gratifying that the results obtained by conventional neutron

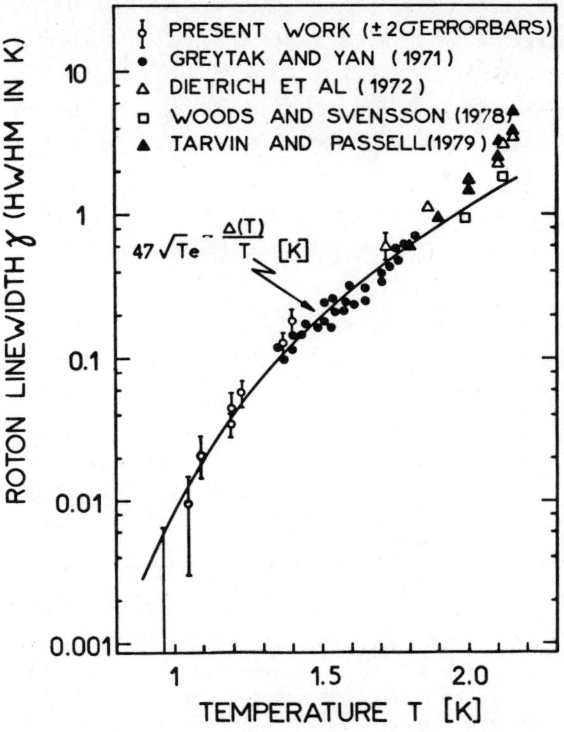

Fig. 5. Present and previous results on the temperature dependence of the roton line-width in ^{4}He at saturated vapor pressure. The previous data were obtained by light scattering (dots, Ref. 7) and conventional neutron scattering (Refs. 1,3,4). The curve represents the theoretical prediction with no fitted parameter (Ref. 8).

scattering at temperatures above 1.7K /1,3,4/,by Raman scattering on the two roton, bound state /7/ between 1.35 and 1.82K and by NSE below 1.4K join smoothly together into a common curve which follows well the theoretical prediction /8/ (continuous line) over two orders of magnitude in $\gamma(T)$ below 1.8K. The scatter of the conventional neutron scattering data above 1.8K, where the full line-width $2\gamma(T)$ and the roton energy $\Delta(T) \simeq 8K$ are comparable, was shown /4/ to be due to ambiguities in the data fitting procedure (1 meV ≃ 11.6K). One of the physical interest of the present results is that with the about 40 times improved resolution the direct neutron scattering measurements could now be extended to the more relevant temperature domain $T < 1.7K$ where $\gamma(T) \ll \Delta(T)$.

The NSE method can also be used for the high resolution investigation of excitation energy shifts. It is obvious from Eq. (21) of I. that a shift $\delta\omega_d$ of $\omega_d(\kappa)$ results in a change $\delta\bar{\varphi}$ of the mean Larmor precession angle $\bar{\varphi}$ which is given as

$$\delta\bar{\varphi} = t\,\delta\omega_d \tag{2}$$

Thus the change of the phase of the oscillations in the NSE group provides a measure of the excitation energy change. In the present case the temperature dependence of the roton energy has been determined between 0.96 and 1.4K. The results are shown in Fig. 6. Note that in fact $\Delta(T) - \Delta(0.96K)$ has been directly measured and the extrapolation to $\Delta(T = 0)$ has been made using the theoretical curve /9/ fitted to the

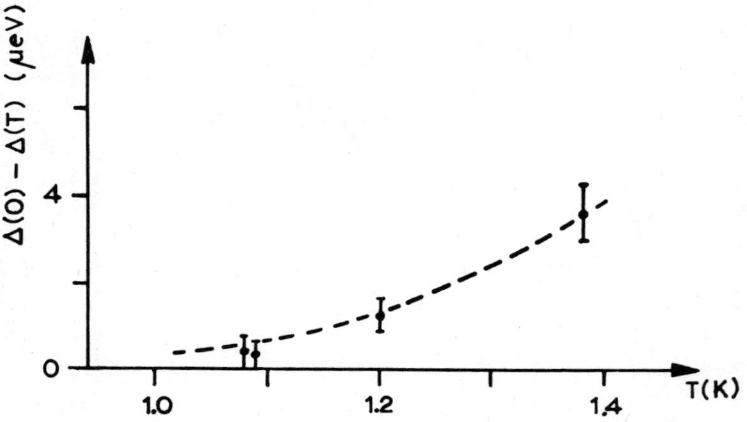

Fig. 6. Present results on the temperature dependence of the roton energy $\Delta(T)$. The dashed curve represents the theoretical prediction (Ref. 9): $\Delta(0) - \Delta(T) = A\sqrt{T}\exp[-\Delta(T)/T]$ with $A = 19 \pm 3$ K, fitted to the present data (both Δ and T are expressed in K)

present data (continuous line) which gave $\Delta(0) - \Delta(0.96\text{ K}) = 0.22\ \mu\text{eV} \simeq 2.5$ mK. Previous data /1,3,4/ on $\Delta(0) - \Delta(T)$ suffered seriously of the above mentioned ambiguities, and the present experiment provides the first model-independent results. The physical significance of the present findings will be discussed in more detail elsewhere /10/.

The other phonons listed in Table I were less extensively investigated, and the following results were obtained. The linewidths of the $\kappa = 1.1$ and 1.72 Å^{-1} phonons at 1.29 and 1.36K, respectively have been found to be identical to the roton linewidth at the same temperatures within the ±10 % error, similarly to previous ovservations made for $T > 1.8$K /3/. The linewidth of the $\kappa = 2.1$, 2.2, 2.3 and 2.4 Å^{-1} phonons has been measured at T = 1K, giving results not significantly different from zero, and I have only been able to establish upper limits viz. $\gamma = 10$, 20, 20 and 10 mK, respectively, with 70 % confidence level. Thus there was no evidence found for the onset of the suggested three phonon decay broadening /11/ at $\kappa \simeq 2.25$ Å^{-1}. This problem has been previously investigated with an order of magnitude worse resolution /5/

3. CONCLUSION

The present experiment is the first in which energy resolutions around 1 μeV (1 part in 1000 in energy transfer) have been attained in neutron scattering study of elementary excitations. Compared with previous work, 10-40 fold resolution improvement has been obtained by the application of the NSE method in its most general form, which offers, in addition, the interesting feature that it is selective for the slope of the dispersion curve. The present results convincingly demonstrate both the feasibility of the NSE method for the high resolution study of the lifetime and energy shifts of elementary excitations, and the possible physical interest of such experiments.

I wish to acknowledge stimulating discussions with numerous colleagues, in particular with Fred Zawadowski and Roger Pynn, and the invaluable technical assistance of the cryogenics group of the I.L.L.

REFERENCES

1. O.W. Dietrich, E.H. Graf, C.H. Huang, and L. Passell, Phys. Rev. A5, 1377 (1972).
2. e.g. A.D.B. Woods, and R.A. Cowley, Rep. Prog. Phys. 36, 1135 (1973).
3. A.D.B. Woods and E.C. Svensson, Phys. Rev. Lett. 41, 974 (1978).
4. J.A. Tarvin and L. Passell, Phys. Rev. B19, 1458 (1979).
5. E.H. Graf, V.J. Minkiewicz, H. Bjerrum Møller, and L. Passell, Phys. Rev. A10, 1748 (1974).
6. K. Abrahams, O. Steinsvoll, P.J.M. Bongaarts and P.W. de Lange, Rev. Sci. Instrum. 33, 524 (1962).
7. T.J. Greytak, and J. Yan, in Proc. 12th Int. Conf. on Low Temperature Phys., ed. E. Kanda (Academic Press, Kyoto, 1971) p. 89.
8. L.D. Landau, and I.M. Khalatnikov, Zh.'ETF 19, 637 (1949).
9. J. Ruvalds, Phys. Rev. Lett. 27, 1769 (1971).
10. F. Mezei, to be published.
11. L.P. Pitaevskii, Sov. Phys. JETP 9, 830 (1959).

COMPARISON OF THE PERFORMANCE OF THE BACKSCATTERING SPECTROMETER IN10 AND THE NEUTRON SPIN-ECHO SPECTROMETER IN11 ON THE BASIS OF EXPERIMENTAL RESULTS

A. Heidemann[x], W.S. Howells[+], G. Jenkin[x]

[x]Institut Laue-Langevin, 156X, 38042 Grenoble Cédex, France

[+]S.R.C. Rutherford Laboratory, Chilton Didcot, Oxfordshire OX11 0QX, U.K.

INTRODUCTION

The aim of the paper is to present a fair comparison of the two high resolution neutron spectrometers at the ILL : The Backscattering Spectrometer IN10 and the Neutron Spin-Echo Spectrometer IN11 /7/. The comparison is performed using the results of experiments carried out on both machines in the fields of polymer dynamics, tunneling spectroscopy and the study of phase transitions. The comparison is made easier by the application of Fourier Transformation as a method for the evaluation of IN10 data.

1. FOURIER ANALYSIS OF IN10 DATA

In any inelastic neutron scattering experiment on a "conventional" spectrometer i.e. triple axis-, TOF- or backscattering spectrometer the measured spectrum $J(\vec{Q},\omega)$ is proportional to the fourdimensional convolution of the scattering law $S(\vec{Q},\omega)$ with the resolution function $R(\vec{Q},\omega)$ of the spectrometer /1/ :

$$J(\vec{Q},\omega) = S(\vec{Q},\omega) * R(\vec{Q},\omega) \tag{1}$$

So $S(\vec{Q},\omega)$ can be obtained in principle by a deconvolution in $\vec{Q}$ and ω. It is well known that a true deconvolution is a very difficult task. Therefore data evaluation is normally carried out in the following way : A model for the system is chosen so that the scattering law $S(\vec{Q},\omega)$ is an analytical function with adjustable parameters. This function is folded with the resolution function and the result is fitted to the data.

Another possible method which is based on an old and well known idea is Fourier transformation of neutron data /2/ : One obtains from eq. (1) :

$$J(\vec{Q},t) = R(\vec{Q},t) \cdot S(\vec{Q},t) \tag{2}$$

where

$$A(t) \equiv \frac{1}{2\pi} \int_{-\infty}^{+\infty} e^{-i\omega t} A(\omega) d\omega \tag{3}$$

The convolution in ω-space is replaced by a multiplication in t-space. The intermediate scattering law $S(\vec{Q},t)$ is then obtained by a simple division :

$$S(\vec{Q},t) = \frac{J(\vec{Q},t)}{R(\vec{Q},t)} \qquad (4)$$

Since this equation looks so very attractive due its simplicity it is surprising that only a few users of neutron scattering have adopted this technique until recently /3/.

Interest in Fourier transformation of neutron data has been stimulated at the ILL in the last three years by the development of the neutron spin-echo spectrometry (NSE). The competition between the standard high resolution techniques like TOF and backscattering on one hand and NSE on the other forced the scientists working in both areas to find a common language. As NSE determines $\overline{S(\vec{Q},t)}$ the idea of using Fourier transformed conventional neutron data for the purpose of comparison was obvious. The complementary way of Fourier transforming NSE data to ω-space is less attractive if one campares the complexity of eq. (1) with the simplicity of eq. (2). (The bar on $S(\vec{Q},t)$ indicates that a wavelength averaged intermediate scattering law is obtained with NSE /3a/ :

$$\overline{S(\vec{Q},t)} = \int_0^\infty f(\lambda)\, S(\vec{Q}(\lambda),t(\lambda))d\lambda \qquad (5)$$

where $t(\lambda) = t_o\lambda^3$ and $f(\lambda)$ is the wavelength distribution of the incident spectrum).

A computer program has therefore been written /4/ which transforms IN10 data into t-space using the FAST FOURIER TRANSFORM technique. The program is now available at both the Rutherford Laboratory /4/ and the ILL /5/. The Fourier Transform program used is the subroutine FT01A of the Harwell library.

The program corrects IN10 data for absorption and self-shielding in the usual way, Fourier transforms the resolution and the sample spectra and calculates $S(\vec{Q},t)$ using eq. (4). It also computes the standard deviation of the Fourier components /6/ arising from the statistical noise on the "raw" data.

Fig. 1 shows two characteristic energy resolution curves R(t) for IN10 in t-space. The full circles represent the best resolution curve (HWHM = 0.17 μeV) of IN10 obtained with polished Silicon (111) monochromator and analyser crystals in exact backscattering. The time scale extends out to 15 nanosec. The dashed line corresponds to a resolution curve with 0.85 μeV (HWHM). This resolution is available for standard small angle scattering experiments in the Q-range $0.07 < Q < 0.3$ Å^{-1}. The time scale now extends to less than 3 nanosec. (wavelength in both cases : 6.27 Å).

Fig. 2 shows the situation on IN11. For 6 Å neutrons the time scale ends at about 4 nanosec while for 8 Å neutrons it ends at 8 nanosec (the time scale is proportional to λ^3). The limit is set by the maximum size of the precession field which is restricted by magnetic field inhomogeneities.

On IN10 the limit is set by the statistical noise which can in principle be reduced by using longer measuring times.

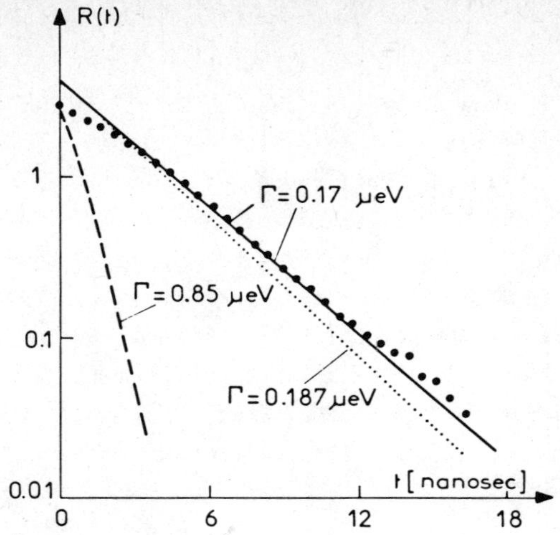

Fig. 1 Semi-logarithmic plot of two resolution curves of IN10 in t-space. Γ is the HWHM in units of μeV. The dotted line indicates the effect of a 10 % broadening.

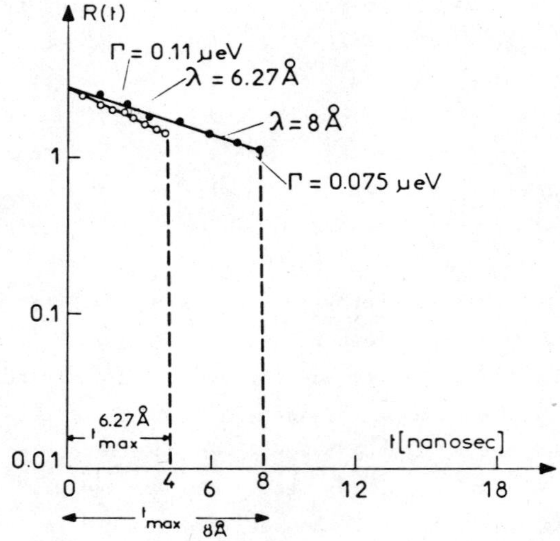

Fig. 2 Semilogarithmic plot of two resolution curves of IN11 for two different wavelengths. The dashed lines indicate the limits of the available time region.

We can conclude from Figs. 1 and 2 that the two instruments show rather similar performances. IN11 is better in resolution for small-angle scattering experiments. The longest times are, however, obtained on IN10 when it is used in its best configuration.

There is now a discrepancy between this conclusion and the rather wide spread view that IN10 has "μeV"-resolution, whereas IN11 has a "spectral" resolution down to several nano eV. In order to investigate this problem we have to look at the definition of energy- and spectral resolution in ω- and t-space more thoroughly. In the "yellow book" /7/ describing the ILL instruments we find that for IN10 the *energy* resolution (FWHM) is given as between 0.3 and 2 μeV while for IN11 the *spectral* resolution at λ = 4 Å is quoted as between 60 nanoeV and 600 μeV and at 8 Å between 8 nanoeV and 80 μeV.

In the IN10 case the energy resolution is defined as the FWHM of the resolution curve. In the IN11 case spectral resolution means the range of measurable energy transfers. The lower limit of the spectral resolution Γ_{Res} is defined on IN11 as the inelasticity which gives rise to a 5 % change in echo polarisation at the maximum time t_{max} obtainable, i.e. assuming a Lorentzian scattering law /7a/

$$\Delta P/P = \Delta S/S = 1 - e^{-\Gamma_{Res} \cdot t_{max}} \approx 0.05 \tag{6}$$

or $$\Gamma_{Res} \approx \frac{1}{t_{max}} \cdot 0.05 \tag{7}$$

From Fig. 2 we get :

$$\Gamma_{Res}^{IN11} = \begin{cases} 4 \text{ nanoeV at } \lambda = 8 \text{ Å} \\ 8 \text{ nanoeV at } \lambda = 6.27 \text{ Å} \end{cases}$$

We now adopt the same definition of the spectral resolution for IN10. t_{max} is here the longest time at which data can be obtained within a reasonable time of data collection (i.e. less than a day). From Fig. 1 we see that t_{max} is smaller than 15 nanosec. Therefore :

$$\Gamma_{Res}^{IN10} \approx 2.5 \text{ nanosec.}$$

Apparently IN10 has a better spectral resolution at 6.27 Å than IN11 at 8 Å. We know, however, from experience that it is quite difficult to measure energy changes of less than 10 neV on IN10. Therefore the definition of the *spectral* resolution seems to be quite artificial. The definition of the energy resolution as the HWHM of the resolution curve is more reasonable. This definition translated into the language in t-space means that the energy resolution is represented by the slope of R(t) on a logarithmic scale. From Figs. 1 and 2 we obtain :

$$\Gamma^{IN10} = 170 \text{ neV at } \lambda = 6.27 \text{ Å}$$

$$\Gamma^{IN11} = \begin{cases} 110 \text{ neV} & \text{at } \lambda = 6.27 \text{ Å} \\ 75 \text{ neV} & \text{at } \lambda = 8 \text{ Å} \end{cases}$$

Conclusion

Defining the energy resolution in ω-space as the HWHM of the energy resolution curve and in t-space as the slope of the logarithm of the resolution curve we obtain the result that IN11 has an energy resolution at 6.27 Å which is slightly better than that of IN10. At 8 Å the energy resolution of IN11 is more than 2 times better than that of IN10. If we define the *spectral* resolution as in /7a/ than IN10 has a more than 2 times better spectral resolution than IN11.

This conclusion is valid at the end of 1979. It is more or less sure that the energy resolution of IN11 will be improved in the future. The backscattering technique, however, has already reached a state where it is difficult to improve the energy resolution.

2. COMPARISON OF EXPERIMENTAL DATA

2.1. Introduction

In order to become familiar with scattering laws in t-space we show in fig. 3 three typical cases :

a) a single Lorentzian centered at $\omega = 0$ with a width Γ (HWHM).

b) a delta function $\delta(\omega)$ at $\omega = 0$

c) a pair of inelastic lines centered at $+\omega_o$ and $-\omega_o$ with a width Γ.

The corresponding intermediate scattering laws are :

α) an exponential $e^{-\Gamma t}$

β) a t-independent horizontal line

γ) a damped cosine function

2.2 Results

We will compare the performance of IN10 and IN11 using three "typical" experiments :

1. Coherent quasielastic small-angle scattering
2. Spin incoherent inelastic scattering
3. Coherent critical scattering in molecular crystals.

2.2.1

Coherent quasielastic small-angle scattering was observed by Richter et al. /8,9/ investigating the dynamics of polymer solutions.

Fig. 4 shows a typical result obtained on IN10. The sample was PDMS in deuterated benzene at 6°C /8/.

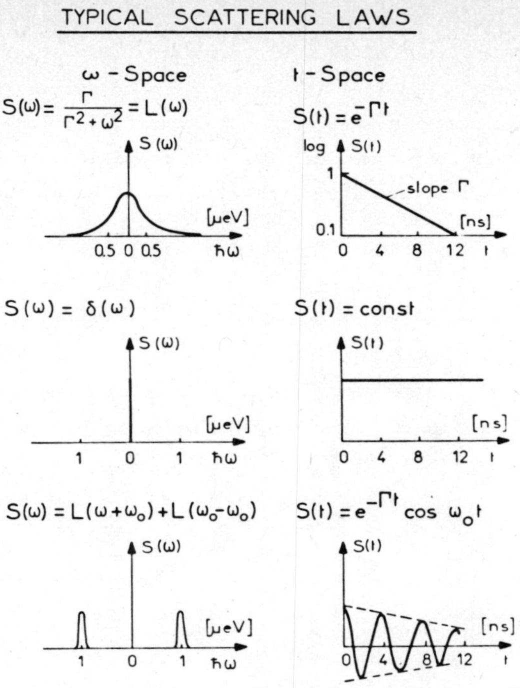

Fig. 3 **Three typical scattering laws in ω- and t-space.**

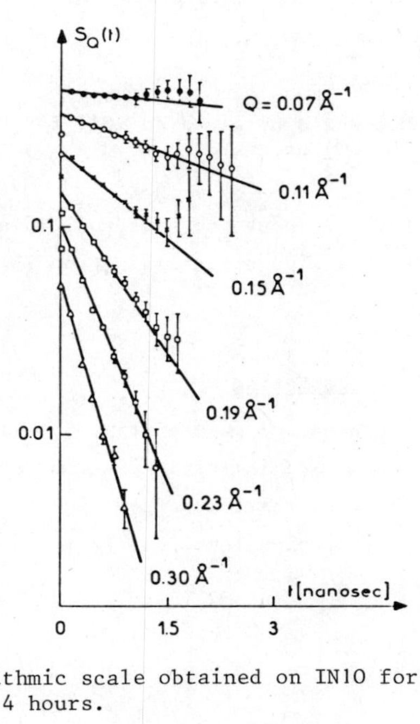

Fig. 4 S(Q,t) on a logarithmic scale obtained on IN10 for PDMS in deuterated benzene, T = 6°C, measuring time: 4 hours.

A set of 8 spectra with different Q-values between 0.07 and 1 $Å^{-1}$ was measured simultaneously. The increasing slope of the curves with increasing Q is obvious. The time scale ends at less than 2 nanosec. Data from IN11 with the same sample (T = 75°C) can be seen in Fig. 5 /9/. The time scale extends out to 6 nanosec, 3 times longer than on IN10. The availability of this long time scale made it possible to observe deviations from a single Lorentzian scattering law which were predicted by the theory.

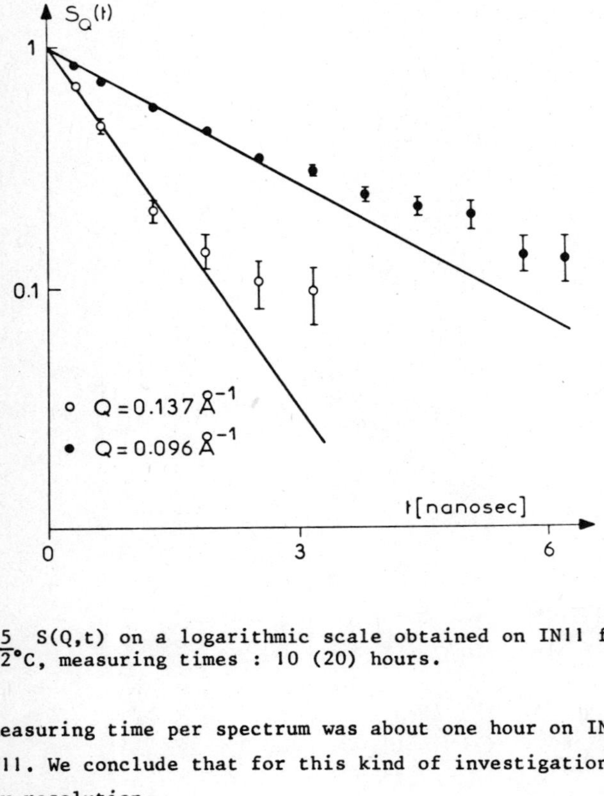

Fig. 5 S(Q,t) on a logarithmic scale obtained on IN11 for PDMS in deuterated benzene. T = 72°C, measuring times : 10 (20) hours.

The measuring time per spectrum was about one hour on IN10 and between 2 and 20 hours on IN11. We conclude that for this kind of investigation IN11 is better than IN10 in energy resolution.

2.2.2 Spin incoherent inelastic scattering

Tunnelling motions of molecules and molecular groups like NH_4^+ and CH_3 are now routinely studied with high resolution inelastic neutron scattering /10/. Fig. 6 shows an energy spectrum of neutrons scattered from Ni-acetate at 4 K on IN10 /11/. The CH_3 groups perform uniaxial rotational motions in a weak potential with three fold symmetry.

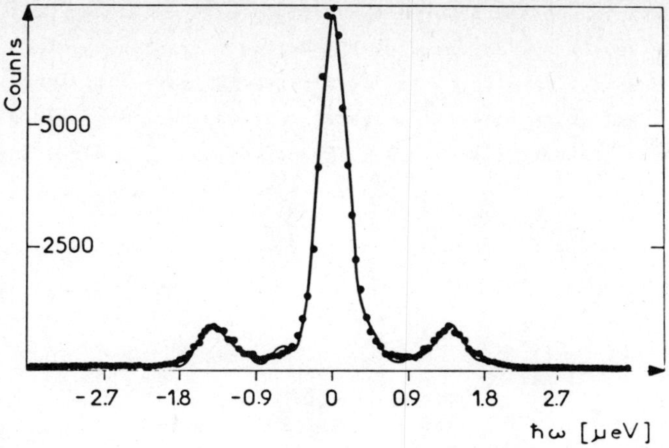

Fig. 6 Energy spectrum of Ni-acetate at 4 K on IN10. Measuring time : 32 hours. Q = 1.8 $Å^{-1}$.

Fig. 7 is a plot of S(Q,t). Four periods of the cosine function are visible with the time scale of IN10 extending out to about 12 nanosec. From the damping of the oscillations the line width of the tunnel peaks can be determined. The amplitude of the

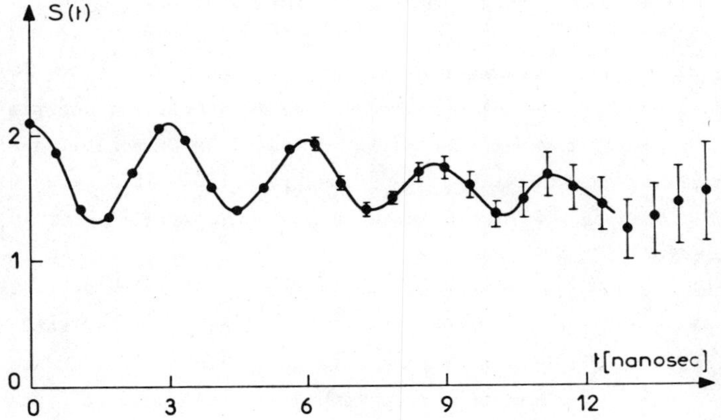

Fig. 7 S(Q,t) of Ni-acetate at 4 K on IN10.

oscillation relative to the average value of S(Q,t) is equal to the ratio of the integrated inelastic to the integrated elastic intensity. This ratio increases with Q^2 for QR < 1 (R is the radius of the molecule). Therefore it is preferable that the experiments are performed at rather high Q values (Q > 1.5 $Å^{-1}$). The error bars of the high Fourier coefficients blow up enormously with increasing time (see Fig. 7). The reason for this is that the absolute errors of the Fourier coefficients of the unnormalised spectra are roughly time independent, whereas the size of the Fourier coefficients decreases strongly with increasing time (see Fig. 1) /6/.

Fig. 8 shows data obtained on IN11 for a similar system (dimethylacetylene) with nearly the same tunnel splitting as in Ni-acetate. Neutrons with a wavelength of 6 Å were used. The momentum transfer was 1.8 $Å^{-1}$. The time scale has already ended by 3 nanosec.

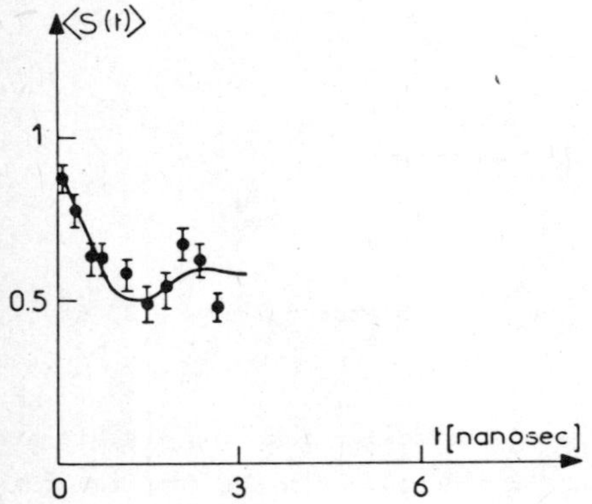

Fig. 8 S(Q,t) of dimethylacetylene on IN11. Measuring time 24 hours.

The reason for the large error bars on the IN11 data is twofold :

1. The scattering (incoherent !) is isotropic and the IN11 analyser accepts only a small solid angle. The ratio of the analyser solid angles of IN10 and IN11 as they were set up for the tunnelling experiment was : $\Omega_{IN10}/\Omega_{IN11} \cong 1500$. This gives rise to intensity problems on IN11 despite the fact that the flux at the sample position of IN11 is more than 100 times higher than on IN10.

2. Two thirds of the neutrons are scattered with spin flip. This is valid for both elastically and inelastically scattered neutrons. The signal to noise ratio is therefore reduced considerably. The beam time necessary to obtain reasonable statistics is about nine times longer compared to a case without partial spin flip scattering /12/.

From a quantitative comparison of the error bars shown in Figs. 7 and 8 we come to the conclusion that the same quality of data on spinincoherent inelastic scattering can be obtained on IN10 with about a *100 times* shorter measuring time than on IN11. This shows very clearly the superiority of IN10 over IN11 in the investigation of phenomena which are studied by spin incoherent inelastic and quasielastic scattering such as tunnelling and proton translational and rotational diffusion. Therefore the statement in a recent paper of F. Mezei about the application of NSE on pulsed neutron sources /12/ that "the effective gain of NSE compared to standard methods (i.e. TOF) is more like 100 for problems like proton diffusion" has to be treated with some caution. The statement is based on the unrealistic assumption of equal analyser solid angles for both types of spectrometers, NSE and conventional TOF. The continuous line in Fig. 8 was calculated using eq. (5) together with the assumption that the width of the inelastic lines if very narrow. The very strong damping is entirely produced by the averaging of S(Q,t) over λ. This averaging kills the oscillation after one period.

A quite impressive demonstration of the power of the Fourier method for IN10 data evaluation is given in the Figs. 9 and 10. Fig. 9 shows the resolution function of IN10 and a spectrum of ferromagnetic UH_2 /13/. The difference is hardly visible.

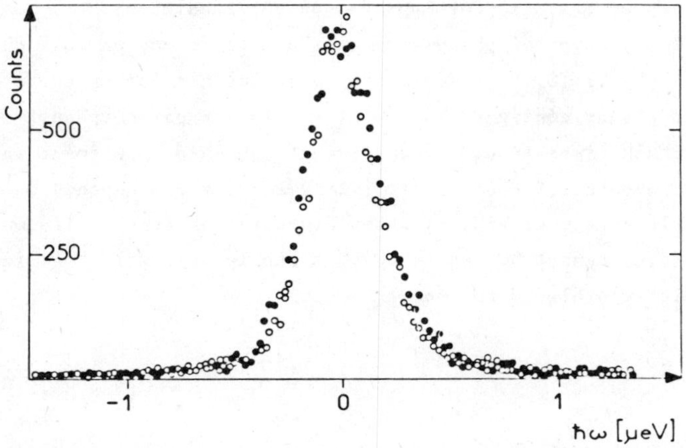

Fig. 9 Energy spectra of UH_2 on IN10 at two temperatures. o T = 200 K
• T = 77 K

In Fig. 10, where S(Q,t) is plotted, the inelasticity shows up very clearly. It is apparently not quasielastic but truly inelastic as seen from the curvature of S(Q,t).

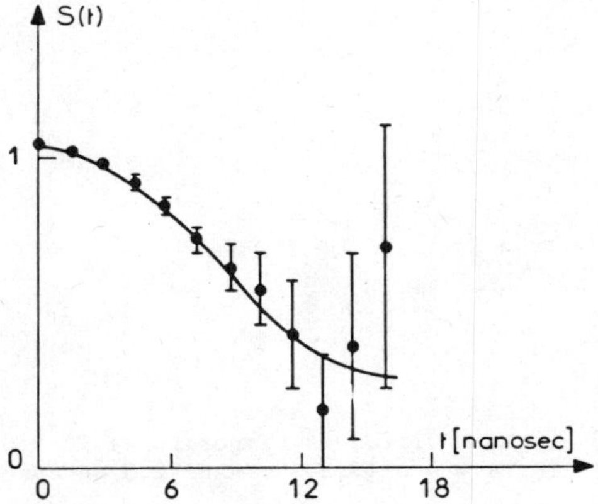

Fig. 10 S(Q,t) of UH_2 for T = 77 K

A fit of a cosine function yields a value of 0.1 μeV for the position of the inelastic peaks. The inelasticity is produced by the transferred hyperfine fields at the proton sites. The size of the hyperfine field is about 4 kG /13/.

2.2.3 Critical coherent scattering near phase transitions

Critical quasielastic scattering has been observed in systems with order - disorder phase transitions. It originates from the formation of clusters of the ordered structure within the disordered phase /14,15/. The energy width of this quasielastic component is proportional to the inverse of the cluster relaxation time τ_c. The intensity and τ_c diverge at the phase transition. This corresponds to a critical slowing down of fluctuations which occurs also for magnetic phase transitions.

An investigation of this type of phenomenon has been performed on IN10 on the molecular crystal paraterphenyl ($C_{14}H_{18}$) /15/. The molecules consist of three phenyl rings which are in a non planar configuration. At T_c = 179.5 K paraterphenyl undergoes an antiferrodistortive phase transition. The purpose of the IN10 experiment was to study the temperature dependence of the cluster lifetime τ_c, which depends on $\vec{q} = \vec{\tau}-\vec{Q}$ ($\vec{\tau}$: superlattice vector), at q = 0. Fig. 11 shows S(q=0,t) for three different temperatures. Apparently S(o,t) cannot be described by a single Lorentzian. An indication of this phenomenon is also visible on the energy spectra.

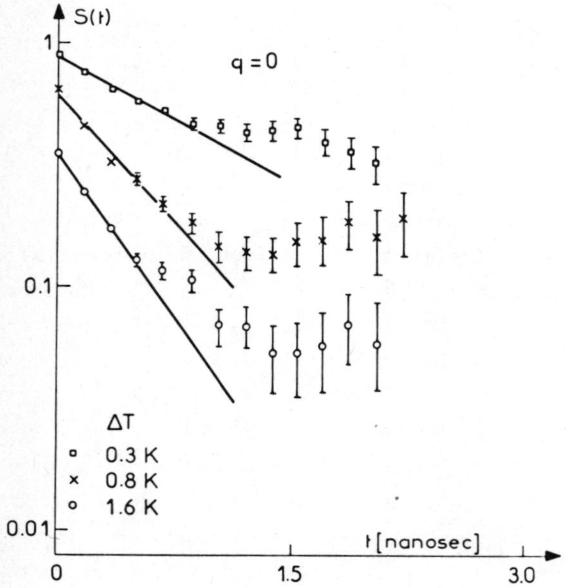

Fig. 11 S(o,t) of para-terphenyl on IN10 at three different temperatures T_c+0.3, T_c+0.8, T_c+1.6 K. The lifetimes τ_c of the clusters are 11.5, 5.7 and 2.5 nanosec.

Recently the experiment was repeated on IN11 /16/. The sample (size 0.1 cm^3) was three times smaller than that studied on IN10. An incident wavelength of 5.5 Å was chosen in order to reach a Q value of 2.08 $Å^{-1}$. A graphite analyser was applied to improve the q-resolution, which was of the order of $5 \cdot 10^{-3}$ $Å^{-1}$ /17/. Fig. 12 shows S(o,t) for several temperatures. Again the non Lorentzian behaviour is observed.

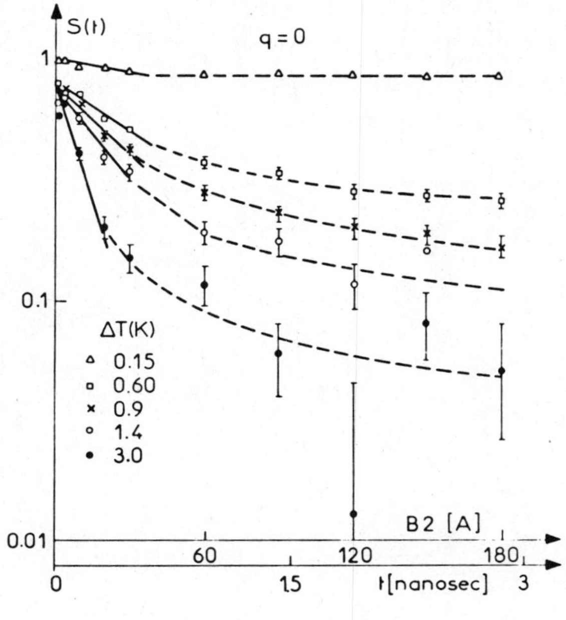

Fig. 12 S(o,t) of para-terphenyl on IN11 at five different temperatures.

Despite the very small sample the statistics of the measured points is better than on IN10 and the time scale extends out to 3 nanosec (on IN10 only to 2 nanosec). This shows clearly the superiority of IN11 over IN10 for this kind of investigation where the signal is concentrated arround points in reciprocal space and where no partial spin flip scattering occurs.

Fitting the initial slope of S(o,t) with an exponential $e^{-\Gamma t}$ yields $\Gamma(T)$ which is plotted against $T - T_c$ in Fig. 13. A power law is observed with an exponent arround 1. This is what one expects from mean field theory.

Due to the small sample size and the limited amount of beamtime only one spectrum at $q \neq 0$ (Fig. 14) could be measured on IN11. A q value of 0.014 $Å^{-1}$ was selected by rotating the sample by an angle equal to the HWHM of the "rocking curve" at $T - T_c$ = 0.6 K. Γ is increased by about a factor 2 compared to the value obtained with q = 0 at the same temperature. The extra elastic intensity which is given by $S(q,t \to \infty)$ decreased.

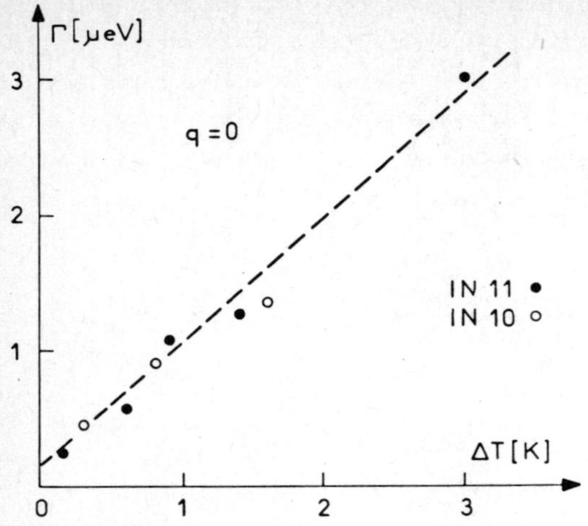

Fig. 13 Linewidth Γ (HWHM) at q = 0 as a function of $T - T_c$.

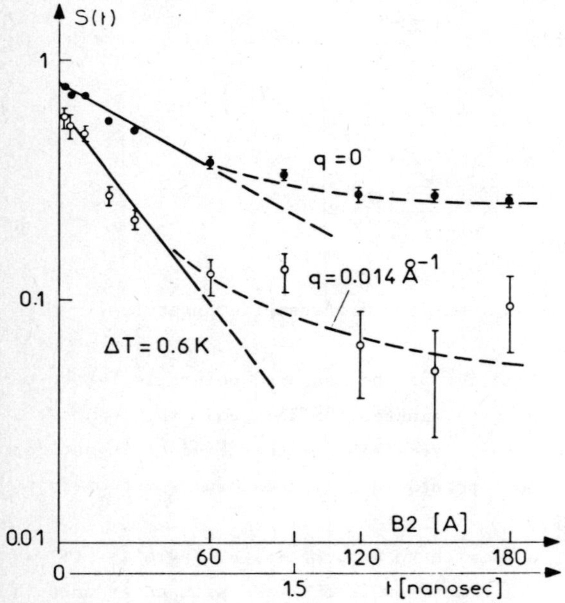

Fig. 14 S(q,t) of paraterphenyl on IN11 at T_c +0.6 K for q = 0(●) and q = 0.014 $Å^{-1}$ (o)

The occurrence of an extra "elastic" intensity, i.e. another *central peak*, was not expected and is a very interesting phenomenon. Further investigations on IN11 in this direction and a careful study of the $\vec{q}$-dependence of $S(\vec{q},t)$ are planned.

Conclusion

The two spectrometers IN10 and IN11 are comparable in their performance and complementary in their applications. Future improvements on both machines are possible and envisaged. For example, on IN11 the solid angle of the analyser can be increased and the time scale extended by having better control of magnetic field inhomogeneities. Indeed, very recently (Dec. 1979) the energy resolution of IN11 has been improved considerably to Γ = 14 neV at λ = 8 Å /18/. On IN10, the energy resolution for low Q experiments can be improved and the energy transfer range be extended.

Literature

/1/ M.J. Cooper, R. Nathans, Acta Cryst. 23, 357 (1967)
/2/ J.D. Bregman, F.F.M. de Mul, Nucl. Instr. Methods 93, 109 (1971)
/3/ C. Steenbergen, Thesis (1979), Delft
/3a/ J.B. Hayter, Z. Physik, in print (1979)
/4/ S. Howells, Internal Report, Rutherford Lab. (1979)
/5/ S. Howells, A. Heidemann, G. Jenkin, Internal Report, ILL (1979)
/6/ U.P. Wild, R. Holzwarth, H.P. Good, Rev. Sci. Instruments 48, 1621 (1977)
/7/ B. Maier, ILL neutron beam facilities at the HFR available for users. Copies can be obtained from B. Maier, Institut Laue Langevin, 156 X, 38042 Grenoble, France.
/7a/ ILL Annual Report, 1977, page 24
/8/ B. Ewen, D. Richter, to be published.
/9/ D. Richter, J.B. Hayter, F. Mezei, B. Ewen, Phys. Rev. Lett. 41, 1484 (1978)
/10/ S. Clough, A. Heidemann, J. Phys. C : Solid State Phys. 12, 761 (1979)
/11/ S. Clough, A. Heidemann, M. Paley, to be published.
/12/ F. Mezei, Nucl. Instr. Meth. 164, 153 (1979)
/13/ B. Gmal, A. Heidemann, A. Steyerl, Phys. Letters 60A, 471 (1977)
/14/ W. Press, A. Hüller, H. Stiller, W. Stirling, R. Currat, Phys. Rev. Lett. 32, 1354 (1974)
/15/ H. Cailleau, A. Heidemann, C.M.E. Zeyen, J. Phys. C, Solid State Phys. 12, L411 (1979)
/16/ H. Cailleau, A. Heidemann, F. Mezei, C.M.E. Zeyen, to be published.
/17/ F. Mezei, Proc. 3rd Int. School on Neutron Scattering, Alushta, USSR (1978)
/18/ ILL Annual Report, 1980, in print.

NEUTRON PHASE-ECHO CONCEPT
AND A PROPOSAL FOR A DYNAMICAL NEUTRON POLARISATION METHOD

G. Badurek
Institut für Experimentelle Kernphysik, A-1020 Wien,
and Institute Laue-Langevin, F-38042 Grenoble

H. Rauch *
Institut für Festkörperforschung der KFA, D-517 Jülich

A. Zeilinger
Atominstitut der Österreichischen Universitäten, A-1020 Wien

ABSTRACT

Experiments in the field of neutron interferometry can be interpreted in terms of a neutron phase-echo concept, which is closely related to the spin-echo concept. Phase echo analysis can be done even with unpolarized neutron beams if coherence between separate beams exists. Furthermore a method is proposed for a dynamical polarization of a neutron beam due to the combination of a momentum separation and a spin overlapping part. This system consists of a multistage h.f. flipper and one half of a spin-echo spectrometer and allows an optimal use of available neutrons.

1. Phase echo concept

In neutron spin-echo systems /1/ the Larmor precessions of the neutron spin within two separate parts of the instrument are balanced. In analogy to this technique we denote a system where the phases of neutron waves along two paths of an instrument compensate each other as a neutron phase-echo system. Because no phase sensitive reflections are known we need a reference beam to detect phase differences. First we assume an idealized system as sketched in Fig. 1 and plane waves throughout the instrument.

The incident plane wave ($\vec{k} = k_x\hat{x} + k_y\hat{y}$) is split into two waves with different k_y-vectors (k_y and $-k_y$) and mirrors change both y-components by 180 degree. The analyser is assumed to allow the observation of the interference pattern independent of its x-position. The focusing conditions are given by the geometrical optics. The maximum phase difference due to the different y-component ($\chi_y = k_y \cdot D$) is balanced at the analyser position for all wave lengths. A phase shift in the x-directions (χ_x) can be obtained by a phase shifting material with an index of refraction

* On leave of Atominstitut der Österreichischen Universitäten

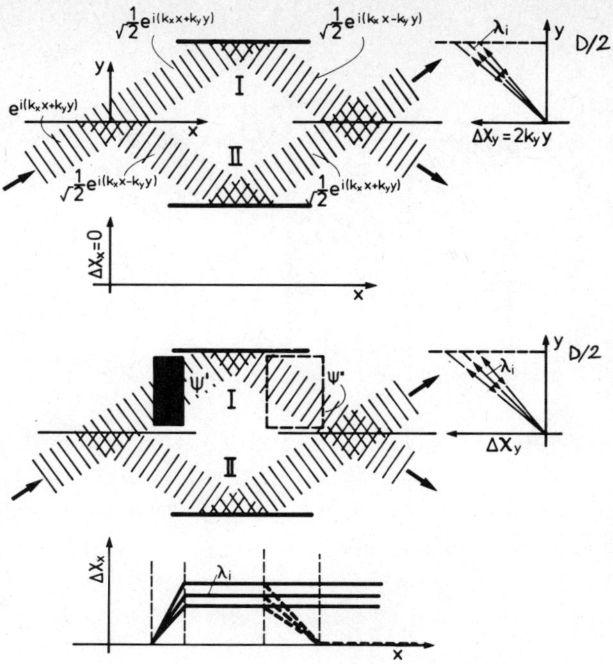

Fig. 1: Sketch of an idealised interferometer composed of conventional optical components without (above) and with samples in the beams (below)

$n = 1 - \lambda^2 N b_c/(2\pi)$ (N ... particle density, b_c ... coherent scattering length, λ ... neutron wave length). This phase shift changes the wave behind the material of thickness D_{eff} along the beam path according to:

$$\psi_o' = \psi_o\, e^{+i(k'-k_o)D_{eff}} = \psi_o\, e^{-ik_o(1-n)D_{eff}} =$$

$$= \psi_o\, e^{-iN_1 b_{c1} D_{eff,1}\lambda} = \psi_o\, e^{i\chi_x} \tag{1}$$

A focusing condition can be obtained also for this x component if another phase shifting material with an opposite scattering length is inserted in a way that

$$\chi = \chi_1 + \chi_2 = 0$$

$$\text{or} \quad b_{c1}\, N_1 D_{eff1} = -\, b_{c2}\, N_2 D_{eff2} \tag{2}$$

One recognizes that there exists a focusing condition which is independent of the wave length - similar to spin-echo systems.

The experimental realization can be achieved with the perfect crystal interferometer /2-5/ (Fig. 2). The relevant transfer functions can be taken from the literature /6/

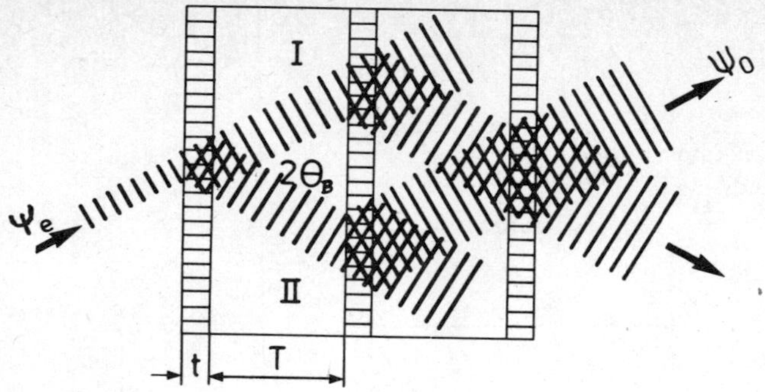

Fig. 2: Sketch of the perfect crystal interferometer

and are given below for the symmetric Si-(220) reflection and the beam in forward direction.

$$\psi_o = \psi_o^I + \psi_o^{II} = \left[v_o(y)\, v_H(y)\, v_{-H}(-y) + v_H(y)\, v_{-H}(-y)\, v_o(y)\right] \cdot$$

$$\cdot \exp\left[-2\pi i\, y\, (T+t)/\Delta o\right] \psi_e \tag{3}$$

with $v_o(y) = \left[\cos A \sqrt{1+y^2} + \frac{iy}{\sqrt{1+y^2}} \sin A \sqrt{1+y^2}\right] \exp(iPt)$

$$v_H(y) = -i \frac{\sin A \sqrt{1+y^2}}{\sqrt{1+y^2}} \exp(iPt)$$

$$P = -\frac{\pi y}{\Delta_o} - \frac{2\pi}{D_\lambda^{Si} \cos\Theta_B}$$

$$y = \frac{(\Theta_B - \Theta)\,\pi \sin 2\,\Theta_B}{\lambda^2 N_c^{Si} b_c^{Si}}$$

$$\Delta_o = \frac{\pi t}{A} = \frac{D_\lambda \cos\Theta_B}{2} = \frac{\pi \cos\Theta_B}{b_c^{Si} N^{Si}{}_\lambda}$$

For the balanced interferometer this relation reduces to

$$\psi_o^I = \psi_o^{II} \tag{4}$$

which remains still valid if spherical wave theory is used /7,8/. Equ. (4) shows the important feature that no y (or λ) dependence exists and therefore a divergent incident beam can be used.

A phase shifting material introduced into beam I changes the wave function to $\psi_o^{I'}$ (equ. 1) and one obtains from equ. (3) the intensity behind the interferometer as:

$$I = |\psi_o^{I'} + \psi_o^{II}|^2 = \frac{I_o}{2} (1 + \cos\chi) \tag{5}$$

Due to the wave length spread a dephasing effect (D_ϕ) occurs ($\exp(i\chi) \rightarrow \exp[i(\chi + \Delta\chi)]$. For a Gaussian wave length distribution centered around λ_o and with a half width (fwhm) of $\Delta\lambda$ we get /9/:

$$I = \frac{I_o}{2} (1 + D_\phi \cos\chi_o) \tag{6}$$

with
$$D_\phi = \exp\left[- A \chi_o^2 (\Delta\lambda/\lambda_o)^2\right]$$
$$A^{-1} = 16 \ln 2$$

This dephasing effect is shown in Fig. 3 for a beam with a half width of $\Delta\lambda/\lambda_o$ = 5 % and is experimentally observed up to high order interferences /10/. From such measurements the wave length distribution $f(\lambda)$ of the incident beam can be obtained by a simple Fourier transformation of the measured intensity modulation ($c = \chi/\lambda$)

$$f(\lambda) \propto \int_0^\infty I(c) \cos c\lambda \, dc \tag{7}$$

which is again very similar to the spin-echo procedure /11/. The dephasing effect vanishes (D_ϕ = 1) if the phase echo condition (equ. 2) is fulfilled.

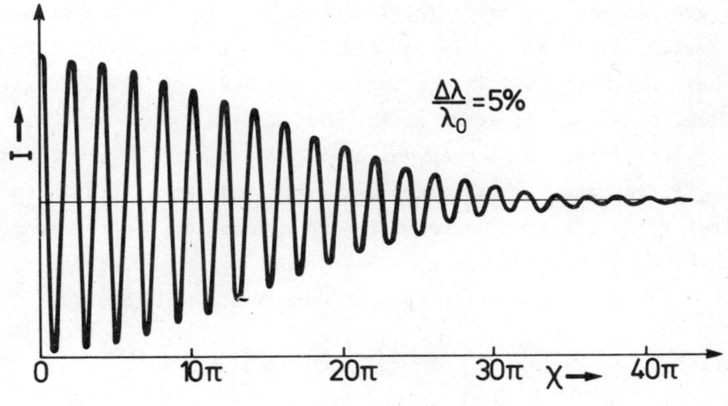

Fig. 3: Loss of contrast due to the dephasing effect of a broad incident wave length spectrum

The magnetic interaction of the neutron introduces an additional phase shift $\vec{\sigma}\,\vec{\alpha}/2$ where $\vec{\sigma}$ are the Pauli spin matrices and $\vec{\alpha}$ a rotation angle ($\alpha = \gamma BD/v$, γ ... gyromagnetic ratio, v ... neutron velocity) around the direction of the magnetic induction B. Characteristic beat effects occur if nuclear (χ) and magnetic phase shifts ($\alpha/2$) exist /12, 13/. Including the dephasing effect we get for unpolarized neutrons

$$I = \frac{I_o}{2} \{1 + \frac{1}{2} \left[D_{\phi_1} \cos(\chi_o + \frac{\alpha_o}{2}) + D_{\phi_2} \cos (\chi_o - \frac{\alpha_o}{2})\right]\} \tag{8}$$

with

$$D_{\phi_1} = \exp \left[-A \, (\chi_o + \frac{\alpha_o}{2})^2 \, (\Delta\lambda/\lambda_o)^2\right]$$

$$D_{\phi_2} = \exp \left[-A \, (\chi_o - \frac{\alpha_o}{2})^2 \, (\Delta\lambda/\lambda_o)^2\right]$$

and for polarised neutrons we obtain

$$I = \frac{I_o}{2} \left[1 + D_{\phi_1} \cos(\chi_o + \frac{\alpha_o}{2})\right] \tag{9}$$

These relations show that the focusing condition for unpolarized neutrons requires a separate compensation of the nuclear and magnetic phase shifts ($\chi_o = 0$ and $\alpha_o = 0$) while for polarized neutrons nuclear and magnetic phase shifts can compensate each other ($\chi_o + \alpha_o/2 = 0$). Similar effects can be discussed concerning gravitational interaction.

So far strictly elastic processes are discussed because experimental experience is available in this regime only. In future experiments with coherent beams may become feasible even in the inelastic (e.g. phonon) regime. In this case a generalized pair correlation function is related to this effect /14/ and Pendellösungseffects may become visible in the inelastic peaks, too /15/. In principle the large phase shift ($\sim 10^9$) inherent in the y-direction can be used with polylithic perfect crystal interferometers, which are tested up till now for X-rays only /16-18/. A disadvantage for a spectrometric application exists in the complicated response of the relevant parts of the interfermeter due to dynamical diffraction theory. No such difficulties should occur for interferometers for ultracold neutrons because nondispersive optical components can be used.

2. Dynamical neutron polarization

The rather low intensity of polarized neutrons limits their general use for solid state physics application. Various techniques are available to polarise thermal and cold neutrons /19/. In any case only neutrons with the desired spin direction are selected and these often with a rather low efficiency. Therefore the possibility of a dynamical polarization method is discussed. For the proposed system a momentum separation and a spin overlapping part is desired, which means a combination of flippers and of components of a spin-echo system /20/.

Many types of neutron spin-turners are used in polarised neutron physics /19/. Here we discriminate between flippers with a time dependent and a time independent magnetic field $\vec{B}$. The behaviour of the first ones are described by the time dependent Schrödinger equation

$$H\,\Psi(\vec{r},t) = i\,\hbar\,\frac{\partial\Psi}{\partial t} \tag{10}$$

with

$$H = -\frac{\hbar^2}{2m}\Delta - \mu\,\vec{\sigma}\,\vec{B}\,(\vec{r},t)$$

while for static flippers the time independent Schrödinger equation can be used

$$\left(-\frac{\hbar^2}{2m}\Delta - \mu\,\vec{\sigma}\,B(\vec{r}) - E\right)\Psi(\vec{r}) = 0 \tag{11}$$

μ is the magnetic moment, m the mass of the neutron and $\vec{\sigma}$ are the Pauli spin matrices. In most cases where the Zeeman splitting is small compared to the energy spread of the incident beam it is allowed to describe the spin precession in a field $\vec{B}\,(\vec{r},t)$ by the classical Bloch equation

$$\frac{d\vec{s}}{dt} = \gamma\,\vec{s}(t) \times \vec{B}(\vec{r}\,(t),\,t) \tag{12}$$

Due to the interference of the two energy eigenstates ($\pm\mu B_o$) the wellknown Larmor precession with an angular frequency ω_L appears

$$\omega_L = 2\,|\mu|\,B_o/\,\hbar \tag{13}$$

Existing spin-echo systems are described by equ. (12) /1, 21, 22/. The behaviour of the spin within a rotating (or oscillating) field superimposed to a constant guide field is calculated within the rotating frame and the resonance conditions for a complete spin turn are fulfilled if the rotational frequency $\omega = \omega_L$ and the amplitude B_1 of the rotating field is related to the length l of the rotating field and the neutron velocity v_o as

$$\gamma B_1\,l/v_o = \pi(2\,n + 1) \tag{14}$$

Oscillating fields are seen as a superposition of two rotating fields. If we take B_1 as the amplitude of the oscillating field we get for $B_1 << B_o$

$$\gamma B_1 \, l/v_o = 2\pi \,(2\,n + 1) \qquad (13\,a)$$

Small deviations from this value are discussed in the literature /23/.

In the case of time dependent fields the total energy is not conserved and at resonance the energy $\Delta E = \hbar\,\omega_L = \pm\, 2\,|\mu|\,B_o$ is transferred to or from the neutrons to change their potential energy according to the related spin turn. In that situation the change of the kinetic energy of the neutron ($E \rightarrow E \pm \mu B_o$) at the entrance into the guide field is enlarged at the exit to $E \pm 2\,\mu\,B_o$, which was first mentioned by Drabkin and Zhitnikov /24/. This energy change is rather small compared to the energy of thermal neutrons ($\Delta E \sim 0{,}12\ \mu eV$ for $B_o = 1$ T) and has not been observed experimentally up to now. A multiplication of this effect ($\Delta E_n = n\Delta E$) can be achieved by a multistage system as shown in Fig. 4. Some "nonadiabatic" spin turns between the stages serve for a situation where a spin com-

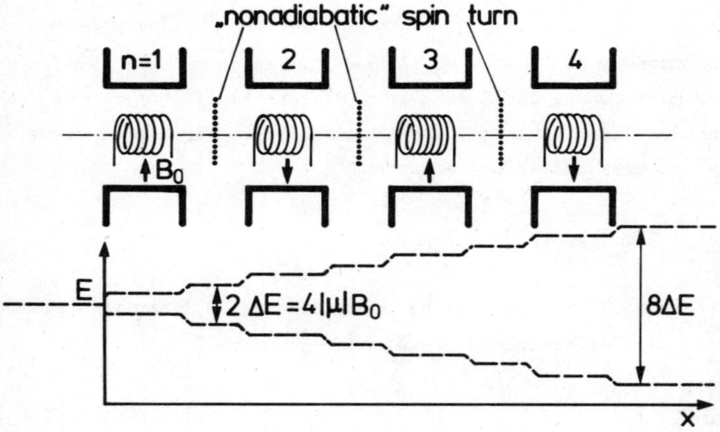

Fig. 4: Multistage system to increase the Zeeman splitting

ponent enters every stage with the same direction to the field. Guide tubes can be used to avoid any loss of luminosity.

A more quantum mechanical treatment describes the neutron by a spinor wave function throughout the device /25-28/. Using some results and a notation similar to that used by Krüger /27/ we write the wave function within the field B_o but before the h.f. field region as

$$\Psi = \begin{pmatrix} \cos\frac{\Theta}{2}\, e^{-i\phi/2}\, e^{ik^+x} \\ \sin\frac{\Theta}{2}\, e^{i\phi/2}\, e^{ik^-x} \end{pmatrix} e^{-iE\,t/\hbar} \tag{15}$$

where we have neglected small additional contributions of reflected waves. Behind the h.f. field ($\omega = \omega_L$) but still within the B_o field we have

$$\Psi = \begin{pmatrix} \cos\frac{\Theta}{2}\, e^{-i\phi/2}\, e^{ik^+x} \\ \sin\frac{\Theta}{2}\, e^{i\phi/2}\, e^{ik^-x} \end{pmatrix} \cos\left(\frac{\mu B_1 l}{\hbar v}\right) e^{-iE\,t/\hbar}$$
$$-i \begin{pmatrix} \sin\frac{\Theta}{2}\, e^{i\phi/2}\, e^{ik^-x}\, e^{-i(E+\Delta E/2)\,t/\hbar} \\ \cos\frac{\Theta}{2}\, e^{-i\phi/2}\, e^{ik^+x}\, e^{-i(E-\Delta E/2)\,t/\hbar} \end{pmatrix} \sin\left(\frac{\mu B_1 l}{\hbar v}\right) \tag{16}$$

In this notation Θ and ϕ may be interpreted as the polar angles of the spinor in the coordinate system related to the B field, which is chosen as z-direction. $k^{\pm}$ are given by the index of refraction $k^{\pm} = k\, n^{\pm} = k\,(1 \mp |\mu| B_o/E)^{1/2}$, where k_o is the wave vector of the neutron outside the magnetic field region and $E = \hbar^2 k^2/2\,m$. If the resonance condition (equ. 14) is fulfilled, only the inelastic contributions remain and during the exit of the neutron from the B field an additional change of the kinetic energy occurs and we write the wave functions behind the whole field arrangement as

$$\Psi = -i \begin{pmatrix} \sin\frac{\Theta}{2}\, e^{i\phi/2}\, e^{ik^{--}x}\, e^{-i(E+\Delta E/2)\,t/\hbar} \\ \cos\frac{\Theta}{2}\, e^{-i\phi/2}\, e^{ik^{++}x}\, e^{-i(E-\Delta E/2)\,t/\hbar} \end{pmatrix} \tag{17}$$

where $k^{\pm\pm} = k^{\pm}\, n^{\pm} \simeq k\,(1 \mp \frac{|\mu| B_o}{E})$. Note that different k-vectors belong to the + and - component and therefore no overlap between the + and - state exist for a strictly monochromatic neutron beam.

In practice a wave packet representation with an amplitude function $f(k-k_o)$ centered around k_o has to be used for the incident wave function. These amplitude functions are transformed to $f^{\pm}(k-k_o^{\pm\pm})$ accordingly. No momentum overlap exist if /25/

$$\gamma = \frac{\left|\int f^+(k-k_o^{++})\; f^-(k-k_o^{--})\, dk\right|}{\left(\int |f^+(k-k_o^{++})|^2\, dk \int |f^-(k-k_o^{--})|^2\, dk\right)^{1/2}} \tag{18}$$

is much smaller than 1. This condition is satisfied if the half width of the incident beam ΔE_o is smaller than the Zeeman splitting $\Delta E_o \leq 2\Delta E = 4\ |\mu| B_o$ (or $4\ n \cdot |\mu| B_o$).

These beams split in momentum and energy space enter the spin overlapping system (Fig. 5) which acts as one half of a spin echo system.

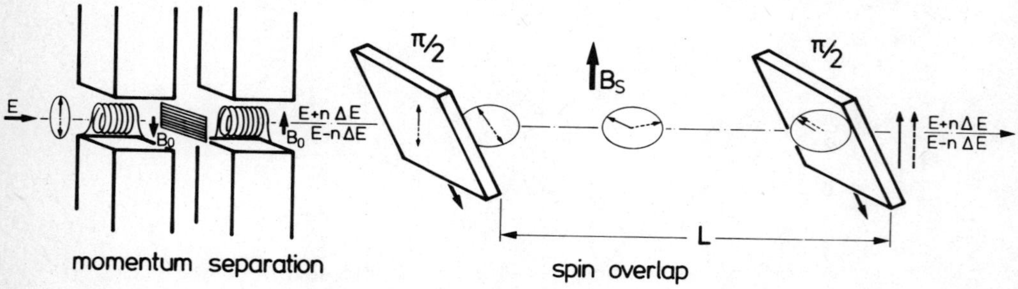

Fig. 5: Proposed arrangement for dynamical neutron polarization consisting of a momentum separation (left) and spin superposition part (right)

The behaviour of the + and - states within the $\pi/2$ turners and the precession field are well described by the rotation operator and are known from spin echo systems /21/. According to the different velocities associated with both states $v \pm \Delta v$ ($\Delta v = n\Delta E/mv$) a condition can be formulated where the spinor points in one direction only. This condition is fulfilled if the difference in the rotational angle of the two states within the precession field B_s reaches a value of π

$$\gamma\ B_s L \left(\frac{1}{v+\Delta v} - \frac{1}{v-\Delta v}\right) = \pi \qquad (19)$$

where L is the length of the precession field. From equ. (19) we get the condition for dynamic neutron polarisation

$$\frac{8\ n\ \mu^2}{m\ \hbar}\ \frac{B_o\ B_s\ L}{v^3} = \pi \qquad (20)$$

which is shown graphically in Fig. 6. It is seen that a multistage Zeeman splitting part allows the polarisation of a broader wave length band and simplifies the spin overlapping part.

It should be mentioned that equ. (20) can be fulfilled for a broad incident spectrum if a defined relation of direction and energy exists and if the precession field is

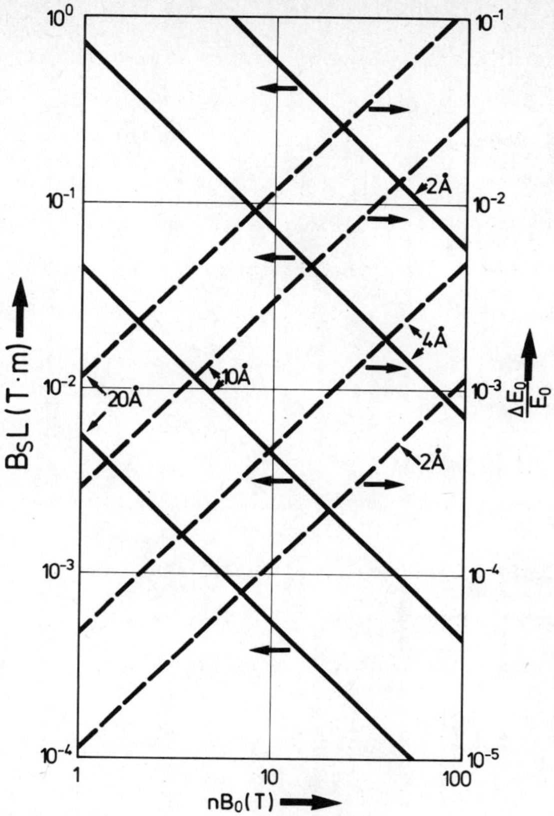

Fig. 6: Condition between the momentum separation and the spin overlapping part for dynamical neutron polarization

shaped to fulfil equ. (20) for any direction. This situation exists for example if perfect crystal reflections are used.

Neutron guide tubes can be used for both parts of the arrangement because the h.f. flippers as well as the $\pi/2$ spin turners are rather insensitive to the neutron velocity or to the effective path length through these devices.

The method described allows a complete polarisation of an incident unpolarised neutron beam with a certain energy width using only electromagnetic interaction and without material in the beam. The method is especially suited for long wave length neutrons and provides an optimal use of available neutrons.

REFERENCES

/1/ F. Mezei; Z. Physik 255 (1972) 146.

/2/ H. Rauch, W. Treimer, U. Bonse; Phys. Lett. A47 (1974) 369.

/3/ U. Bonse, W. Graeff; in "X-Ray Optin" (Edt. H.-J. Queisser), Top. Appl. Phys. 22 (1977) 93, Springer Verlag.

/4/ W. Bauspiess, U. Bonse, H. Rauch; Nucl. Instr. and Meth. 157 (1978) 495.

/5/ U. Bonse, H. Rauch (Edt.) "Neutron Interferometry" Oxford Univ. Press 1979.

/6/ D. Petrascheck; Acta Phys. Austr. 45 (1976) 217.

/7/ W. Bauspiess, U. Bonse, W. Graeff; J. Appl. Cryst. 9 (1976) 68.

/8/ D. Petrascheck, R. Folk; phys. stat. sol. (a) 36 (1976) 147.

/9/ H. Rauch, M. Suda; phys. stat. sol. (a) 25 (1974) 495.

/10/ H. Rauch in "Neutron Interferometry" (Edt. U. Bonse, H. Rauch) Oxford Univ. Press 1979, p. 161.

/11/ J.B. Hayter, J. Penfold; Z. Physik B 35 (1979) 199.

/12/ G. Eder, A. Zeilinger; Nuovo Cimento 34B (1976) 76.

/13/ G. Badurek, H. Rauch, A. Zeilinger, W. Bauspiess, U. Bonse; Phys. Rev. D14 (1976) 1177.

/14/ M.C. Li; Phys. Rev. B12 (1975) 3150.

/15/ S.M. Mendiratta; Phys. Rev. B14 (1976) 155.

/16/ U. Bonse, E. teKaat; Z. Physik 214 (1968) 16.

/17/ R.D. Deslattes, A. Henins; Phys. Rev. Lett. 31 (1973) 972.

/18/ M. Hart; Nature 275 (1978) 45.

/19/ H.B. Hayter; in "Neutron Diffraction" (Edt. H. Dachs) Top. Current Physics 6 (1978) 41, Springer Verlag.

/20/ G. Badurek, H. Rauch, A. Zeilinger; Z. Physik B in print.

/21/ H.B. Hayter; Z. Physik B31 (1978) 117.

/22/ other articles of these proceedings.

/23/ H. Kendrick, J.S. King, S.A. Werner, A. Arrott; Nucl. Instr. and Meth. 79 (1970) 82.

/24/ G.M. Drabkin, R.A. Zhitnikov; Sov. Phys. JETP 11 (1960) 729.

/25/ E. Balcar in "Neutron Interferometry" (Edt. U. Bonse, H. Rauch) Oxford Univ. Press 1979, p. 252.

/26/ F. Mezei in "Coherence and Imaging Processes in Physics" (Edt. M. Schlenker), Springer Verlag, in print.

/27/ E. Krüger, Proc. Int. Conf. Polarized Neutr. in Cond. Matter Research, Zaborów Poland, September 1979, in print at Nukleonika.

/28/ M. Calvo; Phys. Rev. B18 (1978) 5073.

CHAPTER III:

Future Progress and Applications

SEPARATION OF THERMAL DIFFUSE SCATTERING BY NSE IN DIFFRACTION STUDIES

C.M.E. ZEYEN

Institut Laue-Langevin
156X, 38042 Grenoble Cedex, France

ABSTRACT

Thermal diffuse scattering (TDS) is inelastic scattering arising mainly from acoustic phonons and which contaminates the purely elastic Bragg or purely elastic diffuse scattering when intensities are measured by standard diffraction methods. Corrections for the effect of TDS are classically done by more or less approximate calculations. These corrections are in general rather difficult because they suppose a knowledge of the dynamics of the system and of the resolution function of the diffractometer used. At higher temperatures and/or for soft materials these corrections may not be feasible at all. Experimental filtering of TDS is in principle possible by the neutron triple-axis technique and has actually been performed in certain special cases. But in general sufficient energy resolution to guarantee safe filtering can only be achieved by drastic beam collimations rendering the method impracticable intensitywise. Neutron spin echo (NSE) added to a triple-axis neutron diffractometer is shown here to be capable of providing improved resolution preserving good intensities such as to allow an efficient separation of purely elastic and low energy inelastic scattering.

INTRODUCTION

Since the development of triple axis spectroscopy, besides studying dispersion relations in materials one objective has always been to measure purely elastic scattering. The most important case is of course the accurate measurement of pure Bragg intensities aiming at a precise determination of atomic positions and their thermal parameters. It turns out that this is also the most difficult case because superimposed on Bragg peaks one usually finds inelastic diffuse scattering called thermal diffuse scattering (TDS) as it mainly arises from long wavelength acoustic phonons. It is thus particularly strong for soft materials and/or at high temperatures. Thermal parameters are strongly affected by the occurrence of TDS and although representing a very important part of results deduced from diffraction experiments they are often considered to be poorly reliable. Their precise knowledge though is essential in the determination of charge density distributions for the study of chemical bonding, in the study of disorder between near lying sites, of molecular motions and of phase transitions.

The TDS intensity distribution peaks at the reciprocal lattice point just at the positions of the Bragg peaks themselves and is therefore so difficult to separate experimentally. In fact extremely good energy resolution is required. Triple-axis machines have practical resolution limits imposed by the limited neutron source intensities which in relative value are about $\frac{\Delta E}{E_o} \simeq 1\ \%$. To obtain good absolute resolution ΔE one usually lowers the incident neutron energy E_o. Practically used incident neutron energies for good resolution work do not allow crystallographic work because of too small values of incident neutron momentum $k_o = \frac{2\pi}{\lambda_o}$ resulting in too small Ewald spheres for any structure of practical unit cell size to be solved. Of course absorption and extinction as well both rapidly increase with increasing λ_o. Finally TDS has very rarely been separated experimentally in a systematic structural study and the standard procedure of calculating corrections, described in the following paragraph, is adopted for the limited number of cases where it is possible.

One therefore has to look for a method which can provide good energy resolution independent of/or even at large incident momenta still offering acceptable intensity. The neutron spin-echo method in treating each neutron individually before and after scattering will be shown here to be capable of fulfilling this condition. Mezei has applied this method to reach the very best energy resolution (nano-eV) using cold neutrons on IN11, the merits of which are described in other papers of this workshop. For problems where a resolution of a few μeV is convenient NSE with thermal neutrons can be used with the possibility of providing small incident wavelengths and large momentum transfers. In a later section we will show that such a resolution is sufficient to allow a good filtering of TDS (better than 95 %) even for soft materials. Such a machine could thus be used for structural work where TDS is strong and/or difficult to correct for (complicated substances, anisotropy, etc.) and yield reliable temperature factors. A great deal of further applications of such a high energy resolution diffractometer can be thought of such as studies of quasielastic scattering at large momentum transfers (relaxation processes, critical scattering).

In the following we will shortly outline the main properties of TDS. We will then describe the design of an NSE option to the triple axis diffractometer D10 of the I.L.L. and calculate the expected performance. Some future applications will then be discussed as well as expected developments in performance as the described test set-up is far from presenting the best possible resolution with the presented method.

THE TDS-SYNDROM

Generally speaking TDS is inelastic scattering arising from phonons with low energies and small wave vector i.e. likely to be picked-up in the vicinity of Bragg spots. Exceptionally the phonons may be very soft optical branches or magnons, but more commonly TDS is inelastic scattering from acoustic phonons. When squared structure amplitudes are observed in a diffraction experiment the total effect of TDS is usually expressed as /1/

$$|F(\underline{\tau})|_o^2 = |F(\underline{\tau})|^2\ (1 + \alpha) \qquad (1)$$

where $|F(\underline{\tau})|_o$ is the observed structure amplitude corrected for such effects as extinction and absorption at lattice vector $\underline{\tau}$, $|F(\underline{\tau})|$ is the structure amplitude we are aiming at and α gives the TDS correction. Practical estimates of α have been discussed extensively (see for example Willis and Pryor /2/). For a simple discussion of the properties of TDS we will consider one phonon scattering from acoustic modes only in which case α can very simply be written as /1/ :

$$\alpha = \frac{kT}{NV\rho} \sum_{\underline{q}} J(\underline{q}) \tag{2}$$

where k is the Boltzmann constant, T the absolute temperature, N the number of unit cells in volume V and ρ is the density. The sum is over all wavectors $\underline{q}$ of acoustic phonons for which scattering can take place $J(\underline{q})$ can be written as :

$$J(\underline{q}) = \sum_j \frac{|(\underline{q} + \underline{\tau})\ \underline{e}_j(\underline{q})|^2}{\omega_j^2(\underline{q})} \tag{3}$$

where $\underline{e}_j$ is the Eigenvector of mode j with frequency ω_j and the sum is over the acoustic phonons. In the case of an isotropic solid and taking a mean sound velocity v, hence $\omega(q) = v|q|$, the following simple approximation for $J(\underline{q})$ can be used

$$J(q) = \frac{\tau^2}{v^2 q^2} \tag{4}$$

which allows an easy discussion of the main properties of TDS. It is relevant to note that since the TDS contribution is proportional to $\tau^2 = (\frac{4\pi \sin \theta}{\lambda})^2$ it may strongly bias the thermal parameters in the sense of a reduction. This is the most important systematic error, introduced by the occurrence of TDS.

TDS is proportional to temperature so it can be reduced by cooling the sample. This however also reduces the thermal motion parameters themselves so that the relative error in these parameters is not dramatically reduced upon cooling. Furthermore and especially in phase transition work thermal parameters are required at a well defined and not necessarily low temperature.

Another relevant property of α is that it peaks for $\underline{q} \to 0$ so that a singificant part of TDS will be inside the Bragg peak. This is the main reason why a good experimental separation is difficult. It seems therefore logic to calculate the TDS contribution by integrating expression (3) of $J(\underline{q})$ and to correct the measurements. This calculation however requires the knowledge of the dynamic characteristics of the material since ω_j and $\underline{e}_j$ of all acoustic modes are involved in expression (3). In addition the integration of $J(\underline{q})$ presupposes a good knowledge of the experimental resolution function. Thus in practice,calculated TDS corrections are only possible for relatively simple systems and for materials where the TDS contributions is small.

Another approach is to separate the slightly inelastic TDS scattering from the purely elastic scattering experimentally by performing an energy analysis of the scattered beam. Experimental TDS filtering has been achieved for X-ray using Mössbauer resonance techniques /3/. Mössbauer sources however are still too weak to allow systematic structure analysis using this method.

Neutron triple-axis techniques can in principle also be used, as discussed in the early days of this technique /4/, the crystal analyser removing the neutrons scattered inelastically by the sample. Unfortunately sufficient energy resolution can only be obtained through drastic beam collimation which renders the method impractical. Furthermore the wavelength band-pass of the analyser system may then affect the elastic intensities and introduce errors comparable to the TDS eliminated. The triple-axis method though is practical in the case where purely elastic off-Bragg scattering has to be measured in the presence of TDS, for example Huang (distortion)-scattering at high temperatures /5/. In general however, such measurements require further improved energy resolution.

As has been stated earlier during this workshop /6/ the neutron spin echo technique is capable of directly observing changes in the neutron energy independently of momentum resolution. This decoupling of energy resolution and beam collimator is precisely what we require for an efficient TDS filter diffractometer namely independent adjust of momentum resolution for accurate Bragg-peak integration and very good energy resolution so that most of the inelastic scattering can be separated out.

We will show in the following that provided efficient spin polarizer and analyser systems are used the method is feasible. Our approach was to design an NSE option to be added to a classical thermal beam triple axis spectrometer /7/. The NSE option will not alter the classical triple-axis resolution of the instrument which is to be considered as a band-pass. The elastic scattering process is unchanged and the intensity losses are given by the efficiency of the polarizer and analyser systems only. In the following we will describe the principle of this machine and calculate its performance. The option includes the possibility of various modes of operations such as inelastic scattering filter, quasielastic scans with high resolution, inelastic scattering from dispersion free (flat branch) excitations and NSE focussing phonon spectroscopy (this mode is described in detail in the paper by R. Pynn : NSE and triple axis spectrometers, this workshop).

NSE TRIPLE AXIS SPECTROMETER

With the Spin-Echo options added, the experimental arrangement is schematically shown in the Figure. The classical set-up consists of vertically focussing crystal monochromator and analyser, sample table with Eulerian cradle and neutron analyser detector system. Due to the air cushion construction mode, the distances in between the two spectrometer tables are easily variable and can be adjusted to include the spin polarizer/analyser facilities and the two precession electromagnets. Further details can be found in /7/ and /8/. The present design foresees two possibilities for the neutron spin polarizer and analyser systems. In conjunction with vertically

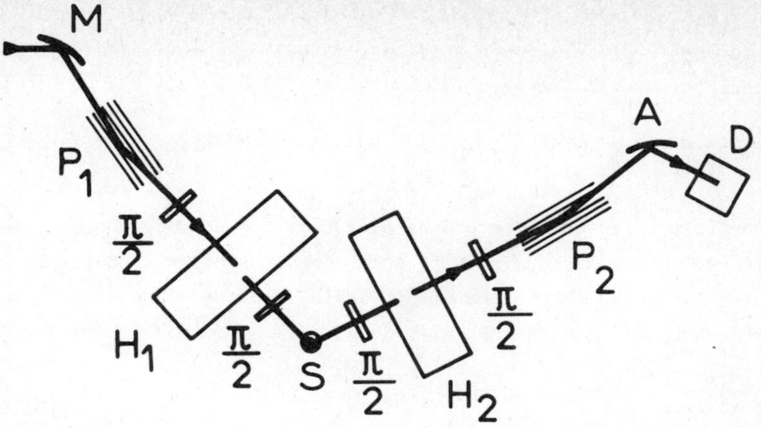

Schematic arrangement of a spin-echo triple-axis diffractometer.
M : focussing crystal monochromator
A : focussing crystal analyser
P_1, P_2 : spin polarizer and analyser (super mirror systems)
D : detector
S : sample
$\frac{\pi}{2}$: spin-turn coils
H_1, H_2 : precession fields.

focussing crystal monochromator/analyser neutron supermirrors systems have been built /9/. They have been shown to operate at low wavelengths but their drawback at short wavelength is the rather small glancing angle and hence the mirror length needed to reach reasonable beam sizes. The second possibility is to use vertically focussing composite Heusler alloy crystals both for monochromatization/analysis and spin polarization/analysis. The technique of composing large area focussing assemblies out of selected bits of crystals has proven to be possible (Copper, Si and Ge assemblies currently used at I.L.L.). As recent experiments with flat Heusler crystals have shown triple-axis NSE spectrometers can be kept rather compact with distances comparable to those of classical triple-axis machines.

ENERGY RESOLUTION

As discussed in previous papers of this workshop the quantity determined experimentally in NSE experiments is the beam average of the final polarization $\langle P_z \rangle$. In our case where the incident wavelength band is rather well defined ($\Delta\lambda/\lambda \simeq 1$ %) and if we consider only quasi-elastic scattering $\langle P_z \rangle$ can be written as

$$\langle P_z \rangle = \frac{1}{K} \int_{-\infty}^{\infty} S(Q, \omega) \cos [\omega t(\lambda_o)] d\omega \tag{5}$$

where $S(Q, \omega)$ is the scattering law, ω the neutron energy change and t the time variable, (K is a normalization prefactor).

$$t = N_o \pi / E_o \quad ,$$

N_o being the number of Larmor precessions, in the first guide field magnet, and E_o the mean incident neutron energy.

$$N_o = \frac{\lambda_o}{135.65} \int_o^d H d\ell \quad ; \quad \int_o^d H d\ell \text{ is the line integral (in Oe.cm)} \qquad (6)$$

of the magnetic precession H, extending over distance d along the neutron path. For elastic scattering and a symmetric spectrometer setting (identical precession fields in the two arms) the number of precessions in the second arm N_1 is equal to N_o. For inelastic scattering N_1 can be written as

$$N_1 = \frac{\lambda_o + \Delta\lambda}{135.65} \int_o^d H d\ell$$

In order to evaluate the magnitude of the precession field line integrals required for the type of resolution we want at a given incident neutron energy E_o (i.e. evaluation of the time variable in the Fourier transform $t = \frac{N_o \pi}{E_o}$) let us consider a simple case of scattering function

$$S(Q,\omega) = \delta(Q - \tau) \; [(1-x)\delta(\omega) + x \, L(\omega)] \qquad (7)$$

as a Bragg peak $\delta(\omega)\ \delta(Q - \tau)$ centered at the end point of the reciprocal lattice vector τ and superimposed on a Lorentzian TDS distribution $L(\omega)$ given by

$$L(\omega) = \pi^{-1} \frac{\omega_o}{\omega_o^2 + \omega^2}$$

where $2\omega_o$ is the FWHM, x is the relative intensity of TDS. With this scattering law the spin echo polarization becomes for the Bragg peak given by τ,

$$\langle P_z \rangle = (1 - x) + x \exp\left[- \frac{\pi N_o}{E_o} \omega_o \right]$$

The first term 1-x, due to purely elastic scattering which we want to measure directly. The second term arises from inelastically scattered neutrons for which the energy resolution of the system is not sufficiently good to be filtered out. In terms of polarization this means that the number of Larmor precessions N_o is not sufficient to completely randomize the spin distribution of inelastically scattered neutrons. We shall now estimate the number N_o in a practical case taking the worst scattering situation where a very weak elastic peak would have to be measured in the presence of a strong inelastic Lorentzian distribution; i.e. we take $x \simeq 1$. This will never be the case really with Bragg scattering and TDS as we have seen that TDS is proportional to the square of the structure amplitude (relation 1). The situation may be different for critical scattering or diffuse inelastic scattering which may be weak and superimposed on relatively larger inelastic backgrounds.

For $x \simeq 1$ we thus require that the inelastic contamination i.e. the exponential term be less than 1 % :

$$\exp\left[-\frac{N_o \pi}{E_o}\,\omega_o\right] \leqslant 0.01$$

so that the measured polarization yields the purely elastic Bragg intensity to within 1 %, which also corresponds to the statistical accuracy standardly obtained in the measurement of the polarization.

This will set a lower limit to the number of spin precessions N_o :

$$N_o \geqslant 1.47 \quad \frac{E_o}{\omega_o} \quad .$$

If we use E_o = 36 meV corresponding to 1.5 Å neutrons and if we aim at an effective filter resolution of 50 μeV we have to design the precession magnets in such a way that N_o can be kept larger than 1060 turns. Following (6) this can be reached with a field integral of $\int_o^d H d\ell = 9.5 \cdot 10^4$ Oe·cm. Such field integrals can easily be achieved with electromagnets producing a mean magnetic field of 3 kOe over a distance of 32 cm for example. These characteristics were chosen for the test magnets. They are far from representating the technically best possible solution. In fact with superconducting coils the field line integral could be increased by at least an order of magnitude and the resolution improved by the same amount.

Our remaining task is to show that the chosen resolution, i.e. HWHM of the Lorentzian inelastic distribution ω_o = 50 μeV is already sufficient in practice to guarantee effective TDS filtering. Let us discuss this point by taking the example of a rather soft organic crystal : naphthalene. The dynamics of this system have been carefully studied and the lowest acoustic phonon has a slope of 12 meV / $\mathring{A}^{-1}$ in the $b^* = 1.05\ \mathring{A}^{-1}$ direction. With a resolution of 50 μeV all scattering from phonons with wave vectors $q_E \leqslant 50 \ .\ 10^{-3}/12 = 4.2 \cdot 10^{-3}\ \mathring{A}^{-1}$ will be detected as elastic. The crystallographic scan to obtain the integrated intensity will be about 10 % of b^* i.e. will extend to about $q_M = 0.1\ \mathring{A}^{-1}$.

The TDS correction α, calculated over a spherical scan volume of radius q tered at the reciprocal lattice point is given by /2/ :

$$\alpha = \frac{|\tau|^2\ kT}{2\pi^2\ C}\, q$$

and C the mean elastic constant (isotropic solid approximation). With ω_o = 50 μeV the NSE filter will be effective for q larger than q_E and the amount of TDS not removed is given by

$$\frac{\alpha(q_E)}{\alpha(q_M)} = \frac{q_E}{q_M} = 0.04$$

so even for soft materials more than 90 % of the TDS can be removed with the 50 μeV resolution which as already said is far from being the possible limit.

ACKNOWLEDGEMENTS

I thank Feri Mezei and John Hayter for patiently introducing me to the N.S.E. principles.

REFERENCES

/1/ W. Cochran, Acta Cryst. A25, 95 (1969).
/2/ B.T.M. Willis and A.W. Prior (1975). Thermal vibrations in Crystallography. Cambridge University Press.
A.W. Prior, Acta Cryst. 20, 138 (1966).
/3/ G. Albanese, C. Ghezzi, A. Merlini and S. Pace, Phys. Rev. B7, 65 (1973).
/4/ G. Caglioti, Acta Cryst. 17, 1202 (1964).
/5/ E. Burkel, B. von Guérard, H. Metzger, J. Peisl and C.M.E. Zeyen, Z. Phys. B53, 227 (1979).
/6/ F. Mezei, first paper, this volume.
/7/ J.B. Hayter, M.S. Lehmann, F. Mezei and C.M.E. Zeyen, Acta Cryst. A35, 333 (1979).
/8/ J.B. Hayter, M.S. Lehmann, F. Mezei, R. Pynn, and C.M.E. Zeyen, I.L.L. 78HA99T Internal Technical Report.
/9/ F. Mezei and P.A. Dagleish, Commun. Phys. 2, 41 (1977).

NEUTRON SPIN ECHO AND THREE-AXIS SPECTROMETERS

by

R. PYNN

Institut Laue-Langevin,
156X Centre de Tri,
38042 GRENOBLE Cédex,
France.

ABSTRACT

The application of the neutron spin echo (NSE) technique to three-axis spectroscopy is discussed. After a brief introduction which recalls the precession-field geometry required in this case, detailed calculations of performance are presented for two typical cases. Effects of detuning the precession fields from their focused conditions are included explicitly in these calculations which should then be a guide to the design and use of such a spectrometer. It is shown that the three-axis NSE technique can be used to measure both phonon line widths and the wave vector distributions of (quasi)-elastic scattering. In the latter case, resolution similar to that available with the best X-ray spectrometers can be achieved. Finally, design calculations for precision magnets suitable for use with three-axis spectrometers are presented.

INTRODUCTION

As other contributions to this workshop have demonstrated, the neutron spin echo (NSE) spectrometer IN11 has been successfully used to measure energy-widths and line-shapes of a certain class of excitations. This class comprises all excitations whose energies are (locally) independent of wavevector. For example, experiments on quasi-elastic scattering and on the roton minimum in ^{4}He have been performed. In the former the nominal energy of the excitation is zero for all wavevectors, $\vec{Q}$, whereas, for the roton, the energy is essentially independent of $\vec{Q}$ in the region of reciprocal space (resolution ellipsoid) sampled by the instrument.

The IN11 geometry does not however permit measurement of the width of excitations with a finite group velocity, such as phonons. To understand the reason for this we refer to the scattering triangle in Fig. 1. $\vec{k}_I$ and $\vec{k}_F$ are the nominal initial and final neutron wavevectors while $\vec{Q}_o$ is the nominal wavevector transfer. Let us suppose for a moment that the distribution of incident wavevectors is infinitely sharp and that only neutrons with wavevector $\vec{k}_I$ are incident on the sample. The most probable scattering process produces neutrons of wavevector $\vec{k}_F$. However, some part of the phonon dispersion surface close to $\vec{Q}_o$ produces scattered neutrons of wavevector $\vec{k}_f$. The locus of $\vec{k}_f$ is generally known as the scattering surface and, in the

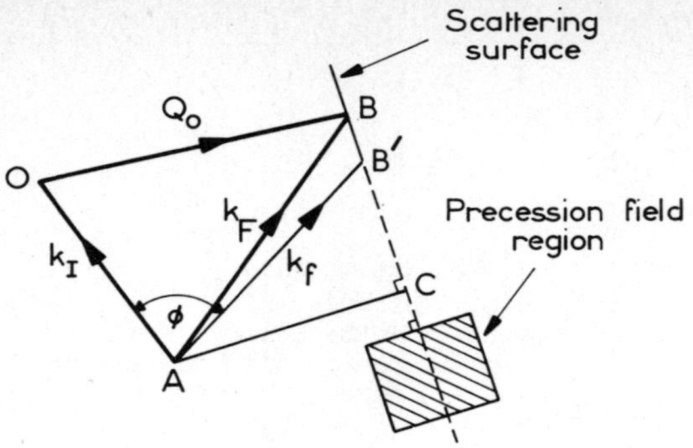

Fig. 1 : Scattering triangle for a typical three-axis measurement, demonstrating tilted magnet geometry.

neighbourhood of $\vec{Q}_o$, it is a straight line in the scattering plane. If the phonon frequency is independent of $\vec{Q}_o$ (flat branch), the scattering surface is perpendicular to $\vec{k}_f$. In the general case, the angle between the scattering surface and $\vec{k}_f$ depends on the gradient of the phonon dispersion surface and the modulus of $\vec{k}_f$ is not constant on the scattering surface.

For the spin echo technique to work we require that all neutrons with wavevectors on the scattering surface BB' in Figure 1 should suffer the same number of spin precessions. This condition can only be met by the IN11 magnet geometry if $\vec{k}_f$ is perpendicular to the scattering surface because the number of precessions is inversely proportional to the modulus $|\vec{k}_f|$ on IN11. It is for this reason that IN11 is suitable for the measurement of linewidths of flat dispersion surfaces. In the general case, the magnet geometry should be chosen so that the number of spin precessions depends on the component of $\vec{k}_f$ perpendicular to the scattering surface, since this component is common to all $\vec{k}_f$. A magnetic field applied over the shaded rectangle of Figure 1 satisfies this condition. Thus in general, the precession magnet in the scattered beam is not symmetrically disposed about the beam but is tilted. A similar argument may be applied to the precession field of the primary spectrometer. This leads to a magnet geometry which is similar to that sketched in Figure 2.

Although the above argument demonstrates simply the reason for the tilted-magnet geometry it does not provide a basis for calculating either the ratio of the magnitudes of the precession fields or their tilt angles. Such a calculation has been carried out both by Mezei /1/ and by the author /2/. Most of this paper will be based on the latter reference which is reproduced in Appendix D of the present volume.

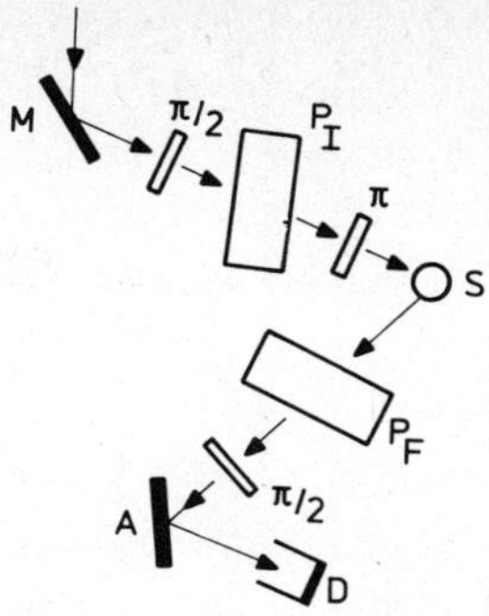

Fig. 2 : Configuration of a three-axis NSE spectrometer.
M : Monochromator, S : Sample, A : Analyser, D : Detector, P_I and P_F : Precession fields.

In this paper, we aim to extend the calculation of Appendix D and to discuss the measurements which might be made on a three-axis NSE spectrometer. In addition we will present some results of magnet-design calculations which have been carried out in conjuction with the proposed test of the three-axis NSE technique at ILL.

I. PHONON LINEWIDTH MEASUREMENTS

In Appendix D the resolution properties of a three-axis NSE spectrometer are considered in detail and the conditions for the focusing of the echo are derived. Expressions (equs (30) & (31) of Appendix D) are obtained for the ratio of the precession field strengths and for the magnet tilt angles in the focussed situation. With these conditions satisfied, the polarisation measured by the spectrometer is found to be

$$P = \exp - \frac{2\pi^2 N_I^2 \Delta^2}{k_I^2 ({}^1/_{\sigma^2} + {}^1/_{\gamma^2})} \quad \cos 2\pi(N_I - N_F) \tag{1}$$

$$\equiv P_o \cos 2\pi(N_I - N_F)$$

Here N_I and N_F are the (mean) numbers of spin precessions in the initial and final beams and Δ is a function of $\vec{k}_I$, $\vec{k}_F$ and the gradient of the phonon dispersion surface. The quantity 2.355 σ is the FWHM of a constant-Q scan through the dispersion surface and γ is the standard deviation of the (assumed) Gaussian phonon lineshape.

If we assume that the focussing conditions for the precession magnets could be set accurately by 'dead reckoning', equ (1) would allow an interpretation of a measurement. As a function of N_I, the measured polarisation ought to be a damped cosine function and, from the damping constant, the intrinsic phonon linewidth γ could be deduced.

However, it is somewhat naive to expect that the precession fields can be set up with sufficient precision to permit the immediate use of equ (1). Rather, we require a more general expression which gives polarisation as a function of N_I, N_F and magnet tilt angles even when the NSE focussing conditions are not satisfied. Such an expression ought to serve as a guide to the use and fine tuning of the spectrometer.

A suitable generalisation of equ (1) may be derived by a conceptually trivial and algebraically boring extension of the arguments given in Appendix D. The general idea is sketched in the Appendix of this paper. Instead of the simple damping term of equ (1) one finds a complicated function which depends on every parameter in the problem ! In this section it is our intention to investigate some features of the general equation for P for two (hopefully !) representative examples.

One example is a transverse [110]T_2 phonon of Nb and the other is a [100]L phonon in the same material. Both are used to demonstrate various points raised in Appendix D. The values of parameters used in the calculations are summarised in table I.

	Q_o (Å^{-1})	ω_o (meV)	E_I (meV)	ε_m	ε_s	ε_a	a' (mev.Å)	b' (meV.Å)
[110] T_2	3.8211	2.38	14.6	-1	-1	-1	0	-12.6
[100] L	3.998	6.6	14.6	1	-1	1	-35	0

Table I : Phonon wavevector (Q_o), phonon energy (ω_o), incident neutron energy (E_I), spectrometer configuration (ε_m, ε_s, ε_a) and components of the gradient of the phonon dispersion surface along and perpendicular to $-\vec{Q}_o$ (a', b'). ε_m, ε_s, ε_a are -1(+1) for clockwise (anticlockwise) scattering at monochromator, sample and analyser respectively. Down scattering ($E_F < E_I$) is used and the monochromator and analyser are taken to be pyrolytic graphite (002) of mosaic (FWHM) 25'.

a) Effect of detuning precession-field magnets

In figure 3 we plot the logarithm of P_o (cf. equ (1)) for the [110]T_2 phonon for linewidths γ = 5 and 50 μeV under the assumption that the NSE focussing conditions are satisfied. Evidently the polarisation is practically undiminished from its maximum value for the γ = 5 μeV excitation in the range of N_I plotted. However the γ = 50 μeV excitation gives a substantial effect even for $N_I \sim 200$. There is very little difference between results with 20'-20'-20'-20' collimations and 40'-40'-40'-40' collimations. The difference arises from the change in the width of a constant-Q scan (σ) which is 0.22 meV for $(20')^4$ collimation and 0.42 meV for $(40')^4$. The more stringent classical resolution (i.e. $(20')^4$) tends to suppress the wings of the phonon lineshape and to yield an apparently smaller linewidth as Fig. 3 shows.

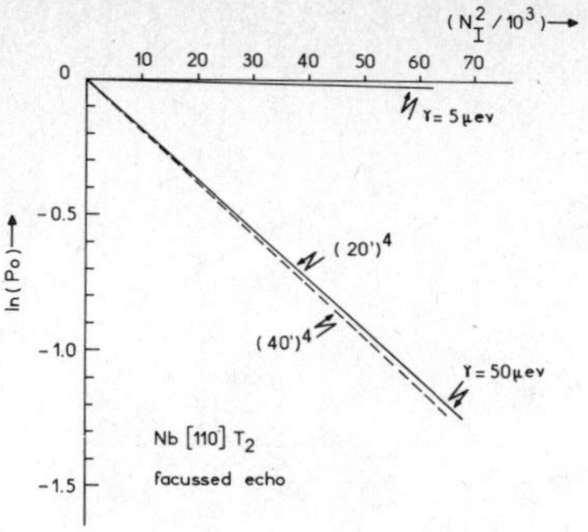

Fig. 3 : Variation of $\ln(P_o)$ with N_I for a typical phonon in the focused configuration.

Figures 4, 5 and 6 show the effect on P_o of the variation of one of the parameters N_F, θ_I or θ_F when the remaining two parameters satisfy the NSE focussing conditions. Here $(\theta_I-\theta_{IO})$ and $(\theta_F-\theta_{FO})$ denote the angles through which the initial and

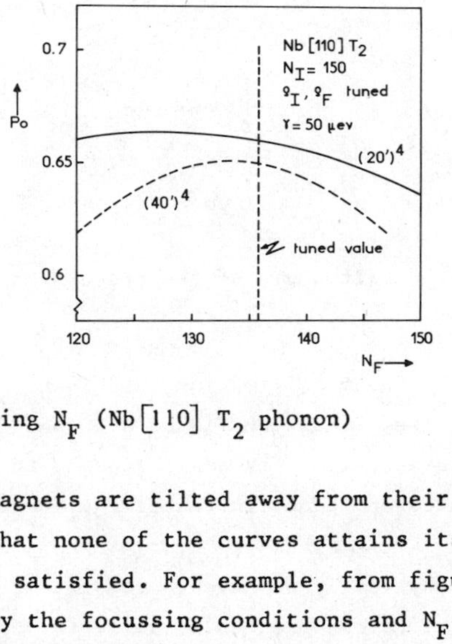

Fig. 4 : Effect of detuning N_F (Nb[110] T_2 phonon)

final precession-field magnets are tilted away from their symmetric positions. It is clear from the figures that none of the curves attains its maximum value when the focussing conditions are satisfied. For example, from figure 4, if the precession-field tilt angles satisfy the focussing conditions and N_F is scanned, the maximum value of P_o does not correspond to the N_F which satisfies the focussing conditions (denoted 'tuned value' in the figures). However, P_o has a greater tendancy to be maximised when the NSE focussing conditions are satisfied if the traditional spectrometer resolution is poor. This is demonstrated by the difference between curves in Figure 4-6 for calculations with 20'-20'-20'-20' collimations and 40'-40'-40'-40' collimations. To understand this result we note that V_o in equ (A8) (see Appendix)

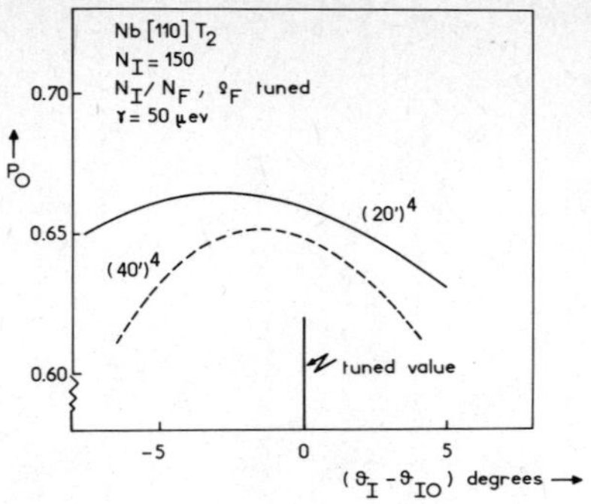

Fig. 5 : Effect of detuning the tilt angle of the precession magnet in the incident beam (Nb [110] T_2 phonon)

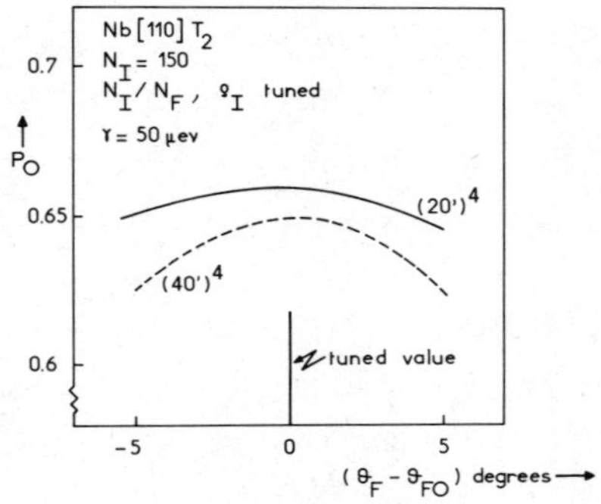

Fig. 6 : Effect of detuning the tilt angle of the precession magnet in the final beam (Nb [110] T_2 phonon)

is smaller for less restrictive beam collimation, and that $\mu = o$ when the focussing conditions are satisfied. As the precession fields are changed from the focussed condition, μ becomes non-zero and, if V_o is small enough, the first term in the exponential of equ (A8) dominates and tends to reduce P_o. The smaller V_o, the greater this reduction will be.

To explain this effect physically we have to reconsider the origin of the parameter μ in equ (A8). The total Larmor precession angle $(N_i - N_f)$ depends in general on the six variables $\vec{k}_i$, $\vec{k}_f$. In order to use the NSE technique we have to arrange for $(N_i - N_f)$ to depend only on the four variables $\vec{Q}$, ω which define the fluctuation spectrum of the scattering sample. This condition, denoted in Appendix D 'obtaining the echo' corresponds to the equality $\mu = o$. Thus if the precession fields are changed in a manner which produces a non-zero value of μ, the echo is lost because the Larmor phase factor depends on variables other than $\vec{Q}$ and ω and averaging over these variables reduces the measured polarisation.

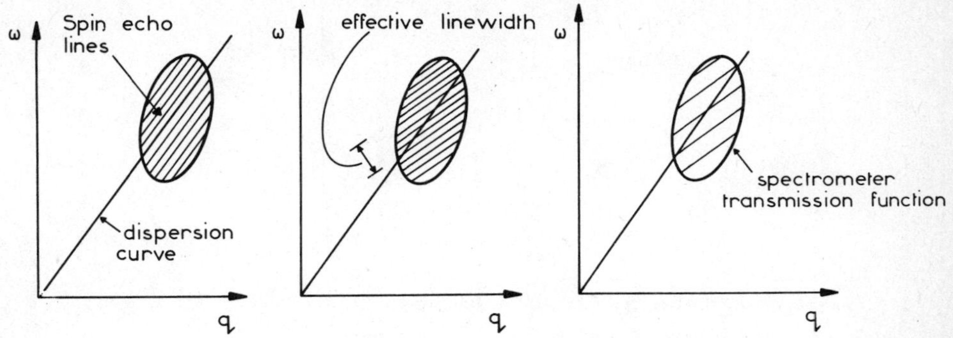

Fig. 7 : a) Spin echo lines and resolution ellipse for focused case
b) c) possible configurations of spin echo lines resulting from detuning of precession fields

If the echo has been obtained and $\mu = o$, the effect of a spin echo measurement may be summarised as in Figure 7. Lines of constant Larmor phase, denoted spin echo lines, are straight lines in $\vec{Q}$, ω space to a first approximation. These lines represent the nodes of a cosine function which Fourier transforms that part of the sample fluctuation spectrum which lies within the transmission function of the spectrometer. When the NSE focussing conditions are satisfied the spin echo lines are parallel to the phonon dispersion surface (Fig. 7a). If the precession fields are changed, with $\mu = o$ and N_I constant, the slope and separation of the spin echo lines are altered. A simple change of slope (Fig. 7b) causes a reduction of P_o because the 'effective' phonon linewidth, measured perpendicular to the spin echo lines, is increased. However, if the change of slope is accompanied by an increase in the separation of the spin echo lines (Fig. 7c), there will be a tendancy for P_o to increase. Thus, even when $\mu = o$, it is possible that P_o does not achieve its maximum value when the NSE focussing conditions are satisfied.

Figures 4-6 demonstrate that for the Nb [110]T_2 phonon the satisfaction of the focussing conditions is not critical. Even if both magnet tilt angles were wrong by 5° and N_I/N_F were in error by 10 % the measured P_o would be 0.56 instead of 0.65 for the case with 40' collimations. If equ (1) were then used to deduce the value of γ, 58 μeV would be found instead of the nominal value of $\gamma = 50$ μeV.

In order to check that the results displayed in Figures 4-6 are not an artifact of the phonon chosen as an example, we display in Figures 8 and 9 corresponding results for the [100]L phonon in Nb. Evidently the results are similar but, in this case, the focussing depends much more critically on the magnet tilt angles. The latter can however be sufficiently well adjusted by scanning the tilt angle until the maximum value of P_o is achieved.

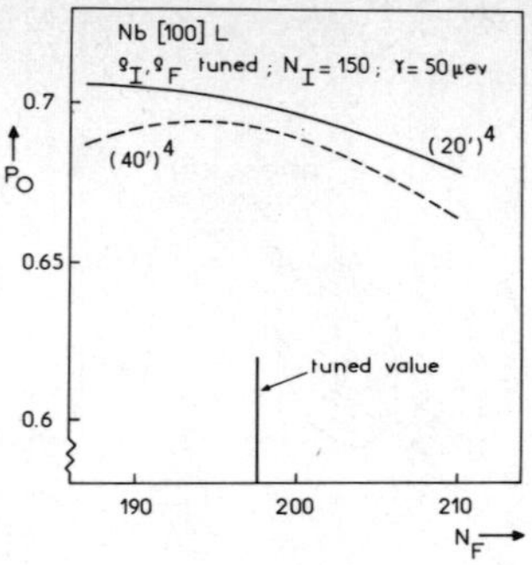

Fig. 8 : Effect of detuning N_F (Nb [100] L phonon)

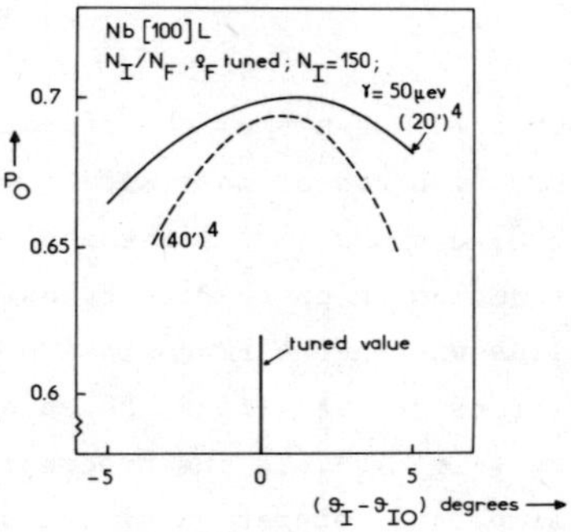

Fig. 9 : Effect of detuning the tilt angle of the precision magnet in the incident beam (Nb [100] L phonon)

One should be cautious of assuming that all wisdom is condensed into figures 3-9. Figures 10 and 11 show results for the same phonon as figures 3-6 but for an assumed phonon width of γ = 10 μeV and N_I = 600. Notice that because the phonon line-width is now considerably less than the classical spectrometer resolution, P_o is essentially independent of collimation at the echo point. Further, since μ is proportional to N_I, moving away from the echo point by changing the magnet strengths or tilt angles causes a rapid loss of echo for this case of N_I = 600. So much so that P_o now depends very sensitively on the magnet tilt angles (Fig. 11) and peaks at or very close to the echo point.

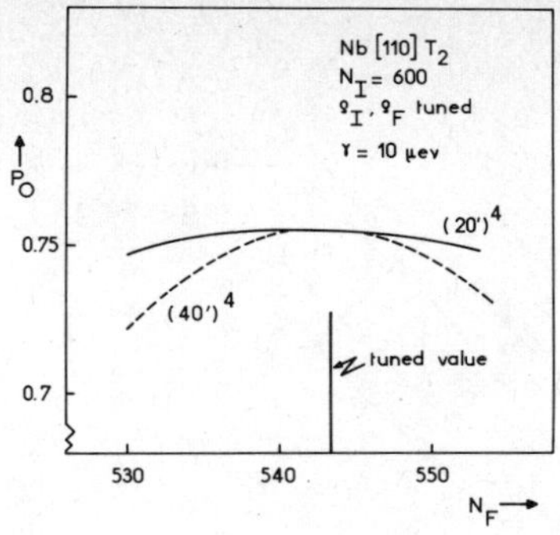

Fig. 10 : As Fig. 4 but with larger value of N_I and smaller intrinsic phonon line width, γ.

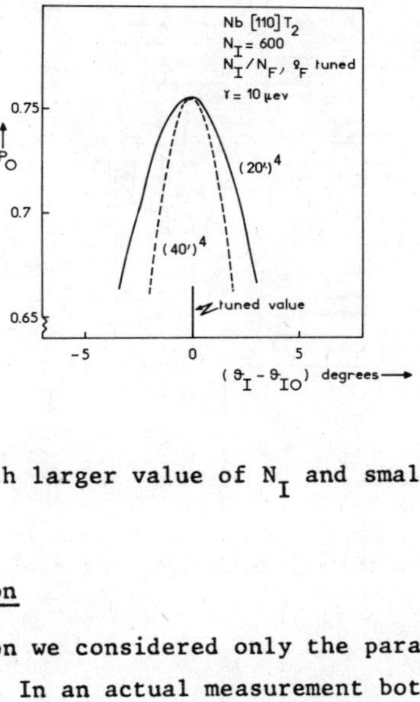

Fig. 11 : As Fig. 5 but with larger value of N_I and smaller intrinsic phonon line width, γ.

b) The measured polarisation

In the previous section we considered only the parameter P_o in equ (1) and ignored the modulating cosine term. In an actual measurement both terms contribute. This means that the measured polarisation (P in equ (1)) looks like a piece of corrugated iron in the N_I-N_F plane. As shown in Fig. 12a the loci of the maxima of the cosine modulation are straight lines equally inclined to the N_I and N_F axes. However, the locus of focussed echoes does not in general obey the relation $N_I = N_F$ (cf. equ (31) of Appendix D). If we imagine the polarisation to be plotted out of the plane of Fig. 12a, an isometric plot of the rectangular region of this figure has the appearance of Fig. 12b ; hence the earlier designation 'corrugated iron'. The region plotted in Fig. 12b is too small to show any significant effects from the variation of P_o.

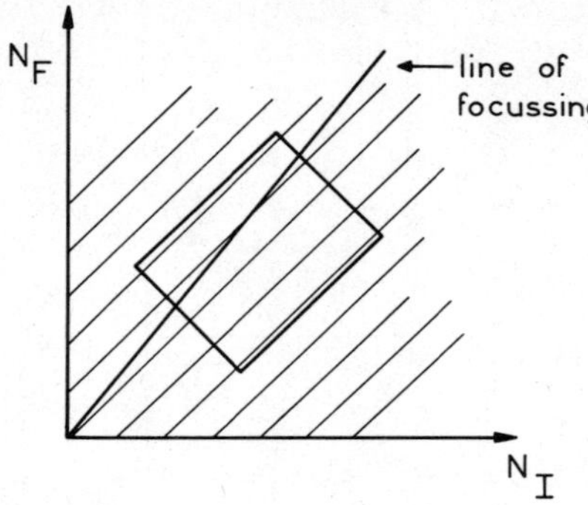

a)

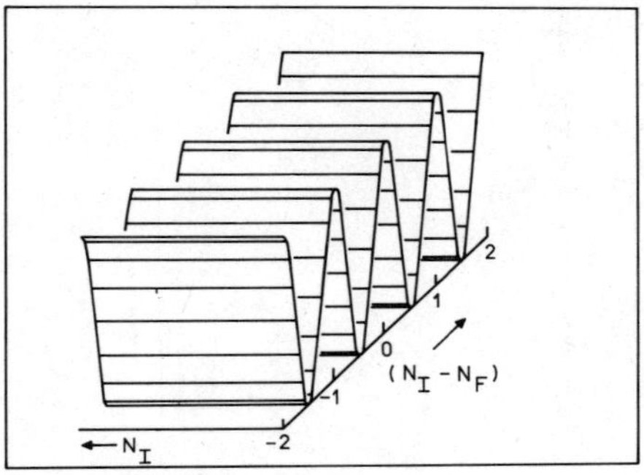

b)

Fig. 12 : a) loci of maxima of cos $2\pi(N_I-N_F)$ in the N_I-N_F plane and the line of NSE focusing
b) isometric plot of measured intensity as a function of N_I and N_I-N_F over the region corresponding to the rectangle of figure 12a.

c) Making a measurement

We are now in a position to combine the information presented in the two previous sections and to suggest how a measurement could be made. For this purpose we consider the phonon used for Figs. 10 and 11. Suppose that we first set all magnet parameters by dead reckoning. We may then change, for example, N_F by a small amount to maximise the cosine modulation term. The quantity measured is then P_o. According to Fig. 11 we may next tune the magnet tilt angles. Finally a further scan of N_F must be carried out to verify that the maximum attainable polarisation (tuned echo) has been obtained.

d) Caveats

The most important caveat actor is that none of the foregoing discussion is based on practical experience ! Nevertheless there are a number of obvious defects even at the calculational level which ought to be mentioned. How, for example, having obtained a maximum echo, do we deduce γ ? Even within the context of the calculation presented here the answer is not obvious because there are a number of parameters, such as N_I, which are not well known a priori and for which no real calibration is possible. Further, the calculation presented here is correct only to lowest order. It takes no account of curvature of either the phonon dispersion surface or of the spin echo lines. Clearly no effects of the inhomogeneities of the magnetic fields are included. All these effects tend to reduce the measured polarisation and are generally corrected for on IN11 by making a comparison with a known elastic scatterer. While such a calibration is also possible in the 3-axis case it furnishes incomplete information. For an elastic scatterer the precession magnets do not need to be tilted and the change of field homogeneity with tilt angle remains uncalibrated. Although we do not have a complete panacea for this problem it is possible the measurement of Bragg peaks, as discussed in the following section, will allow a sufficient calibration.

II a) Elastic scattering

In the previous section the essential concern was the use of the NSE technique for the measurement of phonon linewidths. However, the tilted magnet geometry has other applications, one of which we will now discuss. Suppose we consider cases in which scattering is elastic but is distributed over a region of reciprocal space. Examples range from Huang scattering to Landau-Peierls singularities and include Bragg scattering as a special case. The measured polarisation is still given by equations (A2), (A3) and (A4) of the Appendix but X_4 is to be set identically to zero. We may now choose the magnet parameters so that the echo condition is satisfied (μ = o in equ (A4)) and the spin echo lines have any desired orientation in the scattering plane of the sample. If we wish to measure the width in $|\vec{Q}|$ of a Landau-Peierls singularity for example, we require the Larmor phase factor (equ A3), to be independent of X_2 in addition to μ = o. The constraints on the magnet parameters may then be written as

$$\frac{N_I}{N_F} = \frac{(a-b\rho_F)}{(a+b\rho_I)} \quad (2)$$

$$a(\rho_I - \rho_F) = b(1+b)\rho_I\rho_F \quad (3)$$

where $a = \varepsilon_s \cos \Theta_s$ and $b = \sin \Theta_s$

Here $2\Theta_s$ is the scattering angle at the sample, ε_s is + 1 (-1) for anticlockwise (clockwise) scattering at the sample and ρ_I, ρ_F are the tangents of the magnet tilt angles. When equs (2) and (3) are satisfied the measured polarisation is given by :

$$P = \frac{\int dQ\ F(Q)\ \exp - Q^2/2\sigma^2 . \cos 2\pi\ (N_I - N_F + \delta Q)}{\int dQ\ F\ (Q)\ \exp - Q^2/2\sigma^2} \tag{4}$$

Here F(Q) is the lineshape to be measured, with Q = o at the set point of the spectrometer, and σ is the width of a classical theta-two-theta scan through F(Q) (providing σ is much larger than the width of F(Q)). The quantity δ appearing in equ (4) is given by :

$$\delta = \frac{N_I}{2k} \frac{(\rho_I + \rho_F)}{(a - b\rho_F)} \tag{5}$$

Thus P is essentially the Fourier transform of the lineshape F(Q) multiplied by the ubiquitous cos $2\pi\ (N_I - N_F)$ term. A rough value for the resolution of such measurement is given by the inverse of δ in equ (5).

The advantages of using the spin echo technique to measure a lineshape in Q are threefold. Firstly, the spectrometer does not move during the measurement and thus errors are not related to systematic problems of spectrometer alignment. Secondly, the Fourier transform technique is sensitive to line shape and one ought to be able to distinguish clearly, for example, between Lorentz and Gaussian lineshapes. In the latter case ln (P_o) is independent of δ at small δ whereas, for the Lorentzian, ln (P_o) is proportional to δ. Thirdly, the resolution can be made extremely fine and it can be calibrated directly by measurement of a Bragg peak.

As an example we consider the Landau-Peierls anomaly measured by Als-Nielsen et. al /3/ by X-ray techniques in the smectic-A phase of 80 CB. This anomaly occurs at a wavevector $Q_o = 0.2\ Å^{-1}$ and the X-ray experiment had a resolution (FWHM) of $2 \times 10^{-3}\ Å^{-1}$. For a classical three axis measurement with $k_I = 2.662\ Å^{-1}$ (the graphite filter wavevector) we would expect a resolution of at best (10' collimation), $0.01\ Å^{-1}$. For the spin echo measurement we find that with $\rho_F = 0.5$ (final magnet tilt angle of 26.57°) we require $\rho_I = 0.51$, $N_I/N_F = 0.9627$ and $\delta = 0.5153 \times N_I/k_I$. Thus with $N_I = 10^3$, the resolution ought to be of order $0.005\ Å^{-1}$. This value would be essentially halved by increasing the magnet tilt angles to 45°. It is thus relatively easy to achieve resolution better than that available with a classical 3-axis machine and possible to compete realistically with that obtainable on the best X-ray spectrometers.

b) Critical scattering

The wavevector distribution of critical scattering close to a phase transition may also be measured by the technique described above. In this case there is an additional constraint on the magnet parameters derived from the fact that the coefficient of X_4 in equ (A3) is required to vanish. Provided the spectrometer transmission function is larger than the energy-width of the critical scattering the latter may be callipered in any desired direction in Q space. In fact the NSE technique will probably only be useful in this case if the critical scattering is sufficiently anisotropic for the Q-width in the direction of measurement to be constant within the wavevector transmission function of the spectrometer.

III MAGNET DESIGN

In this section we present a few details of the magnet-design calculations which have been carried out in connection with the proposed test of the 3-axis NSE technique at ILL. The specification of magnetic fields is without doubt the major problem involved in turning the calculations of this paper into a practical reality. If the necessary magnetic fields can be produced there is no reason to doubt that the calculated performance could be obtained.

The magnetic fields required for a 3-axis NSE spectrometer fall into three groups. There are guide fields which simply maintain the neutron polarisation, spin-turn coils which cause well defined rotations of the polarisation direction and precession fields which impart the necessary number of Larmor precessions. Guide fields are well understood from classical, polarised-neutron experiments. Spin-turn coils have been used extensively on IN11 and their properties have been measured and calculated in detail /4/. Mounting such coils on a 3-axis instrument may involve some problems because the fringing fields of the precession magnets, seen by the spin turn coils, vary both with the strength of the precession fields and with the tilt angles of the precession magnets.

The precession magnets themselves present an unusual problem of magnet design. The number of Larmor precessions, N, obtained with a precession magnet is related to the integral, $L\bar{B}$; of the magnet field strength along the neutron trajectory by :

$$N = (L\bar{B})\ \lambda/135.2 \qquad (6)$$

where λ is the neutron wavelength in Å and $L\bar{B}$ is in Gauss. cm. For example, with $\lambda = 2.5$ Å and $N = 100$, we require $L\bar{B} = 5.4 \times 10^3$ Gauss. cm and it is this line integral (rather than B itself) which should be homogeneous across the neutron beam. In order to achieve NSE focussing, both initial and final neutron beams have to be passed through precession fields at angles up to $\sim 45°$. Wide magnets are therefore required.

The basic design of the precession magnets is summarised in figure 13a. The magnetic field in the median (y=o) plane is in the y direction while the neutron trajectory (for zero magnet tilt) is in the z direction. Design calculations for such a magnet have been carried out using the 'finite-element' program /5/ GFUN developped at the Rutherford Laboratory. This program is known to yield line integrals with a relative precision of about 10^{-3}.

The magnet envisaged for the NSE spectrometer is an electromagnet and three possible forms for the windings of the energising coils have been considered.

Bedstead windings : the core of the magnet is as in Fig. 13 (window frame type) and the windings run in the z direction within the magnet and are turned up, out of the beam, at the entrance and exit to the magnet. The windings thus pass above and below the beam at the entrance and exit in the x direction.

H-frame : the magnet core is as in Fig. 13 but with the addition of pole pieces which extend into the magnet from the top and bottom faces. The windings are 'rectangular' and run around the pole pieces in the x - z plane. Such windings are easier and cheaper to make than bedstead windings and pole-face (shimming) windings can be incorporated easily. However, the field homogeneity is not as good as for the bed-stead-wound, window-frame magnet.

Racetrack windings : The iron core is as in Fig. 13 and the coils are wound round the magnet side limbs in the x - z plane. This geometry is easier to produce than bedstead windings, gives similar field homogeneity to the bedstead case and a smoother field variation at the magnet entrances than the bedstead coils. The latter point is an important consideration for the variation of the field line-integral as a function of y (see below). The disadvantage of this geometry is that the magnet width is somewhat more limited for a given length (z direction) of iron core.

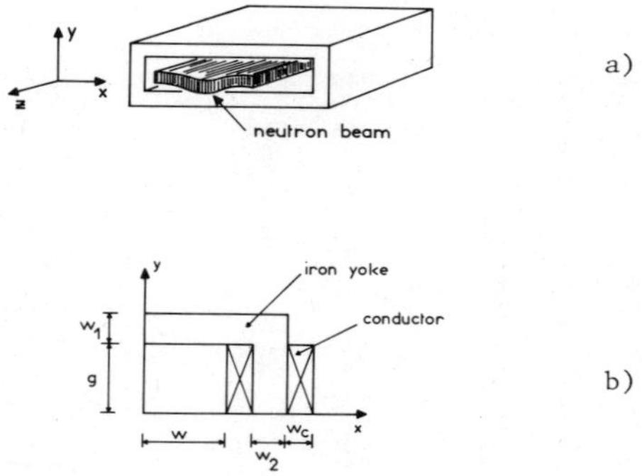

Fig. 13 : a) Typical precision field magnet
b) cross-section of precision field magnet ; dimension chosen for the results of table II are w = 30 cm ; g = 8 cm ; $w_1=w_2=10$ cm ; w_c = 3 cm ; magnet length = 20 cm ; Power consumption = 2.2 kW.

After fairly extensive calculations with GFUN the racetrack windings were chosen. To represent the magnet circuit 200 elements with a B-H curve of high quality magnet steel were used in the program. The dimensions (thicknesses) of the iron core were chosen so that magnetic saturation did not occur even for the maximum envisaged energising current. Conductor dimensions were chosen to yield a maximum current density of 400 A. cm^2. This represents an acceptable compromise between requirements of power density (cooling capacity) and conductor size. Final magnet dimensions were chosen as in Fig. 13b which represents a cross-section of 1/4 of the magnet displayed in Fig. 13a. The performance of the magnet is summarised in table II. In this table

#	x (cm)	y (cm)	θ (deg)	($L\bar{B}$) KG.cm	
1	0	0	0	58.58	Line integral between $z = \pm 70$ cm
2	0	2	0	58.98	
3	0	3.5	0	59.90	
4	3	0	0	58.54	
5	3	2	0	58.94	
6	3	3.5	0	59.80	
7	0	0	30 (1.13)	66.14	
8	3	0	30	66.12	
9	1	0	30	67.08	Line integral between $z = \pm 105$ cm
10	0	0	45 (1.36)	79.92	
11	1	0	45	79.92	
12	0	1	45	80.08	
13	1	1	45	80.06	

Table II : Line integrals of $|B|$ for the racetrack-wound, window-frame magnet shown in figure 13b. The values of x and y are the coordinates of the integration path at z = o (cf. fig. 13a) and θ is the angle of the integration path with respect to z. The dimensions of the magnet are those given in the caption of Figure 13.

the x and y coordinates are those of the beam at z = o and θ is the (tilt) angle of the neutron flight path with respect to the z direction. Values of the line integral of the field modulus ($L\bar{B}$) are given in KG.cm. For the first 8 rows of the table, this line integral was evaluated between $z = \pm 70$ cm while for the final 5 rows the integration limits were $z = \pm 105$ cm. An idea of the degree of convergence of the calculations can be obtained from a comparison of rows 8 and 9. The values in parenthesis in rows 7 and 10 are the ratios of field integrals at θ = 30° and θ = 45° to that at θ = 0. Ideally, if lines of constant field strength were parallel to the x axis throughout the magnet, these ratios should be 1.15 and 1.41 instead of 1.13 and 1.36 as calculated by GFUN. The latter values are acceptable and should pose no experimental problem.

From table II one sees that the variation of the field integrals across the neutron beam in the x-direction is, for $-3 < x < 3$ cm, in the noise of the calculation and that ($L\bar{B}$) should vary by less than 10^{-3} in this direction. However, the variation of ($L\bar{B}$) with y is greater. Empirically /6/, the variation of ($L\bar{B}$) with y is given by the relation :

$$\Delta\int|B|dz = C \frac{y^2}{g(\ell + 1.4\ g)} \int|B|dz \qquad (7)$$

where ℓ is the magnet length (distance between entrance and exit faces) and C is related to the quantity $[\partial B_y/\partial z]_{y=0}$ and is in the range $0.4 < C < 0.7$. Our calculations with GFUN indicate a value of C = 0.42 and this can probably not be improved either by change of magnet design or by shimming. To improve the homogeneity of $(L\bar{B})$ in the y direction, the only solution /7/ is to increase the magnet gap g. For the tests of NSE technique envisaged at ILL this solution would be prohibitively expensive. Thus we are restricted to a neutron beam of total height 2 cm over which the number of Larmor precession, N, varies by about $\pm$ 1.5 x 10^{-3} N. It is not at all certain how serious a limitation this will prove to be however. For any neutron which does not change momentum perpendicular to the scattering plane, the effects of vertical (y) beam inhomogeneity in the initial and final magnets cancel. We have performed a rough calculation for a point scattering sample placed symmetrically between two precession magnets of the design displayed in Fig. 13b. The magnet centres were chosen to be 80 cm apart. Using the empirical relation equ (7) we find that, for $N_I = N_F = 1000$, field inhomogeneity along y ought to reduce the measured polarisation by 90 %, 30 % and 15 % for vertical beam divergences (FWHM) of 6°, 4° and 3° respectively. Since sample-size effects are not included these numbers are indicative rather than reliable.

ACKNOWLEDGEMENTS

I would like to thank Feri Mezei for several stimulating discussions about the interpretation of the results presented in this paper and John Hayter for showing me some of their practical ramifications. Thanks are due also to Jeff Penfold and John Hayter for allowing me to use their computer-graphics programs to produce Figure 12b. The support of the directors of the ILL for the project designed to test a 3-axis NSE spectrometer is also gratefully acknowledged.

REFERENCES

/1/ F. MEZEI, Proc. Int. Symp. on Neutron Scattering (IAEA, Vienna 1977) (see Appendix C of this volume)

/2/ R. PYNN, J. Phys. E 11, 1133 (1978) (see Appendix D of this volume)

/3/ J. ALS-NIELSEN, J.D. LITSTER, R.J. BIRGENAU, A. LINDEGAARD-ANDERSEN and S. MATHIESEN, "Order in Strongly Fluctuating Systems" (Plenum, 1979)

/4/ J.B. HAYTER, Z. Physik B31, 117 (1978) (see Appendix B of this volume) ; O. SCHÄRPF, this volume

/5/ G F U N 3 D User Guide (Rutherford Laboratory report RL-76-029/A)

/6/ Magnetic Fields for Transporting Charged Beams by G. PARZEN (Brookhaven National Laboratory report BNL 50536).

/7/ see however, the Fresnel correction coils considered by F. MEZEI in the last paper of this volume.

APPENDIX

In this appendix we sketch the derivation of a general expression for the polarisation measured by a three-axis NSE spectrometer. The expressions obtained are valid for any values of precession fields and magnet tilt angles.

As explained in Appendix D of this volume the polarisation measured by a 3-axis NSE spectrometer is given by (eqn. (4b) of Appendix D)

$$P = \frac{1}{n_o} \int d^3k_i \int d^3k_f S(\vec{Q},\omega)\ p(\vec{k}_i,\vec{k}_f)\ \cos 2\pi(N_i - N_f) \tag{A1}$$

In this equation $S(\vec{Q},\omega)$ is the scattering law of the sample and $p(\vec{k}_i,\vec{k}_f)$ is the probability that a neutron, incident with wavevector $\vec{k}_i$ and scattered with wavevector $\vec{k}_f$, will be transmitted through the spectrometer. N_i and N_f are the number of spin precessions in the initial and final precession field regions.

To evaluate (A1) we substitute for $p(\vec{k}_i,\vec{k}_f)$ from eqns. (14) and (15) of Appendix D and for $N_i - N_f$ from eqn. (28) of Appendix D and carry out the integrations over y_f and z_f (see Appendix D for definitions). The result is

$$P = \frac{1}{n_o} \exp-\left[2\pi^2\mu^2/V_o^{\,2}\right] \int dX_1 \int dx_2 \int dX_4 \exp-\frac{1}{2}\Sigma\, M_{k\ell}X_kX_\ell$$
$$\times\ F(\omega-\omega_Q)S(\vec{Q})\cos\left[2\pi f(X_1,S_2,X_4)\right] \tag{A2}$$

where the argument of the cosine (cf. eqn (28) of Appendix D) is given by

$$\begin{aligned} f(X_1,X_2,X_4) = N_I - N_F - \frac{N_I}{k_I}\{ & bv_1X_1 + (bu_2-a)X_2 + bu_4X_4 \\ & - \rho_I\left[av_1X_1 + (au_2+b)X_2 + au_4X_4\right]\} \\ + \frac{N_F}{k_F}\{ & Bu_1X_1 + Bu_2X_2 + Bu_4X_4 \\ & - \rho_F\left[Au_1X_1 + Au_2X_2 + Au_4X_4\right]\} \\ & - \mu\Sigma_\ell\ \vec{V}_o\cdot\vec{V}_\ell X_\ell/V_o^{\,2} \end{aligned} \tag{A3}$$

with $\rho_I = n_{I2}/n_{I1}$ and $\rho_F = n_{F2}/n_{F1}$.

Except for μ, all quantities which occur in the above equations are defined in Appendix D. μ is the coefficient of y_f in eqn (28) of Appendix D and is thus given by

$$\mu = -N_I\ (a + \rho_I b)/k_I + N_F(A + \rho_F B)/k_F \tag{A4}$$

The condition denoted in Appendix D as 'obtaining the echo' is satisfied if $\mu = 0$ and the calculation presented in Appendix D is valid for this case. Here we allow μ to be finite in order to calculate P as a general function of precession fields (N_I and N_F) and magnet tilt angles (ρ_I and ρ_F).

To evaluate (A2) we note that there is a change of variable

$$\omega-\omega_o \equiv X_4 = X'_4 + \alpha X_1 + \beta X_2 \tag{A5}$$

which removes all dependence of $f(X_1,X_2,X_4)$ (eqn (A3)) on X_1 and X_2. If we make this substitution and, in addition, assume a Gaussian form for the phonon lineshape

$$F(\omega-\omega_Q) = \exp-(\omega-\omega_Q)^2/2\gamma^2 \tag{A6}$$

where ω_Q is given by eqn (24) of Appendix D (with c' = 0) all the integrals in (A1) may be performed straightforwardly provided $S(\vec{Q})$ is independent of $\vec{Q}$ within the transmission volume of the spectrometer. The result is :

$$P = P_o \cos\left[2\pi(N_I-N_F)\right] \tag{A7}$$

where $$P_o = \exp-\left[2\pi^2(\mu^2/V_o{}^2 + \delta^2\Gamma^2)\right. \tag{A8}$$

$$\frac{1}{\Gamma^2} = M'_{44} + \frac{2M'_{12}\,M'_{14}\,M'_{24} - M'_{11}\,M'^2_{24} - M'_{22}\,M'^2_{14}}{M'_{11}\,M'_{22} - M'^2_{12}} \tag{A9}$$

$$\begin{aligned}
M'_{11} &= M_{11} + \alpha^2 M_{44} + 2\alpha M_{14} + (\alpha - a')^2/\gamma^2 \\
M'_{22} &= M_{22} + \beta^2 M_{44} + 2\beta M_{24} + (\beta - b')^2/\gamma^2 \\
M'_{44} &= M_{44} + 1/\gamma^2 \\
M'_{12} &= M_{12} + \alpha\beta M_{44} + \beta M_{14} + \alpha M_{24} + (\alpha - a')\,(\beta - b')/\gamma^2 \\
M'_{14} &= M_{14} + \alpha M_{44} + (\alpha - a')/\gamma^2 \\
M'_{24} &= M_{24} + \beta M_{44} + (\beta - b')/\gamma^2
\end{aligned} \tag{A10}$$

$$\alpha = \frac{\mu\vec{V}_o\cdot\vec{V}_1/V_o{}^2 + N_I A_1 v_1/k_I - N_F B_1 u_1/k_F}{-\mu\vec{V}_o\cdot\vec{V}_4/V_o{}^2 - N_I A_1 u_4/k_I + N_F B_1 u_4/k_F} \tag{A11}$$

$$\beta = \frac{\mu\vec{V}_o\cdot\vec{V}_2/V_o{}^2 + N_I(A_1 u_2 - A_2)/k_I - N_F B_1 u_2/k_F}{-\mu\vec{V}_o\cdot\vec{V}_4/V_o{}^2 - N_I A_1 u_4/k_I + N_F B_1 u_4/k_F} \tag{A12}$$

$$\delta = -N_I A_1 u_4/k_I + N_F B_1 u_4/k_F - \mu\vec{V}_o\cdot\vec{V}_4/V_o{}^2 \tag{A13}$$

$$A_1 = b - \rho_I a \qquad A_2 = a + \rho_I b$$

$$B_1 = B - \rho_F A \qquad B_2 = A + \rho_F B$$

These expressions reduce to those given in Appendix D when $\mu = 0$ and the magnet tilt angles are given by eqn (30) of Appendix D. In this case α and β are identical to the components a' and b' of the gradient of the phonon dispersion surface and the definition of δ given by eqn (A13) is the same as that in eqn (32) of Appendix D.

CONCLUSION : CRITICAL POINTS AND FUTURE PROGRESS

F. MEZEI

Institut Laue-Langevin, 156X, 38042 Grenoble Cedex, France

and

Central Research Institute for Physics, Pf. 49, Budapest 1525, Hungary

INTRODUCTION

In this paper, the last of the volume, I will try to summarize those points which, to the best of my understanding, are the most relevant to future work with Neutron Spin Echo (NSE). The first of these points that I am going to discuss in detail is related to the question : how reliable are the NSE data ? i.e. how well does the NSE scheme for instrumental resolution corrections really work ? And how can spurious effects occur ? It will be shown that these problems are now satisfactorily understood and experimentally checked. Next I will consider the main technical limitations of NSE, viz. the inhomogeneities of the precession fields and the intensity losses due to the polarization analysis involved. In the following section of this concluding paper I will then try to pin down the types of experiments in which the NSE results are likely to most usefully complement the information provided by other methods while the conclusion is meant to put NSE into the context of inelastic neutron scattering as a whole.

1. A CRITICAL VIEW ON NSE DATA REDUCTION

In the previous papers the effective, measurable NSE polarization has been given using various approximations, which are most useful in the practice. In order to assess the possibility of errors, however, we have to reconsider the NSE spectrometer response with no approximations at all. The expression which will be obtained this way, is much too complex for any practical use and contains in addition many unknown functions, but it will permit the precise definition of those conditions under which simple practical procedures can be used. As far as I can tell, the following analysis offers a plausible explanation for the few controversial results I am aware of.

The problem of data reduction is most relevant for quasielastic scattering experiments making use of the highest available energy resolution (10^{-4} - 10^{-6}). Therefore the calculations will be done for this case, but the conclusions are of more general use. To start with, we have to make the following observations :

a) Due to the inhomogeneities of the precession fields, and also due to trivial geometrical factors, such as finite sample volumes, the neutron spin Larmor precession angles depend not only on the neutron velocities, but also on their trajectories.

b) For both the polarizer and the analyser neither the polarization efficiency nor the transmitted neutron flux are perfectly constant over the whole beam and they also depend to some extent on the neutron velocity.

Each neutron scattering event can be fully characterized by a) the neutron trajectory from the polarizer exit window via the scattering point in the sample to the analyser entrance window, and b) the incoming and outgoing velocities. Let us refer to a particular neutron trajectory by the symbolic vector quantity $\vec{T}$, which in fact stands for 7 parameters viz. the coordinates at the polarizer exit window (2), at the sample scattering point (3) and at the analyser entrance window (2). The total Larmor precession angle ϕ can be written as

$$\phi(\vec{T}, v_o, v_1) = \phi_o - \phi_1 = \gamma_L \left(\frac{h_o}{v_o} + \frac{\Delta h_o(\vec{T})}{v_o} - \frac{h_1}{v_1} - \frac{\Delta h_1(\vec{T})}{v_1} \right) \tag{1}$$

where γ_L is the Larmor constant, v_o and v_1 the incoming and outgoing neutron velocities and

$$h_i = \left< \int H_i(\vec{T}) ds \right> \qquad \Delta h_i(\vec{T}) = \int H_i(\vec{T}) ds - h_i$$

with H_i (i = 0, 1) being the precession fields. Thus h_o and h_1 represent the average values of the line integrals of the magnetic fields from the first $\frac{\pi}{2}$ coil to the sample scattering point and from there to the second $\frac{\pi}{2}$ coil, respectively. More precisely, e.g. for normal NSE with the flipper close to and just before the sample, the small part of the field integral from the flipper to the sample has to be taken with a negative sign in calculating h_o. The Δh's are the deviations of the line integrals from the corresponding average values for a particular trajectory $\vec{T}$. A given experimental configuration can be characterized by the probability distribution function $F(\vec{T}, v_o, v_1)$ which describes both the spatial and velocity distribution of the incoming beam, and the probability for the scattered beam to be detected. (This function would describe the actual $\vec{T}$, v_o, v_1 distribution for a hypothetical sample of the considered geometry which scatters the incoming neutrons with equal probability into any outgoing direction and velocity state). The combined inhomogeneous polarization efficiency of the polarizer, the analyser and the $\frac{\pi}{2} - \pi - \frac{\pi}{2}$ flips can be described by the function $P(\vec{T}, v_o, v_1)$. Finally, the measured NSE polarization signal will be given, by the use of the familiar complex notation, as

$$P_{NSE}^{eff} = \frac{\int d\vec{T} dv_o dv_1 F(\vec{T}, v_o, v_1)\, P(\vec{T}, v_o, v_1)\, S(\vec{\kappa}, \omega)\, P_S(\vec{\kappa}, \omega) e^{i\phi(\vec{T}, v_o, v_1)}}{\int d\vec{T} dv_o dv_1 F(\vec{T}, v_o, v_1)\, S(\vec{\kappa}, \omega)} \tag{2}$$

where the momentum transfer is given as a function $\vec{\kappa} = \vec{\kappa}(\vec{T}, v_o, v_1)$ and the energy transfer as $\hbar\omega = \frac{1}{2} mv_1^2 - \frac{1}{2} mv_o^2$.

The sample is characterized by the scattering function $S(\vec{\kappa},\omega)$ and the NSE polarization factor $P_S(\vec{\kappa},\omega)$ describing the neutron spin change in the scattering, as discussed in my first paper in this volume. (Strictly speaking, for magnetic scattering P_S explicitly depends on $\vec{\kappa}$ - with simple limiting values as given above - and in addition it will depend implicitely on $(\vec{\kappa},\omega)$ if $S(\vec{\kappa},\omega)$ contains contributions with different P_S values, e.g. nuclear spin incoherent scattering, $P_S = -\frac{1}{3}$, and non-magnetic coherent scattering, $P_S = 1$). This exact equation is completely untreatable and most functions in it are not well measurable anyway. It can only be simplified if we can assume that only elastic and very small energy change scattering takes place, i.e. $|v_o - v_1| \lesssim 10^{-4}\, v_o$, and thus v_1 can be practically always replaced by v_o in the argument of the functions F and P, furthermore for the κ variable in the argument of S and P_S. Thus the relevant variable space reduces to $(\vec{T}, v_o)$, and v_1 only occurs in ω and ϕ. This approximation is fundamental in the NSE method, and was in fact pointed out in my introductory paper above : viz. the neutron transmission properties of the spectrometer described by the F and P functions must not be sensitive to the neutron energy differences measured by the NSE. These transmission properties in fact correspond to the classical resolution that the instrument would have without the use of NSE (host-spectrometer resolution). The just excluded contributions of scattering processes with energy changes comparable to the host-spectrometer resolution can be evaluated using classical resolution calculation schemes with all their complexities. In high resolution NSE work these effects generally do not contribute to the nominator in Eq. (2), due to the rapidly oscillating $\exp(i\phi)$ term. Often this is all what we need to know about them.

With these assumptions we can then write

$$P^{eff}_{NSE} = \frac{\int d\vec{T}\, dv_o\, F(\vec{T},v_o)\, P(\vec{T},v_o) \int d\omega\, S(\vec{\kappa},\omega)\, e^{i\phi(\vec{T},v_o,v_1)}}{\int d\vec{T}\, dv_o\, F(\vec{T}, v_o) \int d\omega\, S(\vec{\kappa},\omega)} \tag{3}$$

where for convenience, the integration over v_1 has been transformed into an integration over ω. In Eq. (1) the $\gamma_L \Delta h_i / v_i$ terms should not exceed about 300 degrees, otherwise the Δh_i field inhomogeneities would completely destroy the precessing polarization, thus the

$$\gamma_L \frac{\Delta h_o(\vec{T})}{v_o} - \gamma_L \frac{\Delta h_1(\vec{T})}{v_1} \simeq \gamma_L \frac{\Delta h_o(\vec{T}) - \Delta h_1(\vec{T})}{v_o} \equiv \Delta\phi_h(\vec{T},v_o) \tag{4}$$

approximation is valid to a fraction of a degree in ϕ, i.e. it is fully justified for evaluating Eq.(3). Finally, at the experimentally well defined echo point : $h_o = h_1$, we have

$$\gamma_L \left(\frac{h_o}{v_o} - \frac{h_1}{v_1} \right) \simeq \gamma_L \frac{h_o}{v_o^2} (v_1 - v_o) \simeq \gamma_L \frac{\hbar\, h_o}{m\, v_o^3}\, \omega \tag{5}$$

to an accuracy of better than 0.1 deg. in ϕ. Introducing the notation

$$S_P(\vec{\kappa},t) = \frac{\int d\omega\, S(\vec{\kappa},\omega) P_S(\vec{\kappa},\omega)\, e^{i\omega t}}{S(\vec{\kappa})} \tag{6}$$

where

$$S(\vec{\kappa}) = \int d\omega\, S(\vec{\kappa},\omega) \tag{7}$$

is the well known structure factor and considering Eqs. (1), (3), (4) (5) and (6) we arrive at our final result :

$$P_{NSE}^{eff} = \frac{\int d\vec{T}\, dv_o F(\vec{T},v_o)\, P(\vec{T},v_o)\, e^{i\Delta\phi_h(\vec{T},v_o)}\, S(\vec{\kappa})\, S_P(\vec{\kappa}, t)}{\int d\vec{T}\, dv_o\, F(\vec{T},v_o)\, S(\vec{\kappa})} \tag{8}$$

where $t = \gamma_L \hbar \frac{h_o}{mv_o^3}$ and $\vec{\kappa} = \vec{\kappa}(\vec{T},v_o)$, which together with $F = F(\vec{T},v_o)$, $P = P(\vec{T},v_o)$ and $\Delta\phi_h = \Delta\phi_h(\vec{T},v_o)$ will be understood in what follows, unless otherwise stated.

Equation (8) can be written in the shorter form

$$P_{NSE}^{eff} = \langle P\, e^{i\Delta\phi_h}\, S(\vec{\kappa},t)\, \rangle_{FS(\vec{\kappa})} \tag{9}$$

where the angle bracket stands for averaging over the $(\vec{T},v_o)$ variable space using the weight function given in the index (F.S$(\vec{\kappa})$ in this case). For a standard elastic coherent scatterer with $S_o(\vec{\kappa},\omega) = S_o(\vec{\kappa})\,\delta(\omega)$, obviously $S_P^o(\vec{\kappa},t) = 1$ [cf. Eqs : (6) and (7)] and the measured echo signal becomes

$$P_{NSE}^{o} = \langle\, P\ e^{i\Delta\phi_h} \rangle_{F_o S_o(\vec{\kappa})} \tag{10}$$

which is the weighted average of just the inhomogeneity effects, and which can be looked upon as the instrumental resolution function. If the weight functions in the two averages, Eqs. (9) and (10) are the same, FS = $F_o S_o$, we can write

$$P_{NSE} = \frac{P_{NSE}^{eff}}{P_{NSE}^{o}} = \frac{\langle\, Pe^{i\Delta\phi_h}\, S_P(\vec{\kappa},t)\, \rangle_{FS}}{\langle\, Pe^{i\Delta\phi_h} \rangle_{FS}} = \langle S_P(\vec{\kappa},t)\rangle_{FSP\exp(i\Delta\phi_h)} \tag{11}$$

This is the recipe which was given in my introductory paper for the measurement of the NSE signal. (Note that in practice usually only the maximum amplitude of the NSE group is measured, and not the absolute phase of the Larmor precession, i.e. the above equations are only used to determine the modulus of $S_P(\vec{\kappa},t)$ if it happens to be complex).

The measured NSE signal, as given by Eqs. (6) and (11) is inevitably the result of averaging processes, which have to be looked into in detail in each case when interpreting the results (see e.g. the examples in John Hayter's contribution). A major part of the averaging in Eq. (11) occurs in fact over the $\vec{\kappa}$ resolution of the spectrometer configuration, and often it is just a minor effect, if $S_P(\vec{\kappa},t)$ is a smooth function. What is really crucial, on the other hand, is to be sure that Eq. (11) holds, i.e. to check all of the assumptions under which it could be derived. These were

a) The relevant values of $\hbar\omega$ i.e. $|v_o - v_1|$ are small enough so that for the transmission and polarization efficiency of the analyser-detector system and the spin flip coils $v_1 = v_o$ can be taken.

b) $S(\vec{\kappa},\omega)$ depends on $\vec{\kappa}$ slowly enough so that $\vec{\kappa}$ can be taken with v_1 replaced by v_o.

c) The shape of the structure factor $S(\vec{\kappa})$ and the scattering geometry function $F(\vec{T},v_o)$ are identical for the sample and the standard within the relevant variable range. (In fact the identity of the F.S product is sufficient, but not obvious to check.) Note that when considering sample geometry, we have to take into account both the absorption self-screening and, in particular, the multiple scattering effects which have been neglected up to now. It is vital to check these latter effects, since $S(\vec{\kappa},\omega)$ only refers to single scattering and, in addition, multiple scattering in a voluminous sample corresponds to an $F(\vec{T}, v_o)$ trajectory distribution drastically different from that for single scattering.

In high resolution work, conditions a, and b, are usually automatically and largely met. Condition c, on the other hand, has to be carefully checked. The effect of the shape of $S(\vec{\kappa})$ can be studied by comparing different elastic standard samples. Well measurable differences have been observed e.g. between standars displaying isotropic diffuse scattering and small angle scattering strongly peaking in the forward direction, showing that it can be essential to use standards matched to the sample structure factors. Yolande Alpert and myself have investigated in addition the effect of the scattering geometry function $F(\vec{T},v_o)$ which in practice only varies with the sample geometry. We have determined P_{NSE} via Eq.(11) using two different polarizer-analyser configurations in order to have a cross check. In the first experiment (Fig. 1a) the effective cross-section of the sample (22 x 35 mm rectangle) was not identical to that of the standard (Ø 28 mm circle), and indeed systematic deviations show up in the data. In the second run (Fig. 1b) with identical sample

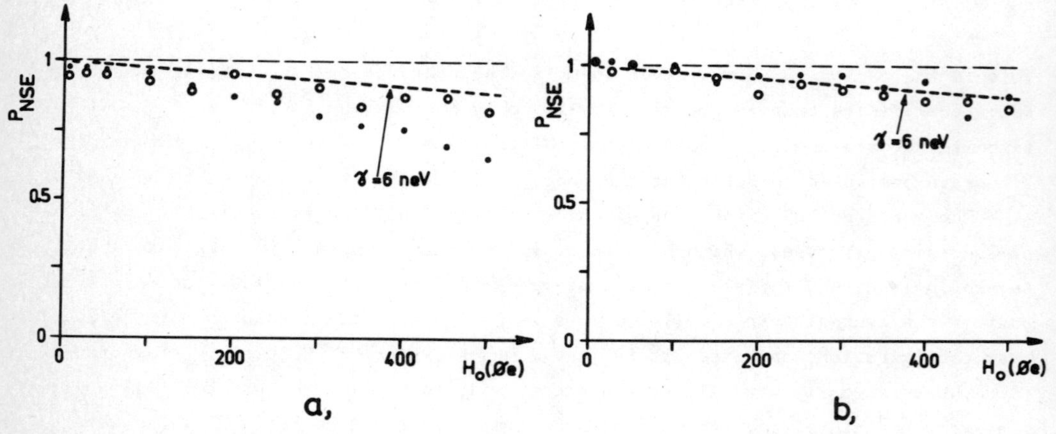

Fig. 1. Reduced NSE spectra obtained using the full beam (Ø 30 mm) cross section (dots) and with Ø 17 mm diaphragms at the polarizer and the analyser (open circles) for different (a,) and rigorously identical (b,) sample and standard geometries (see text). The dashed lines correspond to the expected Lorentzian of $\gamma = 6$ neV half width. (IN11 data)

geometry for sample and standard (∅ 25 mm circle), the results nicely reproduce. The dashed line corresponds to the expected NSE spectrum, i.e. Lorentzian line with γ = 6 neV (∿ 1.5 MHz), which value is about 10 % of the instrumental resolution "broadening". (P^{o}_{NSE} has the shape shown in Fig. 5 of the first paper of the volume). Thus we conclude that the no-deconvolution NSE resolution correction formula Eq. (11) can reliably eliminate as much as 95 % of the instrumental resolution effects, if conditions a-c, are respected.

Note that if lesser resolution is required, the inhomogeneities can become negligible, since for lesser precision $P(\vec{T},v_o) = P_o$ = const. can be taken, and $\Delta\phi_h \sim 0$ due to its proportionality to $t \propto H_o$, Eq. (9) immediately reduces to

$$P_{NSE} = P_o < S_P(\vec{\kappa},t) >_{FS} \qquad (12)$$

i.e. the standard is only necessary to determine P_o, for each H_o value used, and condition c., can be dropped.

2. TECHNICAL LIMITATIONS OF NSE

2.1. Polarizer-analyser systems

The main technical difficulties in building a NSE spectrometer are related to the polarizer and analyser systems and to the inhomogeneities of the precession fields, Δh_o and Δh_1[(cf.Eq.(1)]. Essentially the polarizer and the analyser determine the neutron luminosity of the instrument, and this represents the key problem in the use of the method. Neutron supermirrors /1/ now offer a satisfactory solution for neutron wavelengths above 3-4 Å. High intensity can be achieved by the use of large detector banks, essentially the same way as in time-of-flight instruments. In such an NSE set-up the time-of-flight method could logically provide the $\vec{\kappa}$ resolution, and it could be used both on a pulsed neutron source, as described in Ref. 2 (see Appendix) or with a chopper on a continuous reactor. The instrument would be best suited for the study of large scattering angle diffuse scattering processes, which require high luminosity.

For shorter wavelengths (1.5 - 3 Å) which are most often required for the study of elementary excitations, there is no really satisfactory solution for the polarizer and analyser problem. Both present day supermirrors and polarizing crystals are expected to provide a 1-10 μeV resolution NSE-triple axis spectrometer with a mere 3-5 % of the luminosity of a medium resolution (100-200 μeV) conventional instruments. This is a serious limitation in many cases.

2.2. Homogeneity of precession fields

The inhomogeneities of the magnetic precession fields determine the limits of resolution achievable with NSE, and to a minor extent they have an indirect effect on the neutron economy too. The NSE requirements of the precession fields are rather unusual : the fields have to change along the neutron path from low values (1-10 Øe) at the sample and the $\frac{\pi}{2}$ coils to values as high as possible in between, in order to produce a large number of precessions. On the other hand, the fields must be homogeneous across the beam in order to avoid the dephasing of the precession. These requirements are contradictory, and considering the div H = 0 and rot H = 0 field equations it is easy to show /3/ that for a field configuration cylindrically symmetrical around the z axis, the field H at a general point in space is given as

$$H_z(r,z) = H(z) - \frac{1}{4}\frac{d^2H(z)}{dz^2}r^2 + \ldots$$
$$H_r(r,z) = -\frac{1}{2}\frac{dH(z)}{dz}r + \ldots \tag{13}$$

where obviously H(z) is the field at the axis, and r radial coordinate stands for the distance from the axis. In the case of IN11 the z axis is the mean neutron trajectory. It is seen that the very fact that H(z) has to change along z creates an inhomogeneity in the perpendicular direction, across the beam

$$|H(r,z)| = \left[\left(H(z) - \frac{1}{4}\frac{d^2H(z)}{dz^2}r^2\right)^2 + \frac{1}{4}r^2\left(\frac{dH(z)}{dZ}\right)^2 + \ldots\right]^{1/2}$$
$$\simeq H(z) - \frac{r^2}{4}\left[\frac{d^2H}{dz^2} - \frac{1}{2H(z)}\left(\frac{dH}{dz}\right)^2\right] \tag{14}$$

The Larmor precessions are related to the field integral along the trajectory, and thus the inhomogneity can be characterized by the integral

$$\int|H(r,z)|dz = \int H(z)dz + \frac{r^2}{8}\int\frac{1}{H(z)}\left(\frac{dH(z)}{dz}\right)^2 dz \tag{15}$$

taken between the field minima at the sample and at the $\frac{\pi}{2}$ coils. (At these limits dH/dz = 0, therefore the d^2H/dz^2 term in Eq. (14) integrates to zero). Note that for iron yoke magnets, having a symmetry plane rather than a symmetry axis, (considered in detail in the contribution of Roger Pynn) Eq. (15) applies if the inhomogeneity term is taken with 4 times bigger coefficient : $r^2/2$). The solution of the variational problem of minimizing the ratio of the two integrals on the right hand side of Eq.(15) is not known to me, but it is easy to see what it should look like. For a solenoid with maximum field H_o at the centre, length ℓ and diameter D, the most inhomogeneous field regions are those at both ends, where $H \simeq \frac{H_o}{2}$ and over a distance of the order D the field gradient is about H_o/D. Thus Eq. (15) can be roughly evaluated, viz. :

$$\int|H(r,z)|dz \simeq H_o\ell + 2\frac{r^2}{8}\frac{2}{H_o}\left(\frac{H_o}{D}\right)^2 D =$$
$$= H_o\ell + H_o\frac{r^2}{2D} \tag{16}$$

and the relative inhomogeneity is given as

$$\eta \simeq \frac{r^2}{2D\ell} \tag{17}$$

What Eq. (16) implies is that the field should change over the longest possible distances D, and in fact I found in numerical calculations that the best homogeneity corresponds to a Lorentzian-like shape of H(z) with the maximum half way between the

$\frac{\pi}{2}$ coil and the sample. For the IN11 solenoids with $\ell = 200$ cm $D = 25$ cm, Eq. (17) gives for $r = 1.5$ cm, which corresponds to the usual beam diameter, $\eta \simeq 2.25 \cdot 10^{-4}$, which is in very good agreement with the exact figure ($1.9 \cdot 10^{-4}$). This value is only two times bigger than the ideal one, obtained numerically for the Lorentzian shaped field. Thus there is not too much to be gained by optimum solenoid design, i.e. using non-uniform winding density simulating the Lorentzian field pattern.

Substantial precession field homogeneity improvements can only be achieved by introducing current loops into the beam volume, i.e. by using field configurations with rot $H \neq 0$, so that Eq. (14) does not apply. Fig. 2 illustrates the effect of a current loop inside the relatively strong precession field $\vec{H} \parallel z$. The magnetic field

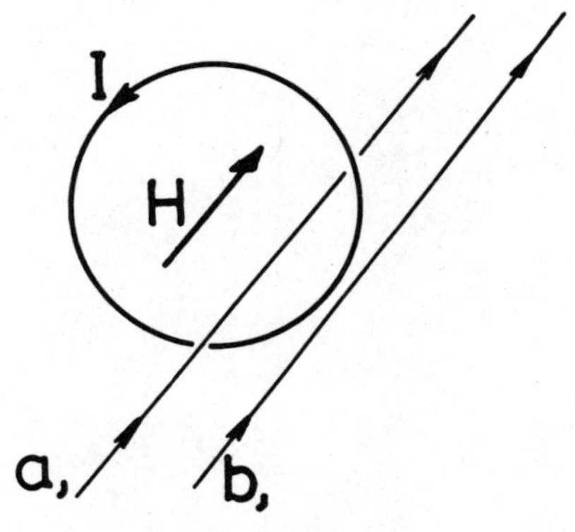

Fig. 2. Modification of the line integral of the precession field for neutron trajectories going through (a,) and outside (b,) a current loop I. The strong magnetic field H, external to the loop, is parallel to the mean neutron beam direction.

integral along a trajectory parallel to z is given as

$$\int \left| \vec{H} + \vec{L} \right| dz = \int \left[(H + L_z)^2 + (L_x + L_y)^2 \right]^{1/2} dz$$
$$= \int H dz + \int L_z \, dz + \frac{1}{2H} \int (L_x^2 + L_y^2) dz \tag{18}$$

where $L \ll H$ is the field of the loop, and thus the last integral on the right hand side is negligible. The second integral, on the othèr hand, is given by Ampère's law :

$$\int L_z dz = \begin{cases} \frac{4\pi}{c} & \text{if the trajectory passes inside the loop (a,)} \\ 0 & \text{if it passes outside (b,)} \end{cases}$$

Thus by an appropriate array of current loops the precession field integral can be spatially modulated in any desired fashion. This principle was the one used to synthetize the field asymmetry ("tilt") necessary for the study of phonons as described in my contribution on superfluid He in this volume. To compensate the inhomogeneity term, proportional to r^2 (cf. Eq. (15)) we will need an array of concentric loops carrying the same current I and having radii $r \propto \sqrt{n}$, where n is the serial number of the loop with n = 1 being the smallest one. Thus the loops form just an optical Fresnel plate pattern, which is by no means surprising, since the physics of focussing light and/or dephased Larmor precessions are basically the same. In practice this array can be best approximated by two adjacent $r \propto \sqrt{\phi}$ spiral coils (we will call them Fresnel spirals) one of which carries the current inward from the rim to the center where the two spirals are connected and the other carries it back to the rim while going around in the same sense. In Fig. 3 the original etching mask is reproduced, which was used to produce this double Fresnel spiral on both

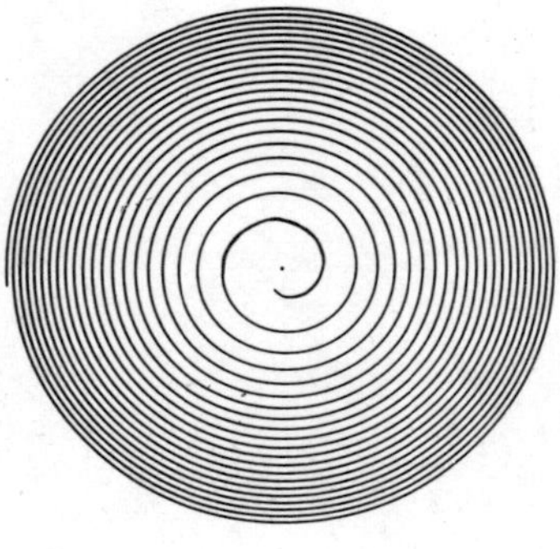

Fig. 3. The mask used for producing the Fresnel spiral coils by photo-lithography. The original computer drawing was photographically reduced to this actual size.

sides of a 0.1 mm thick integrated circuit board. The spirals thus obtained cause only 5 % neutron intensity loss, and,applied (cf. Fig. 2) at the ends of the precession field solenoids on IN11, they proved to be highly efficient indeed. This is shown in Fig. 4 where the two oscillation periods a., and b., are the NSE signals measured for strongly inhomogeneous (i.e. higher than usual) precession fields without and with Fresnel correction coils, respectively. $N_{\uparrow}$ and $N_{\downarrow}$ represent the beam polarization, i.e. the ideal NSE signal. It is seen, that this way some 80 % of the instrumental resolution "broadening" can be eliminated on IN11.

It is worth noting that instead of the Fresnel spirals real "Larmor precession lenses" could be in principle used, viz. concave ones made from diamagnetic materials and convex ones from paramagnets. These have the advantage that their strength is automatically proportional to the field i.e. to the inhomogeneity. The thickness of

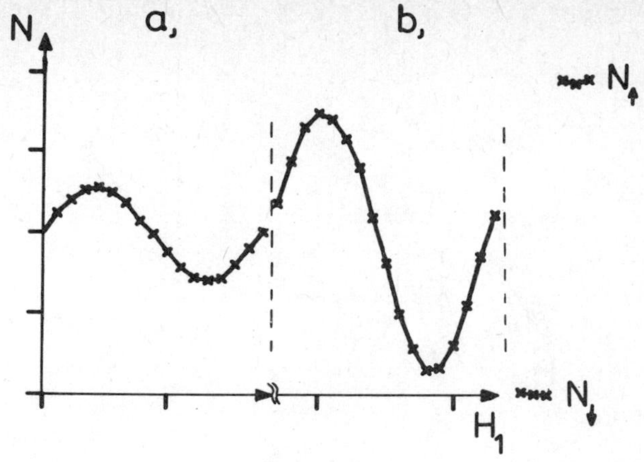

Fig. 4. Neutron spin echo signals measured on IN11 at 900 Øe precession fields (1.8 times the usual maximum value) without (a,) and with (b,) using the Fresnel spiral coils in Fig. 3 for the correction of the precession field inhomogeneity. N↑ and N↓ give the ideal signal. (Neutron wavelength 6.2 Å, raw data)

such lenses would unfortunately be a few centimeters for the best available materials (e.g. wuestite, FeO) except for ferromagnets some 10 degrees above the Curie point (e.g. ∿ 1 mm for the non absorbing ^{160}Gd isotope).

Besides this inherent quadratic ($\propto r^2$) inhomogeneity of the precession fields, we have to consider the effect of external magnetic fields superposed on the precession fields. Even if these external fields are themselves fairly homogeneous they disturb the cylindrical symmetry of the precession fields,which can lead to serious inhomogeneities of linear character (e.g. proportional to the $x \perp z$ direction across the beam) which was observed in early experiments. By numerical analysis I was able to identify these effects as being due to the appearance at the solenoid ends of an overlapping external field component perpendicular to the solenoid axis. Fortunately, such linear inhomogeneities are relatively easy to correct for. If they occur at the $\frac{\pi}{2}$ coil end of the precession field, it is sufficient to tilt the $\frac{\pi}{2}$ flipper coil with respect to the beam direction, thus adding some extra precessions onone side of the beam with respect to the other. On the sample end of the precession fields this can not be used except for ferromagnetic NSE. (Tilting the π flipper coil would be inefficient anyway, because it is in a very low precession field region). The correction method which can be applied here consists of adding another transversal field around the end of the precession field coils, as shown in Fig. 5a where the stray fields of the H_o and H_1 precession field solenoids act as transversal external fields for

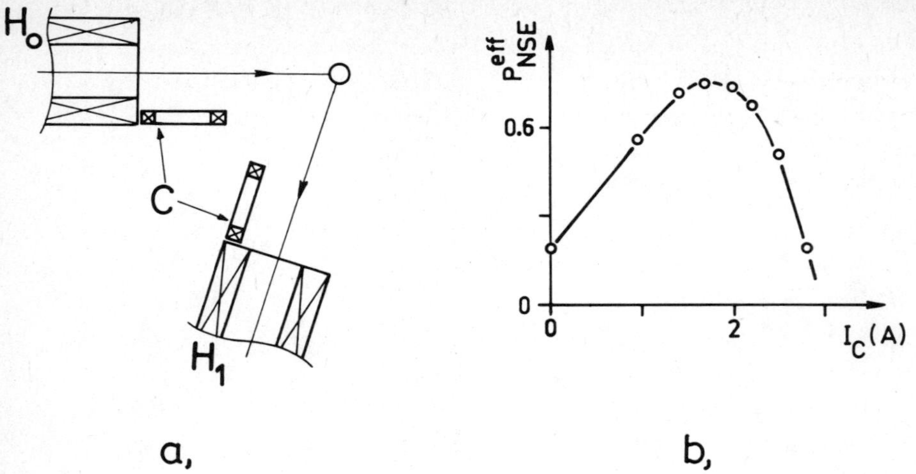

Fig. 5. Ring shaped coils C for correction of the linear field inhomogeneities produced by the mutual overlap of the precession fields H_o and H_1 (a,) and the NSE signal measured as a function of the C coil activating current I_C (b,). H_o and H_1 was 300 Øe at the center of the precession field solenoids, the C coils with 256 turns each were operated in series. (IN11 data)

each other. The ring shape correction coils C produce transversal fields opposite to these stray fields with change roughly the same way across the beam. Since the linear inhomogeneities add algebraically, there is no need to off-set the external field everywhere along the beam, but it is sufficient to use a correction rather locally at the solenoid ends. With Paul Dagleish I have verified this point in an extensive trial-and-error experiment by testing more elaborate correction configurations, but which gave just the same results. Fig. 5b illustrates the excellent efficiency of the correction method, where at the optimal correction coil current I_c the NSE signal is practically as big as for small scattering angles, where the effect is absent. This type of correction is necessary on IN11 above 100-110° scattering angles, the data in Fig. 5b were taken at 130°. Interestly, as depicted in Fig. 24 of Otto Schärpf's contribution, the corrected field configuration is locally less homogeneous than the uncorrected one, but the field integrals along the neutron trajectories are more uniform.

Precession fields produced by iron yoke magnets and perpendicular to the mean beam direction are, for the same dimensions, some 4 times less homogeneous than the solenoidal fields, cf.Eq.(17) as pointed out above. Taking a practical example, with 60cm gap and 40cm polepiece dimension we get an effective length about 130 cm and about $6 \cdot 10^{-4}$ homogeneity for 3 cm beam height. The advantage of such a field is that it can be used with a detector bank covering several scattering angles at the same time, and thus providing higher detector solid angle at the expense of resolution. There is no equivalent of the Fresnel correction coil in this case, but there should be other comparable inhomogeneity correction methods (I know of one).

3. FUTURE APPLICATIONS

In this section I will try to set up a list of the types of future NSE instruments and their most probable applications one can think of as today. The characteristics of the NSE spectrometers mainly depend on the host spectrometer with which NSE is combined, and which provides the momentum resolution ("transmission function"). In what follows this classification principle will be used. The spectral resolution values quoted below correspond to the energy equivalent (HWHM) of the instrumental "broadening" (i.e. dephasing) effects, assuming that the precession field magnet design reflects best up-to-date knowledge within reasonable practical limits. By this I mean that at present day prices typically \$ 100.000 is needed for the NSE rig itself (magnets and power supplies including $\frac{\pi}{2}$, π, correction coils etc.) in addition to the classical host spectrometer set-up (\$ 200.000 - 500.000) and polarizer and analyser systems, whose costs mostly depend on the detector area used, and which for supermirrors will probably amount to the costs of the detectors + amplifiers + scalers themselves.

3.1. NSE-small angle scattering spectrometers

The existing IN11 instrument is of this type. It is characterized by a relatively high best resolution : $\sim$ 20 neV, small detector solid angle : 0.5° x 0.5°, and a momentum transfer range of 0.02 - 2.5 $Å^{-1}$. This type of instrument could be improved in two directions viz. higher resolution by using longer wavelengths, as was actually first thought of (1 neV $\simeq$ 250 KHz at λ $\sim$ 30 Å), and higher solid angle (1.5° x 1.5°) with somewhat less resolution (60 neV at λ = 8 Å and 3 neV at λ = 30 Å). In the latter case a small 10 x 10 cell multidetector could be envisaged too.

The very high resolution version would be most useful in extending the already widespread application of IN11 for the study of macro-molecular dynamics (polymers, liquid crystals, biological materials, mycelles etc.) and κ = 0 phase transitions in which cases good momentum resolution is required in a limited range of 0.005-0.2 $Å^{-1}$.

The higher solid angle version would be best adapted for the study relaxation processes at higher momentum transfers (0.04 - 2.5 $Å^{-1}$),which include spin dynamics (spin glasses, low dimensional systems, etc.) and diffusion in crystals (e.g. superionic conductors). It is of particular importance that NSE can now significantly complement the extremely sensitive but often difficult-to-interpret resonance experiments (NMR, ESR, PAC, muon decay).

3.2. NSE-time-of-flight spectrometers

The particularity of such a set-up is the fortunate and elegant combination of the NSE information which is essentially $t_1 - t_o$, (where t_1 and t_o are outgoing and incoming neutron time-of-flights) and the total time-of-flight $t_1 + t_o$. Thus it provides complete description of the scattering event using simple means. It is the natural way to use high NSE resolution at pulsed neutron sources. On continuous reactors this machine should be aimed at the highest possible detector solid angle, by the use of wide acceptance iron yoke (transversal field) magnets and extended detector banks. It can be used either with a course resolution chopper or with just a velocity selector, eventually using the neutron counting rate modulation corresponding to the NSE signal for statistical spin flip time-of-flight analysis /4/. The best resolution in the 4-8 Å wavelength range is limited to 0.1 to 1 μeV, and the detector solid angle can be as much as (10° - 40°) x (2° - 3°). The instrument is best suited to the lesser resolution study of weakly dispersive diffuse scattering effects, both quasielastic and inelastic (e.g. crystal field levels, optical phonons, singlet ground states, etc.).

This type of instrument could be advantageous from the sole point of view of the neutron intensity , too. In NSE the neutron intensity is determined by the host spectrometer configuration, which defines the momentum resolution, and by the polariser-analyser losses, amounting to a factor of 20 around 3 Å wavelength and 10 above 4 Å. In experiments where only a poor κ resolution but good energy resolution is required, there is no need for good beam monochromatization in NSE, and polarization losses may be recovered. Thus the NSE-time-of-flight type instrument could provide excellent neutron flux for inelastic diffuse scattering experiments with 5-10 % momentum and 0.02-0.2 % energy resolution.

3.3. NSE-triple axis spectrometers

They can basically be of two types. The high momentum resolution version (cf. Claude Zeyen's contribution) is designed for crystallographic problems, like separation of the thermal diffuse scattering in structure analysis, and the study of critical fluctuations in structural phase transition . Experiments of this latter type were successfully made on IN11. The nearly complete D10 NSE instrument will work at shorter wavelengths, 1.5 - 3 Å, with a κ range extended to 8 $Å^{-1}$.

Finally the use of NSE on a poor or medium resolution triple axis host spectrometer is aimed at the study of any kind of dispersive elementary excitations (cf. Roger Pynn's paper). The spectrometer should work at 2 - 4 Å wavelength and achieve 1 - 10 μeV resolution. Until further improvements of neutron polarizers in this wavelength range, the instrument will only achieve 3-5 % of the neutron intensity of a medium resolution triple axis machine. It is the only method, however, which can make available the μeV resolution domain for the study of elementary excitations. This is well illustrated by the results I have presented above on superfluid ^{4}He, which were obtained at about two orders of magnitude worse than usual neutron counting rates, since IN11 is not ideally adapted to this type of work. In spite of this, various phonon groups could be studied within reasonable time with 10-40 times better resolution than it was previously possible.

Furthermore, the NSE has the particular feature of being selective for the slope of the dispersion relation, i.e. it can single out a particular excitation branch by its slope, even when crossing other branches. Therefore, besides lifetime and energy shift studies, it is particularly promising for the investigation of crossing branches and hybridization. These problems are related to electron-phonon, phonon-phonon, magnon-phonon, magnon-magnon etc. interactions, impurity effects, superconductivity, phase transitions etc.

4. CONCLUSION

High resolution in inelastic neutron scattering can now in 1979 be defined as better than 10 μeV (or better than 10^{-3} with respect to the incoming neutron energy). Not quite 10 years ago these values were at least 10 times bigger. The main problem in achieving high resolution is not so much that of distinguishing neutrons with very little energy differences, but more that of overcoming the neutron intensity losses produced by the fine monochromatizations (Liouville theorem). The two known high resolution methods, backscattering and NSE, get around this problem in very different ways. In backscattering (cf. the contribution by Tony Heidemann et al.) huge detector solid angles (up to 1 sterad) are used to win back intensity, and therefore it is suited to the study of practically momentum-independent (non-dispersive) scattering processes. In the conceptually new NSE the whole problem of intensity is circumvented by measuring the neutron velocity changes directly. (On the other hand, NSE requires polarization analysis which leads in a number of cases to well known difficulties of neutron intensity independently of the resolution required).

These two high resolution methods, quite understandably, are in many respects complementary in the investigation of non-dispersive scattering effects. The two spectrometers at the ILL (IN10 and IN11) are on the whole comparable in both neutron intensity and resolution, although for the NSE method the detector solid angle can be substantially increased and considerably better resolutions can be obtained by going to longer neutron wavelengths. However, due to the special character of the information collection in both cases, the two methods perform very differently for different types of scattering spectra. For the study of spectra with a lot of structure or several peaks within a small energy range, the direct energy scan in backscattering is much more preferable to the time Fourier transformed NSE scans. On the other hand, for the study of line shapes the time domain NSE data are better and the unusually large dynamic range (about three orders of magnitude) of NSE is particularly valuable. Similarly, the polarization analysis inherently involved in NSE leads to a 3 fold loss of signal for nuclear spin incoherent scattering, but proves to be extremely useful for the investigation of magnetic effects, such as spin relaxation processes.

In the study of elementary excitations using Neutron Spin Echo a breakthrough of more than an order of magnitude in resolution can be achieved, as has been demonstrated. In this field NSE is the only known high resolution method whose full scale application could well be one of the most spectacular developments in neutron scattering for the next few years.

REFERENCES

1. F. Mezei, Comm. Phys. 1, 81 (1976)
   F. Mezei and P.A. Dagleish, Comm. Phys. 2, 41 (1977).
2. F. Mezei, Nucl. Instrum. & Methods, 164, 153 (1979).
3. V.V. Batygin and I.N. Toptygin, in Problems in Electrodynamics (Pion and Academic Press, London, 1978) p. 279.
4. K. Sköld, Nucl. Instrum. & Methods 63, 347 (1968).

Z. Physik 255, 146–160 (1972)

Neutron Spin Echo: A New Concept in Polarized Thermal Neutron Techniques

F. Mezei

Institut Laue-Langevin, Grenoble, France* and
Central Research Institute for Physics, Budapest, Hungary

Received July 7, 1972

A simple method to change and keep track of neutron beam polarization non-parallel to the magnetic field is described. It makes possible the establishment of a new focusing effect we call neutron spin echo. The technique developed and tested experimentally can be applied in several novel ways, e.g. for neutron spin flipper of superior characteristics, for a very high resolution spectrometer for direct determination of the Fourier transform of the scattering function, for generalised polarization analysis and for the measurement of neutron particle properties with significantly improved precision.

I. Introduction

In the traditional technique of polarized thermal neutron beam studies one is concerned only with the component of the beam polarization parallel to the applied magnetic field. The polarizer devices are used to produce a neutron beam polarized parallel to the magnetic field applied to the polarizer, adiabatic magnetic guide fields are used along the whole neutron path to keep polarization parallel to the field throughout, and the polarization analysis used measures only the polarization component parallel to the field applied to the analyser. Within this framework neutron beams are often considered to consist of two incoherent parts, one of them polarized in the direction of the magnetic field, and the other one oppositely.

We note that this simplified approach is not necessary. Any spin wave function corresponds to a definite spin direction and by the use of the Schrödinger equation one can keep track of the change of spin direction during passage through an arbitrary inhomogeneous and time dependent magnetic field. Consequently, by properly designed static inhomogeneities in the magnetic field traversed by the neutron beam one can transfer the original polarization direction to any desired direction at a given place. One can also follow the motion of the neutron spin orientation throughout and finally one can determine any particular spin direction by finding a controlled way to turn it back parallel to

* Present address.

the magnetic field, i.e. into the direction of analyser device action. This is the basis of the concept outlined below. On the other hand, this approach led us to propose several novel applications in the field of thermal neutron beam research, the description of which constitutes the main purpose of this paper. The basic ideas together with one of the proposed applications (a new spin flipper device) have been tested experimentally at the Nuclear Reactor of the Central Research Institute for Physics in Budapest.

The technique we use to control neutron spin orientation resembles the turning of nuclear spins through definite angles by the application of pulses of defined duration in the NMR spin echo. Furthermore we have established and utilized an important focusing effect, which is an exact analogue of the NMR spin echo. This is why we have chosen the name "neutron spin echo".

In the next section we describe the elementary action we use to change the neutron beam polarization direction with respect to the magnetic field orientation and the principle of the neutron spin echo effect. In Section III we shall briefly discuss some device applications, and Section IV is devoted to a summary.

Before we proceed we have to mention that conventional rf spin flipper coils can also be used to tip polarization out of the magnetic field direction by changing the rf current[1]. This way, however, one produces a time dependent polarization rotating with the rf frequency the time average of which is parallel to the magnetic field. This method therefore did not become practical except that in principle it is used in the well known split coil rf spin flipper device[2].

II. Fundamentals

Let us first recall the general features of the motion of an individual neutron spin in a magnetic field. We can describe the spin state of the neutron by the quantum mechanical expectation values

$$s_\alpha(t) = \langle t \,|\, \hat{S}_\alpha \,|\, t \rangle$$

where $|t\rangle$ is the neutron spin wave function at the instant t, and $\hat{S}_\alpha$ ($\alpha = x, y, z$) are the operators for the three spin components. The equation of motion of the vector $\boldsymbol{s}(t) = (s_x(t), s_y(t), s_z(t))$ in a magnetic field $\boldsymbol{H}(t)$ is known to be

$$\frac{d}{dt} s_\alpha(t) = \frac{d}{dt} \langle t \,|\, \hat{S}_\alpha \,|\, t \rangle = \frac{2|\gamma|\mu_N}{\hbar} [\boldsymbol{H}(t) \times \boldsymbol{s}(t)]_\alpha \tag{1}$$

1 Shull, C. G.: International Summer School on Neutron Physics, Alushta, 1969. Edited by N. Kroo (Dubna, 1970).

2 Ramsey, N. F.: Phys. Rev. **78**, 695 (1950).

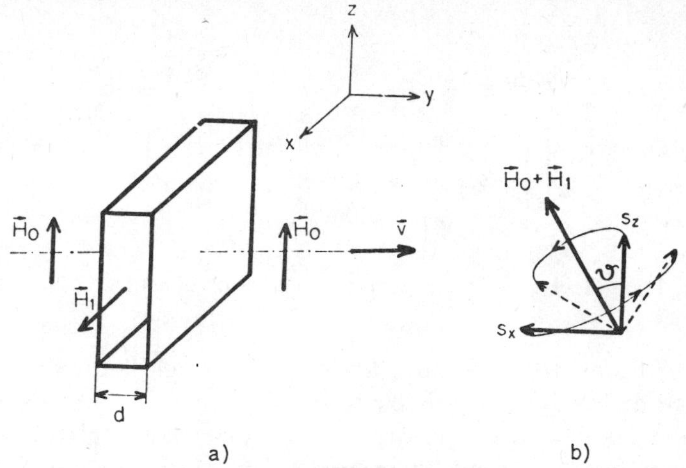

Fig. 1a and b. The coil device to change neutron spin direction with respect to the magnetic field (a) and its action (b)

where $\gamma = -1.91$, and μ_N is the nuclear magneton. In addition we have that $|s(t)| = 1/2$, thus $s(t)$ can be looked at as the spin vector, which evolves according to the equations of motion of a classical magnetic moment.

Considering the linearity of Eq. (1) and the property

$$\frac{d}{dt}(s_1 s_2) = s_1 \frac{d}{dt} s_2 + s_2 \frac{d}{dt} s_1 = \frac{2|\gamma|\mu_N}{h}[s_1[H \times s_2] + s_2[H \times s_1]] \equiv 0 \quad (2)$$

we find that one can follow the motion of each initial spin component s_α independently, in this way arriving at the orthogonal set of final components $s_{\alpha'}$ with $|s_{\alpha'}| = s_\alpha$. For example, if the neutron traverses a sufficiently slowly varying magnetic field (an adiabatic guide field) the spin component originally parallel to the field will follow its change, staying parallel, so that the angle between neutron spin and magnetic field is conserved.

We now describe a simple method for changing the neutron spin direction in a controlled way with respect to the guide field H_0. Let the neutron beam pass a rectangular coil with an axis (x direction) perpendicular to H_0 (z direction), as shown in Fig. 1a. Applying a dc. current to the coil we will have a magnetic field $H_0 + H_1$ inside the coil, and H_0 unchanged outside the coil. Because of the sudden change of the field across the winding sheets the neutrons will enter and leave the interior of the coil with unchanged spin. Let us assume that the time the neutrons spend within tee coil corresponds to one half Larmor precession, that is:

$$\frac{d}{v \sin\beta} = \frac{\frac{1}{2}\pi\hbar}{|\gamma|\mu_N \sqrt{H_0^2 + H_1^2}} \quad (3)$$

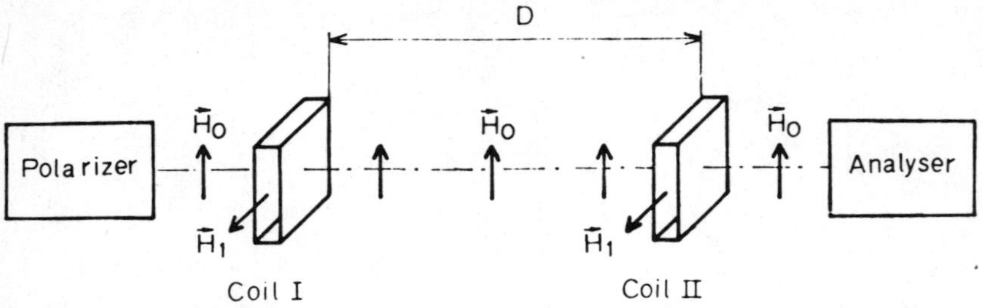

Fig. 2. Experimental set-up for tracing Larmor precession of neutron beam polarization

where v is the neutron velocity, and β is the angle between $\boldsymbol{v}$ and the lateral face of the coil. In this case, as also shown in Fig. 1 b, neutrons will leave the coil with a spin orientation corresponding to the original z and x components turned by an angle of 2ϑ and $2\vartheta-\pi$ within the xz plane, respectively, and to the original y component reversed. For a given coil with a width of d and for a given neutron velocity v we have to tune the magnetic fields H_0 and H_1 in order to fulfill the condtition (3) and to set the value of the angle ϑ between $\boldsymbol{H}_0$ and $\boldsymbol{H}_0+\boldsymbol{H}_1$.

As a simple demonstration of this method we have used it to follow directly the Larmor precession of neutrons between the two coils in the array shown in Fig. 2. For both coils $\vartheta=45^\circ$ has been chosen. Thus coil I turns the polarization of the neutron beam (defined as the average of $2\boldsymbol{s}$ over all neutrons), originally parallel with the direction of $\boldsymbol{H}_0$, into the direction of $\boldsymbol{H}_1$ (x axis). Coil II and the analyzer together act as an analyzer for the polarization component in the x direction P_x at the entrance face of coil II. The experimental determination of P_x as a function of the distance between the coils D showed a nice trace of the Larmor precession of the $\lambda=1.55$ Å wave length neutron beam in the field $H_0=15.5$ Oe (cf. Fig. 3). The slight decrease of the P_x maxima at larger D values observed is due to the velocity scatter of the beam, since neutrons with different velocities spend different precession times between the coils. This was verified by changing the beam monochromatization.

We may, however, get rid of this beam depolarization effect caused by the neutron velocity scatter by utilizing the spin echo focusing technique illustrated in Fig. 4. Here we have the above arrangement modified by dividing it into two symmetrical parts with equal and opposite magnetic field directions separated by a sudden field reversal device (Majorana[3] field-flip) which will be passed by the neutrons without a change in spin direction. Such a sudden field reversal can be achieved by inserting a current sheet perpendicular (or almost perpendicular)

3 Majorana, E.: Nuovo Cimento **9**, 43 (1932).

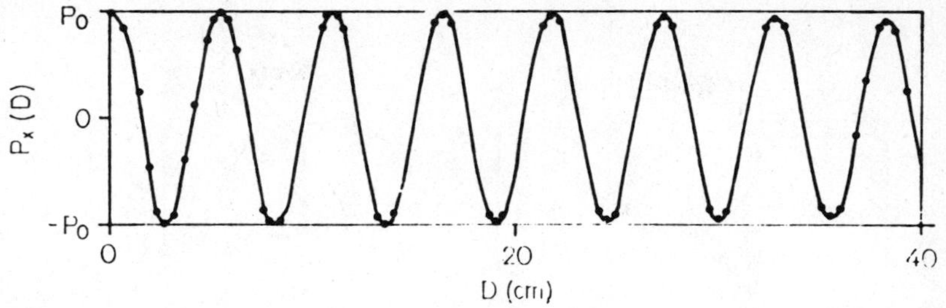

Fig. 3. Larmor precession of neutron beam polarization measured in a magnetic field $H_0 = 15.5$ Oe for $\lambda = 1.55$ Å

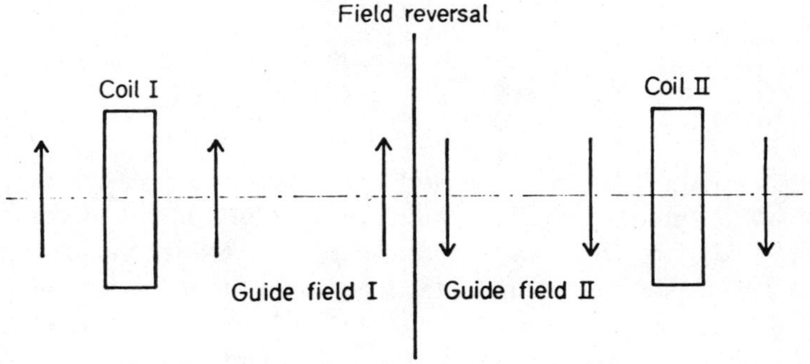

Fig. 4. Neutron spin echo arrangement

to the neutron path[4]. In this case the spin of each neutron in the beam will perform exactly the same number of Larmor precessions between coil I and the field reversal as in the opposite direction between the field reversal and coil II. As a result each neutron will arrive at coil II with the same spin orientation with which it left coil I. To demonstrate this idea we have determined, as above, the P_x component of the neutron beam polarization at the entrance of coil II using the neutron spin echo array shown in Fig. 4 with a distance between the coils of 38 cm, which corresponds to approximately 7 Larmor precession periods for the same value of H_0 and average neutron wave length as above. Within the 0.5% experimental accuracy we recovered the original value of the beam polarization P_0, in contrast to a maximum $P_x = 0.93\ P_0$ found for the same coil distance without the spin echo focusing effect (cf. Fig. 3).

Finally we mention some alternate possibilities. Instead of the coil shown in Fig. 1 one can turn the neutron spin direction with respect to the magnetic field by applying a single current sheet similar to that used for the Majorana field reversal[4], as shown in Fig. 5. If the guide

4 Steinsvoll, O., Abrahams, K., Riste, T.: Kjeller Reports KR-37. Institutt for Atomenergie, Kjeller, Norway (1962).

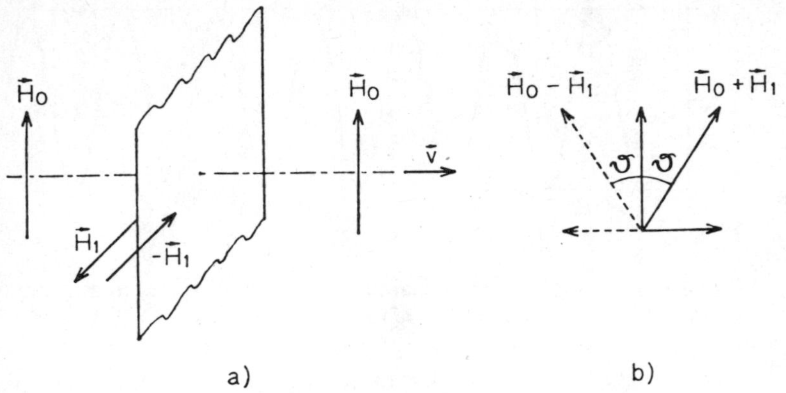

Fig. 5a and b. The sheet device to change neutron spin direction with respect to the magnetic field (a) and its action (b)

field $\boldsymbol{H}_0$ and the perpendicular field $\boldsymbol{H}_1$ produced by the current sheet are large enough, the neutron spin follows the change of the magnetic field from $\boldsymbol{H}_0$ to $\boldsymbol{H}_0+\boldsymbol{H}_1$ adiabatically. Crossing the current sheet there will be no change in spin direction while the magnetic field changes suddenly from $\boldsymbol{H}_0+\boldsymbol{H}_1$ to $\boldsymbol{H}_0-\boldsymbol{H}_1$. The process is completed by a second adiabatic transition from $\boldsymbol{H}_0-\boldsymbol{H}_1$ to $\boldsymbol{H}_0$ again. As a result we have the same action as that of the coil described above (cf. Figs. 1b and 5b). This current sheet device has the advantage as compared to the coil that it works equallv well for a large range of neutron velocities without tuning, and the disadvantage that it produces extensive stray magnetic fields.

There is yet another way to realize neutron spin echo experimentally. Instead of reversing the magnetic guide field half way between the two coils, as described above (Fig. 4) we may apply a neutron spin flipper device at this point and keep the guide field constant throughout. As we shall see later, the neutron spin flipper reverses the x component of the spin, and leaves unchanged the y component (cf. Fig. 6). This action is equivalent to changing the angle of Larmor precession between coil I and the flipper to its negative, which will be canceled by the further Larmor precession between the flipper and coil II. This method is in fact the strict analogue of NMR spin echo.

The choice of a coil or sheet device at any place in an experimental arrangement is a priori independent of similar choices at the other places. Thus there are many possible combinations. In what follows we shall consider only one particular alternative of each experimental set-up described. We have to keep in mind, however, that some other combination may be more advantageous in a given case.

Before we turn to the other applications let us point out that the present direct detection technique might be of considerable experimental interest in determining the Larmor precession of neutron polarization within a material traversed by the beam.

III. Applications

1. Neutron Spin Flipper Device

Let us place the coils with ϑ equal to 45° and to −45° next to each other as shown in Fig. 6. As it is immediately seen, during the passage across the two coils both the x and the z components of the neutron beam polarization will get reversed, and the y component will stay unchanged. We have tested such a device consisting of two identical coils of thickness $d=2$ cm made by winding one tight layer of 0.4 mm diameter enamelled copper wire on an insulating frame. By tuning both the current through the coils connected in series (about 0.5 A) and the guide field H_0 we have achieved a flipping efficiency 100% within the experimental accuracy 0.5% in a range as wide as $14.9 \leqq H_0, H_1 \leqq 15.7$ Oe. These values of H_0 and H_1 agree well with those calculated using Eq. (3) for the $\lambda = 1.55$ Å wave length neutron beam employed. (The flipping efficiency has been measured by comparison to a conventional rf flipper.) The experimentally observed tuning sensitivity has been found to agree well with the result of an elementary geometrical calculation for ideally rectangular coil shape, which reads

$$1-\eta \simeq \frac{\pi^2}{2}\left(\frac{\delta H_1}{2H_1^0}+\frac{\delta H_1'}{2H_1^0}+\frac{\delta H_0}{H_0^0}\right)^2 + 2\left(\frac{\delta H_1}{2H_1^0}+\frac{\delta H_1'}{2H_1^0}-\frac{\delta H_0}{H_0^0}\right)^2 \tag{4}$$

where $H_0^0 = H_1^0$ is the optimal field value obtained from Eq. (3), and $\delta H_1 = H_1 - H_1^0$, etc. We mention that our coils had a thickness uniform

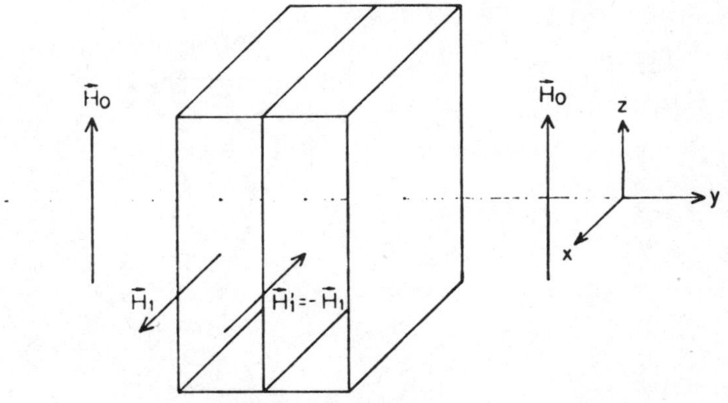

Fig. 6. Neutron spin flipper device

within 0.5 mm, and the passage of the neutron beam across the 4 winding sheets caused a decrease of 8% in intensity.

Besides simplicity, this type of neutron spin flipper has many advantages as compared to the conventional rf resonance flipper[4]:

– it is less sensitive to variations of the magnetic field, irregularities in geometry and scatter in neutron velocity, which makes it possible to achieve very high flipping efficiences with reasonable precision, e.g. with an overall accuracy and beam monochromatization of 1% one will have $\eta \geqq 99.9\%$ (cf. Eq. (4));

– there is essentially no limitation for the neutron beam cross section;

– the flipping action happens within a well defined beam volume with no stray fields whatever outside;

– the length of the neutron path within the flipper is independent of the beam cross section, and can be made very short, which offers the possibility of high repetition rate chopper applications (by delaying the switching of the second coil according to neutron passage time one could easily obtain an effective switching length of 0.5–1 cm).

2. Neutron Spin Echo Spectrometer

One of the most promising applications of neutron spin echo is its utilization for the investigation of extremely small energy change scattering processes. Let us consider the symmetrical arrangement, similar to that in Fig. 4, which is shown in Fig. 7. (The arrows in the figure indicate the magnetic field directions.) It is easily seen that elastically scattered neutrons will arrive at the surface of coil II with a polarization parallel to the coil axis (x direction). (We assume here that the scattering process does not change the neutron polarization direction.

Fig. 7. Neutron spin echo spectrometer arrangement

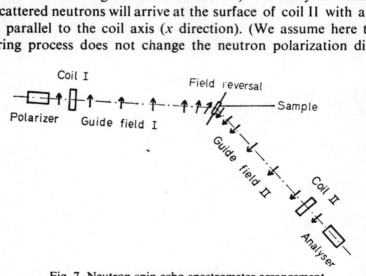

For nuclear spin and magnetic scattering effects such changes must of course be taken into account. However they do not make any fundamental difference.) An inelastically scattered neutron, on the other hand, will have an x polarization component at the coil surface

$$P_x(v, \delta v) = \cos\left(2\pi N \frac{\delta v}{v+\delta v}\right) \tag{5}$$

where v and δv are the incoming velocity and the change in the velocity of the neutron, respectively, and N is the total number of Larmor precessions in guide field II for the original neutron velocity v. The quantity δv is determined by the energy change of the neutron by

$$\hbar\omega = \tfrac{1}{2} m\left((v+\delta v)^2 - v^2\right). \tag{6}$$

To obtain the value of P_x which will be seen by the analyser system we have to integrate Eq. (5) over all incoming neutron velocities and all possible energy changes:

$$P_x = \int dv \int d\omega\, f(v)\, S(\omega) \cos\left(2\pi N \frac{\delta v}{v+\delta v}\right) \tag{7}$$

where $f(v)$ is the velocity distribution of the incoming neutron beam and $S(\omega)$ is the scattering function of the sample in the given detector position.

Now let us consider the case when $f(v)$ is sharply peaked around the average velocity v_0, with a characteristic width of $\Delta v_0 \ll v_0$. We denote the solution of Eq. (6) for $v = v_0$ by $\delta v_0(\omega)$, and we apply the approximations

$$\frac{\delta v}{\delta v + v} \simeq \frac{\delta v_0(\omega)}{\delta v_0(\omega) + v_0 + (v - v_0)} \simeq \frac{\delta v_0(\omega)}{v_0 + \delta v_0(\omega)} \left(1 - \frac{v - v_0}{v_0 + \delta v_0(\omega)}\right)$$

to evaluate the integral in Eq. (7) giving

$$\begin{aligned} P_x &= \int d\omega \int dv\, S(\omega)\, f(v)\, \mathrm{Re}\left\{ e^{2\pi i N \frac{\delta v_0(\omega)}{v_0 + \delta v_0(\omega)}}\, e^{-2\pi i N \frac{\delta v_0(\omega)(v - v_0)}{(v_0 + \delta v_0(\omega))^2}} \right\} \\ &= \int d\omega\, S(\omega)\, \mathrm{Re}\left\{ F\left(2\pi N \frac{\delta v_0(\omega)}{(v_0 + \delta v_0(\omega))^2}\right) \cdot e^{2\pi i N \frac{\delta v_0(\omega)}{v_0 + \delta v_0(\omega)}} \right\} \end{aligned} \tag{8}$$

where

$$F(x) = \int f(v)\, e^{-ix(v - v_0)}\, d(v - v_0)$$

is the Fourier transform of the displaced velocity distribution function $f(v)$ which obviously has the properties

$$F(x) \simeq 1 \quad \text{if } x \ll 1/\Delta v_0$$

$$F(x) \simeq 0 \quad \text{if } x \gg 1/\Delta v_0.$$

We will be mostly interested in the very small energy change scattering, i.e. $|\hbar\omega| \leqq \hbar\overline{\omega} \ll \frac{1}{2} m v_0^2$. To obtain this piece of information we must choose N values such that $N \geqq \overline{N} = v_0/2\pi\,\delta v_0(\overline{\omega})$. In this particular case the dominant contributions to P_x come generally from the part of the integral (8) where $|\omega| \lesssim \overline{\omega}$ for two reasons. Firstly, for large N the exponential function in the integrand oscillates rapidly and for $|\omega| \gg \overline{\omega}$ the $S(\omega)$ is expected to be a smooth function without any sharp structure on the scale of $\overline{\omega}$. Secondly, for large arguments the function F will introduce an additional cut-off. (By comparing the arguments of F and the exponential in Eq. (8) it is easily seen that this cut-off occurs after about $v_0/\Delta v_0 \gg 1$ oscillations of the exponential.) On the other hand, for $|\omega| < \overline{\omega}$ we may take $\delta v_0(\omega) = \dfrac{\hbar\omega}{m v_0}$ and introducing the notation $t = 2\pi N\,\hbar/m v_0^2$ we arrive at the approximate expression

$$P_x = \int_{-\overline{\omega}}^{\overline{\omega}} S(\omega)\,\mathrm{Re}\big(F(\omega t/v_0)\,e^{i\omega t}\big)\,d\omega.$$

Taking into account again that we expect $S(\omega)$ to be a smoothly varying function, we find that if $v_0/\Delta v_0 \gg 1$ the cut off introduced by F does not play any essential role and the integration can be formally extended to infinity, which gives the final result

$$P_x = \int_{-\infty}^{+\infty} S(\omega)\cos(\omega t)\,d\omega = \tilde{S}(t). \tag{9}$$

We conclude that by this method measuring the scattered beam polarization component P_x as a function of $N \propto t$ one directly determines the Fourier transform of the scattering function, which is just the time dependent correlation function itself. Note that the time variable t can be controlled by changing the magnetic fields I and II. We recall that Eq. (9) is valid for $t \gg 2\hbar/m v_0^2$ only. One can, however, extend this method to the general case by applying more complicated mathematical procedures starting from Eqs. (5) and (6).

It has to be pointed out that the measurement of the Fourier transform often does not simply give equivalent information, but can reveal such details which are hardly observable in the direct measurement of the function itself[5].

The basic practical advantage of the neutron spin echo technique is twofold: (1) it provides, via the focusing effect, the possibility of measuring very small energy changes without corresponding monochromatization of the ingoing neutron beam and (2) it offers an easy way to determine very small relative changes of the neutron velocity $\delta v/v$ by converting them into large changes in polarization direction, if we apply guide fields causing a large number of Larmor precessions N (cf. Eq. (5)). For slow neutrons ($\lambda \gtrsim 4$ Å) it is possible to achieve $N = 10^4$–10^5, with a corresponding stability and cross sectional homogeneity of the field. So the construction of a neutron spin echo spectrometer with a sensitivity of 10^{-4} relative energy change $\delta\varepsilon/\varepsilon_0$ should present no difficulty. This resolution requires a modest beam collimation of 0.5° both before and after the sample to keep neutron paths approximately at the same length within the guide fields. Considering the cut-off effect an ingoing beam monochromatization of 1–10% seems to be reasonable in most cases. As compared to the back-scattering spectrometer[6], which is the best ultra high resolution technique proposed up to now, the neutron spin echo principle offers the following advantages:

– the poor ingoing beam monochromatization required results in an enormous gain in neutron intensity for the same resolution (the beam collimations required are exactly the same for the two methods);

– the ingoing neutron energy can be varied freely below 10 meV independent of relative resolution (above this value the resolution will be limited by the unconfortably high value of the guide fields needed);.

Let us briefly consider some practical aspects of the construction of a neutron spin echo spectrometer. There is a special problem of the neutron path differences in the sample, which can be circumvented by keeping the magnetic field (Larmor precession rate) at a minimal value around the sample. In general one has to use a magnetized mirror or a polarized (magnetized) transmission analyser in order to cover an appropriate energy range of the scattered beam. In very small energy change experiments, however, mosaic crystal analysers can be applied as well. By the proper choice of the sign of P_x in the polarization analysis at coil II one can effectively filter out the elastic part of the scattering, if one uses high efficiency polarizing and analysing systems (and if we are concerned with those kinds of elastic scattering processes which

---

5 The case of a broad maximum with small peak intensity, but sizable integrated intensity is an example of this. The author is indebted to Prof. Maier-Leibnitz for calling his attention to this point.

6 Alefeld, B.: Bayer. Akad. Wiss. Math. Nat. Kl. **11**, 109 (1967).

preserve the initial degree of beam polarization[7]). Apart from this, the use of very high polarization efficiencies is unessential in neutron spin echo.

This method offers yet another possibility in high resolution neutron spectroscopy. Changing the magnetic guide fields I and II independently, one may adjust the echo condition (in this case exact only to first order) for a given energy change, and this way one can investigate inelastic line shapes with the high resolution discussed above.

In sum, we have seen that the neutron spin echo technique offers a practical method for very high resolution investigation of inelastic neutron scattering effects. A reasonable limit for velocity resolution seems to be of the order of 10^{-6}, which corresponds to energy changes of 10^{-10} eV for $\lambda=25$ Å wavelength neutrons (cf. the limit for back-scattering method of $\sim 10^{-7}$ eV). The unique property of measuring the Fourier transform of the scattering function directly, and the enormous gain in neutron intensity due to the low monochromatization required seems to make this method ideal for opening up a new generation of neutron measurements of particular interest for organic crystals, biological media, critical phenomena, low temperature effects, etc.

In addition, this principle of determination of extremely small neutron energy changes could be utilized in the search for the neutron charge[8]. One has to apply an electrostatic potential difference between the guide fields I and II, and determine the eventual acceleration of the direct beam. (In such a direct beam experiment one has the further advantage that no collimation is necessary because of the symmetry of the array.) Assuming a cross sectional homogeneity and long term stability of the guide fields of 10^{-6}, the detection of a relative change in the beam average velocity of 10^{-10} seems to be reasonable. For 40 Å wave length neutrons and an electrostatic potential difference of 1 MV this corresponds to a sensitivity in a neutron charge experiment of 10^{-20} of the electronic charge, a figure which is some two orders of magnitude below the upper limit obtained in the best neutron beam experiment up to now[8]. This value, on the other hand, is not necessarily an ultimate limit, and the real difficulty is presented mainly by the short term instabilities of the whole system during the neutron passage time (e.g. thermal expansion effects).

3. *Search for a Neutron Electric Dipole Moment*

The use of neutron spin echo technique also offers a very suitable method for the search for a neutron electric dipole moment which is

7 Marshall, W., Lovesey, S. W.: Theory of thermal neutron scattering. Oxford 1971.
8 See Shull, C. G.: Loc. cit. and references therein.

more adequate than the resonance beam (Ramsey) method[9] which has been the most successful up to now. Using the arrangement shown schematically in Fig. 4 we have to apply in addition an electric field parallel to the magnetic field within guide field I and antiparallel to the magnetic field within guide field II, and determine the change of the Larmor precession at coil II caused by the reversal of this electric field. This can be done most efficiently if we move out of the exact echo position by 90° Larmor angle ($P_x \simeq 0$). Beam collimation is not required in this case.

It is easily seen that the electric dipole and magnetic dipole Larmor precession angles will be approximately additive, if the magnetic guide field applied keeps the Larmor precession axis close to the electric field direction. (By symmetry electric and magnetic dipole moments are expected to be coupled together parallel to each other.) This means that we have to apply a magnetic field H_0 which is at least an order of magnitude larger than the apparent magnetic field

$$\boldsymbol{H}^* = (\boldsymbol{E} \times \boldsymbol{v})/c$$

due to the interaction of the neutron magnetic dipole moment μ_M with the electric field $\boldsymbol{E}$[10]. ($\boldsymbol{v}$ is the neutron velocity.) Consequently the total magnetic Larmor precession angle within each guide field must satisfy the condition

$$\Omega_M = \frac{2}{\hbar} \mu_M H_0 t \gg \frac{2}{\hbar c} \mu_M v E t$$

(where t is the time that the neutrons spend in the guide field). The corresponding electric Larmor precession angle reads

$$\Omega_E = \frac{2}{\hbar} \mu_E E t$$

where μ_E is the eventual neutron electric dipole moment we wish to determine. We find for the ratio of the electric and magnetic Larmor precession angles that

$$(\Omega_E/\Omega_M) = \frac{\mu_E E}{\mu_M H_0} \ll \mu_E \frac{c}{v \mu_M}.$$

For reasonably available slow neutrons with wave length of $\lambda = 30$ Å, this inequality reduces to

$$(\Omega_E/\Omega_M) \ll \frac{\mu_E}{10^{-20} e \cdot \mathrm{cm}}$$

9 Baird, J. K., Dress, W. B., Miller, P. D., Ramsey, N. F.: Phys. Rev. **179**, 1285 (1969).

10 Foldy, L. L.: Rev. Mod. Phys. **30**, 471 (1958).

where e is the electronic charge. Considering that the present experimental upper limit for μ_E is about $10^{-23}\ e \cdot$cm, we see immediately that in order to improve this figure substantially one has to be able to determine the total Larmor precession with an extremely high accuracy.

This is the point where the advantages of neutron spin echo are most prominently shown up. By producing the guide fields I and II by the use of identical Helmholtz coils connected in series, we can make the magnetic Larmor precessions in the two guide fields compensate each other exactly automatically. We emphasize that all we have to do is to keep the current through the Helmholtz coils constant within the neutron passage time, which can be achieved with an accuracy of about 10^{-8}–10^{-10}, and no long term stability is required. It follows that by making use of the spin echo compensation effect one can achieve a system stability corresponding to a value of the neutron electric dipole moment of 10^{-26}–$10^{-28}\ e \cdot$cm for $\lambda = 30$ Å. (The worst problems in this case are short time thermal expansion effects and mechanical stability.) On the other hand, in the classical resonance beam method[9] one is directly concerned with the determination of Ω_E/Ω_M, which means in practice a comparison of two independent systems, i.e. the magnetic guide field and the resonance oscillator. Thus both of these systems must meet extremely high absolute stability conditions, which would seems to be hardly achievable on the above scale. Considering the other experimental aspects (neutron intensity, preferential use of slow neutrons, elimination of the spurious effect of the apparent field H^*) the problems arising are equivalent for both methods.

We conclude that the neutron spin echo technique offers a substantial advantage over the resonance method in that it imposes orders of magnitude easier stability requirements. Therefore we expect no serious difficulty in improving resolution even by orders of magnitude. On this ground it is reasonable to apply brute force methods to increase sensitivity by extending the time that neutrons spend in the electric field (e.g. extremely long guide fields).

4. Generalized Polarization Analysis

It is well known that the polarization of a scattered neutron beam is generally not parallel to the magnetic (guide) field applied to the sample[7]. The method described in the first half of the previous section offers for the first time the possibility of determining the exact direction of the scattered beam polarization. One way of doing this can be deduced from our previous discussion in connection with Fig. 2. As we have pointed out, the coil II and the analyser together act as an analyser for neutron polarization component in the x direction (coil axis). Without activating the coil we have a classical analysis in the z direction (guide field direction), and finally moving the coil by 90° Larmor precession distance from its original position in the beam direction, we will find the original y component of polarization as the new x component. After having determined the polarization at the coil surface, we have only to trace back the Larmor precession to the sample. It is obviously also possible to measure the absolute value of the polarization directly by finding a coil position for which $P_y = 0$, and an appropriate ϑ angle to turn the polarization into z direction.

IV. Summary

We have described a simple, static method to turn neutron beam polarization into directions non-parallel to the magnetic guide field, and also to determine polarization directions non-parallel to the magnetic field. The method can be used to trace the Larmor precession of neutron spins in a magnetic field, which provides an extremely sensitive means of determining neutron velocities. In addition, by the use of the focusing effect we have called neutron spin echo, one can determine average velocity changes of a neutron beam with very high accuracy not limited by the degree of beam monochromatization. Several applications of these ideas, which offer some interesting improvements in current thermal neutron beam experimental techniques, have also been discussed.

The author is grateful to Peter Kleban for his help in preparing this paper and to Prof. R. Mössbauer for critical reading of the manuscript.

F. Mezei
Institut Max von Laue-Paul Langevin
CEDEX 156
F-38-Grenoble-Gare/France

Z. Physik B 31, 117–125 (1978)

Zeitschrift
für Physik B

Matrix Analysis of Neutron Spin-Echo

John B. Hayter
Institut Laue-Langevin, Grenoble, France

Received March 22, 1978

We describe the general action of a neutron spin-turn coil by a rotation matrix, and the nett action of a sequence of spin-turns is then evaluated as the product of the relevant matrices. In the ideal neutron spin-echo configuration, the spin-turn sequence used is shown to have a resultant action described by the unit matrix if there is only elastic scattering, so that an initially polarized beam is transmitted with unchanged polarization. This is spin-echo focussing. Measurement of changes in the final polarization then provides information on the sample dynamics.

Spin-echo focussing permits high resolution to be coupled with a broad incident wavelength spread. We calculate in detail the effect of this polychromaticity on the spin-turn coils involved, and present experimental confirmation. General spin-echo configurations are then considered. Finally, we take into account the effect of spin-flip scattering from spin-incoherent samples, and show that spin-flip effects may be decoupled from dynamic effects in the analysis of spin-echo intensity.

The techniques evolved are relevant to both monochromatic and polychromatic spin-echo spectrometry. Emphasis has been placed on practical aspects of establishing correct spin-turn conditions and analysing spectrometer response in terms of relative transmitted intensity.

1. Introduction

The spatial dependence of the phase of a moving neutron spin, undergoing Larmor precession in a static magnetic field, provides a very sensitive measure of the neutron's velocity. This effect is used in the neutron spin-echo technique [1–3] to analyse small energy changes on scattering, the information being encoded as a phase shift in the precessing polarization of each neutron. Non-monochromatic (white) beams may be used, giving high intensity.

The net phase of the precessing polarization is measured by projecting the appropriate spin-component on to a conventional analyser, as in generalised polarization analysis. Spin-turn coils are used for this purpose, and a variety of applications is discussed in [1–10].

Although it is possible to construct white-beam coils for certain applications [7], most commonly used spin-turn coils operate by causing the neutron spin to undergo a controlled Larmor precession over a fixed distance, and as such are wavelength dependent in their action. We shall develop here a rotation matrix description of the popular Mezei [1] spin-turn coil when it is used as a general spin-turn device under white beam conditions. (The applicability of such a description is discussed in [3].) Practical aspects of coil tuning will also be discussed, and a comparison between theory and experiment presented for the special cases of π and $\pi/2$ spin-turns.

The matrix description associates each coil with an operator connecting the neutron spin vectors before and after passage through the coil. Assemblies of coils are handled simply by successive applications of the appropriate operators to the neutron spin vector. We therefore have a prescription for quickly writing down the

0340-224X/78/0031/0117/$01.80

operator corresponding to the neutron spin-echo spectrometer as a whole, and for including in a natural way the effect of spin-flip scattering in the sample. The technique evolved is of general use in setting up and analysing neutron spin-echo spectrometer configurations, under monochromatic as well as white-beam conditions.

2. The Neutron Spin-Echo Experiment

Neutron spin-echo belongs to the general class of neutron polarization analysis experiments shown schematically in Figure 1a. The polarizer P_1 is characterized by transmissions τ^+ and τ^- for neutrons respectively parallel and anti-parallel to a guide field, which the neutron spin is assumed to follow adiabatically throughout. We shall take the direction of the guide field as the z-direction. Following Kendrick et al. [11], the polarizing efficiency is defined as $P_1=(\tau^+-\tau^-)/(\tau^++\tau^-)$. A similar definition holds for the analyser, P_2.

The operator $\hat{\mathbf{T}}$ is any net spin-turn operation. We specifically exclude spin-projection operators, such as polarizers, from this region. If $\mathbf{P}_i$ and $\mathbf{P}_f$ are the neutron spin-vectors before and after the operation, $\hat{\mathbf{T}}$ is defined by

$$\mathbf{P}_f=\hat{\mathbf{T}}\mathbf{P}_i \qquad (1)$$

In particular, for the cases of incident (+) and (−) neutrons, respectively, $\mathbf{P}_i=(0,0,\pm 1)$ and $\mathbf{P}_f=\pm(T_{13},T_{23},T_{33})$.

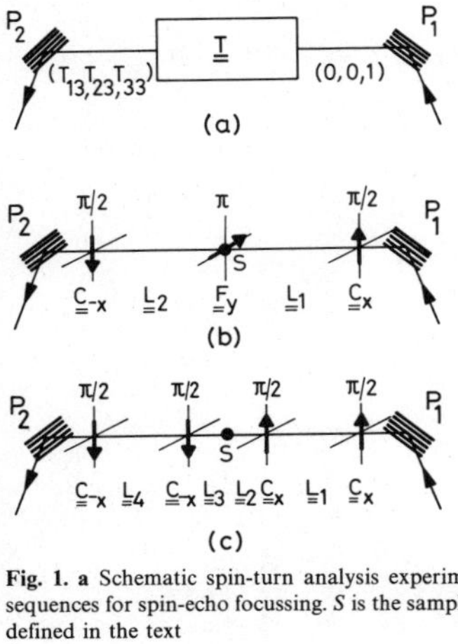

Fig. 1. a Schematic spin-turn analysis experiment. **b, c** Spin-turn sequences for spin-echo focussing. S is the sample. Other notation is defined in the text

The operator $\hat{\mathbf{T}}$ will in general be wavelength dependent. For application to a white beam with spectral density $I(\lambda)$, we define the spectral average for each matrix element:

$$\langle T_{ij}\rangle=\int_0^\infty I(\lambda)\,T_{ij}\,d\lambda \Big/ \int_0^\infty I(\lambda)\,d\lambda \qquad (2)$$

For brevity, the final average z-component $\langle T_{33}\rangle$ will be denoted simply by P_z.

P_z is the fundamental quantity to be measured in any experiment of this type. If the intensities transmitted by the analyser in the presence and absence of $\hat{\mathbf{T}}$ are I_1 and I_0, respectively, the relative intensity is easily shown to be

$$R^{-1}=I_1/I_0=(1+P_1P_2P_z)/(1+P_1P_2) \qquad (3)$$

In the case where $\hat{\mathbf{T}}$ represents a spin-flipper of efficiency f [11], $P_z=1-2f$ and R is the well-known flipping ratio.

The polarizer-analyser efficiency product P_1P_2 may be measured using two flippers, ON-OFF combinations of which give three intensity ratios relative to the intensity with both off. If R_1 and R_2 are the intensity ratios corresponding to the use of f_1 and f_2, respectively, and R_{12} corresponds to their simultaneous use, then

$$P_1P_2=R_{12}(R_1-1)(R_2-1)/(R_1R_2-R_{12})$$
$$f_i=(R_i-1)(1+P_1P_2)/2P_1P_2R_i \quad (i=1,2) \qquad (4)$$

Neutron spin-echo spectrometry seeks to establish a sequence of spin-turn operations before and after scattering such that the net spin-turn $\hat{\mathbf{T}}$ depends only on the energy transfer on scattering, independent of incident wavelength. Measurement of the final polarization by the technique just outlined then provides information on the sample scattering law.

The most commonly proposed spin-echo configuration [1–3] is shown in Figure 1b. The spins are first turned $\pi/2$ with respect to the guide-field, and then undergo Ω_1 radians Larmor precession in the region L_1. A π spin-turn is introduced at the sample, so that after scattering Ω_2 radians Larmor precession take place in the opposite sense in region L_2. A final $\pi/2$ turn then projects a particular component of the polarization back into the analysis direction, and it is this component which is measured as a function of Ω_1.

We shall see below that, if the spin-turns are independent of wavelength, $P_z=\cos(\Omega_2-\Omega_1)$. In the absence of any energy transfer on scattering, we may make $\Omega_2=\Omega_1$, independent of wavelength, by making the field integrals over L_1 and L_2 equal. $\hat{\mathbf{T}}$ is then just the unit matrix, giving the so-called spin-echo focussing condition. If there is now an energy transfer ω on scattering, we have in quasielastic approximation $\Omega_2 - \Omega_1 = \omega t$, where the time-variable t depends linearly on Ω_1. The detected intensity is thus modulated by a factor $\cos(\omega t)$, and since the detector integrates over all energy transfers, the integrated intensity is related to the cosine transform of the scattering probability. Full details of the quasielastic calculation are given in [1]; the more general inelastic calculation has been recently treated by Pynn [12].

The focussing property means that resolution is effectively decoupled from beam monochromatization, and energy changes of less than 10^{-7} eV may be detected using an incident beam wavelength spread of order 40% FWHM [13]. Our concern here is to calculate the effect of this wavelength spread on the action of real spin-turn coils, and thence show how to calculate the spin-echo intensity for any spectrometer configuration.

2.1. Spin-Turn Coil Operators

We restrict discussion to the widely used flat Mezei coil [1–3], set so that the coil field $\mathbf{H}_c$ and the coil normal $\mathbf{n}$ form a plane containing the beam direction (Fig. 2). This slightly restricted geometry covers all cases of practical interest. The beam travels in the z-direction, and for convenience we use a solenoidal guide field $\mathbf{H}_0$ coincident everywhere with this direction. The effective coil thickness in the beam direction is $d = d_0/\cos\beta$.

We assume at the outset that the field outside the coil is $\mathbf{H}_0$ alone, and that inside the coil the nett field $\mathbf{H}_N$ is uniform over the thickness d, so that line integrals through the coil of the form $\int H_N\, dz$ will be replaced by simple products $H_N d$. This assumption has proved completely adequate in practice, with the proviso that an effective coil thickness slightly different from the geometric thickness d must be used. Badurek et al. [9] and Van Laar et al. [10] have discussed the exact calculation. In what follows, however, d is only used to provide a guideline for setting up any practical system, and exact coil tuning may be accomplished without its knowledge.

The dimensionless variable $\Gamma = H_c/H_0$ will be used throughout. The equations describing the coil geometry are then (see Fig. 2)

$$\cos\theta = \Gamma\cos\beta/(1+\Gamma^2-2\Gamma\sin\beta)^{1/2} \tag{5}$$

$$\sin\theta = (1-\Gamma\sin\beta)/(1+\Gamma^2-2\Gamma\sin\beta)^{1/2} \tag{6}$$

$$H_N = H_0(1+\Gamma^2-2\Gamma\sin\beta)^{1/2} \tag{7}$$

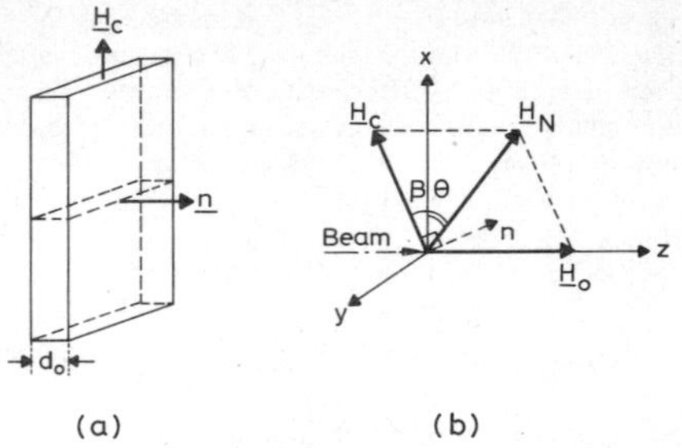

Fig. 2. Coil geometry **a** and axes **b** used in the calculation. d coil thickness; **n** normal to coil; $\mathbf{H}_c$ coil field; $\mathbf{H}_N$ nett field inside coil; $\mathbf{H}_0$ guide field

A neutron of wavelength λ which enters the coil non-adiabatically with its spin not coincident with $\mathbf{H}_N$ will accumulate, during transit through the coil, a Larmor precession angle given by

$$\psi = 4\pi|\gamma|\mu_N m\lambda H_N d/h^2 \tag{8}$$

where $\gamma = -1.913042$ [14], μ_N is the nuclear magneton, m the neutron mass and h Planck's constant. For H_N in oersteds, λ in Å and d_0 in cm,

$$\psi = 4\cdot 632\times 10^{-2}\,\lambda H_N d_0/\cos\beta \tag{9}$$

For a given value of the field ratio Γ, let λ_Γ be the wavelength at which there are just π radians Larmor precession in the coil. We then have in general

$$\psi = \lambda\pi/\lambda_\Gamma \tag{10}$$

where

$$\lambda_\Gamma = 67\cdot 824\cos\beta/H_0 d_0(1+\Gamma^2-2\Gamma\sin\beta)^{1/2} \tag{11}$$

We shall be interested later in finding values of Γ such that $\lambda_\Gamma = \lambda_0$, the central wavelength to which we wish to tune the coil for the particular spin-turn involved.

The relevant coil transfer operator $\hat{\mathbf{T}}$ corresponds to a precession of ψ radians about $\mathbf{H}_N$. Using a coordinate transformation to put z parallel to $\mathbf{H}_N$, rotating about $\mathbf{H}_N$ and transforming back, we find

$$\hat{\mathbf{T}} = \begin{pmatrix} \cos^2\theta+\sin^2\theta\cos\psi & -\sin\theta\sin\psi & \sin\theta\cos\theta(1-\cos\psi) \\ \sin\theta\sin\psi & \cos\psi & -\cos\theta\sin\psi \\ \sin\theta\cos\theta(1-\cos\psi) & \cos\theta\sin\psi & \sin^2\theta+\cos^2\theta\cos\psi \end{pmatrix} \tag{12}$$

It is readily verified that $\hat{\mathbf{T}}$ is orthogonal ($\hat{\mathbf{T}}^{-1} = \hat{\mathbf{T}}^T$). We shall now consider two special cases of particular interest.

(i) For a given $\beta > 0$, we can choose Γ and $H_0 d_0$ so that the z-component of the polarization is turned through

π and just reversed for a particular wavelength. This corresponds to the spin-flipper used in conventional polarization analysis, although the commonly used radio-frequency flipper has no time-independent matrix representation, and is not of interest in neutron spin-echo [3].

(ii) If we set $\beta=0$, Γ may be chosen for any given wavelength such that the z-component of the polarization is projected on to x, and vice versa. This $\pi/2$ turn, which is used to create or analyse precessing polarization, has no simple conventional equivalent.

The transfer operators for these two particular cases will be denoted by $\hat{\mathbf{F}}$ and $\hat{\mathbf{C}}$, respectively.

2.2. *The π Spin Turn*

Exact reversal of the z-component requires precession of $(2n-1)\pi$ radians about an orthogonal axis, say x, where $n=1,2,3\ldots$ is the order of the spin-turn. We shall restrict the present discussion to the widely used case $n=1$; generalisation is easily accomplished.

Since precession in the coil takes place about $\mathbf{H}_N$, reference to Figure 2b shows that we require $\theta=0$, and hence $\beta>0$ for finite fields. The ratio of fields Γ_0 at which $\theta=0$ follows from (5) and (6):

$$\Gamma_0=1/\sin\beta \quad (\beta>0) \tag{13}$$

The coil field must therefore always be larger than the guide field. We call the configuration where $\Gamma=\Gamma_0$ "correctly set".

The coil is perfectly tuned for a π-turn at wavelength λ_0, obtained by setting $\Gamma=\Gamma_0$ in (11):

$$\lambda_0=67\cdot 824\sin\beta/H_0\,d_0 \tag{14}$$

and hence

$$H_c=\Gamma_0 H_0=67\cdot 824/d_0\,\lambda_0 \tag{15}$$

Depending on whether H_N is parallel or anti-parallel to x ($\theta=0$ or π), the transfer operator for a correctly set π-turn coil is then, from (12),

$$\hat{\mathbf{F}}_{\pm x}=\begin{pmatrix}1 & 0 & 0\\ 0 & \cos\psi & \mp\sin\psi\\ 0 & \pm\sin\psi & \cos\psi\end{pmatrix} \tag{16}$$

We shall see later that in some spin-echo configurations it is useful to place the nett field along y, with corresponding operators

$$\hat{\mathbf{F}}_{\pm y}=\begin{pmatrix}\cos\psi & 0 & \mp\sin\psi\\ 0 & 1 & 0\\ \pm\sin\psi & 0 & \cos\psi\end{pmatrix} \tag{17}$$

In this correctly set case, $\lambda_\Gamma=\lambda_0$ and ψ is given by (10) and (14). When $\lambda=\lambda_0$, the matrices (16) and (17) may be described by the schematic transformations

$$\hat{\mathbf{F}}_{\pm x}\colon (x,y,z)\rightarrow(x,-y,-z) \tag{16'}$$

$$\hat{\mathbf{F}}_{\pm y}\colon (x,y,z)\rightarrow(-x,y,-z) \tag{17'}$$

Note that only two of the three components are reversed. This is essential for spin echo, and will be discussed further in Section 3.

2.3. *Efficiency and Tuning of a π Spin-Turn*

The efficiency of a π-turn coil is of general interest, since such a coil provides a cheap and simple flipper. From (16) and (17), the calculation of the efficiency amounts to calculating $\langle\cos\psi\rangle$ using (2). A very good approximation to most practical cases is obtained by taking the spectrum to have the triangular form (normalised to unit area) shown in inset in Figure 3. We then obtain

$$f=0\cdot 5+(1-\cos\pi\,\delta)/\pi^2\,\delta^2 \tag{18}$$

where $\delta=\Delta\lambda/\lambda_0$ and we have used $P_z=1-2f$. The efficiency is independent of the angle β, and any convenient value of β may therefore be used.

A comparison between theory and experiment is shown in Figure 3. The agreement (better than 0.3%) is considered good, especially as the measured velocity selector spectra showed significant deviations from triangular at the long wavelength end. Numerical integration of (2) using the exact spectrum (measured by time-of-flight) shows that in fact $\langle\cos\psi\rangle$ is not highly sensitive to details of $I(\lambda)$, and the triangular

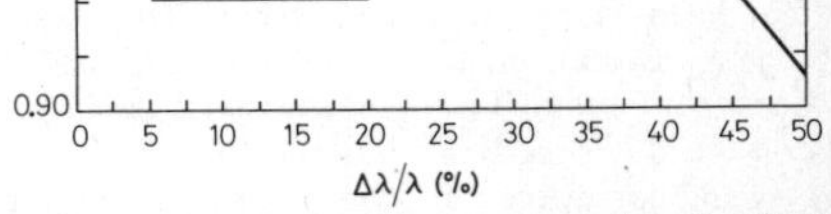

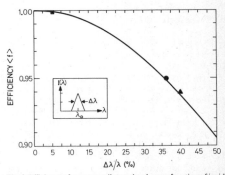

Fig. 3. Efficiency of a π-turn coil, tuned to λ_0, as a function of incident spectral width. ——— calculated from (18) using the spectrum shown in the inset. ■ Measured using graphite monochromator, $\lambda_0=4.3$ Å. ●, ▲ Measured using a helical velocity selector with $\lambda_0=6.2$ Å and 5.5 Å, respectively

approximation is generally good enough. The measured points were obtained using (4). The coil was of 1 mm diameter anodised aluminium wire close wound on a flat aluminium plate with $d_0 = 6$ mm, and surface dimensions $100 \times 150\,\text{mm}^2$. The coil current was 1.62 amps at $\lambda_0 = 4.3$ Å. Measurements were performed on the IN11 spectrometer at I.L.L., Grenoble [13].

Finally, we consider the practical problem of setting, or "tuning", a π-turn coil. If it is desired to tune to a central wavelength λ_0, the relevant guide and coil field strengths are given by (14) and (15), where the angle β is chosen for experimental convenience. (In practice, β should lie in the range 5°–30°.) In the light of our previous remarks about d_0, these values are approximate, and fine tuning is accomplished by varying the fields until P_z is minimised. Equation (3) shows that this corresponds to minimum transmitted intensity. Although this is a 2-parameter optimisation, the minimum is fairly flat and the starting values given by (14) and (15) are close enough for tuning to be easily and quickly established.

To calculate the form of the tuning curve (intensity *versus* coil current) at points other than correct setting, we must use the general form for P_z given by (1) and (12). At a general value of Γ, we may write

$$\psi = \lambda\,\phi/\lambda_0 \tag{19}$$

where, from (9) and (14)

$$\phi = \tan\beta(1 + \Gamma^2 - 2\Gamma\sin\beta)^{1/2}\,\pi \tag{20}$$

The relative intensity is then given by (3) with $P_z = \sin^2\theta + \cos^2\theta\langle\cos\psi\rangle$, the terms in θ being calculated from (5) and (6).

If the triangular spectrum of Figure 3 is again assumed,

$$\langle\sin\psi\rangle = 2\sin\phi(1-\cos\phi\,\delta)/\phi^2\delta^2$$
$$\langle\cos\psi\rangle = 2\cos\phi(1-\cos\phi\,\delta)/\phi^2\delta^2 \tag{21}$$

A series expansion verifies that $\langle\sin\psi\rangle \to \sin\phi$ and $\langle\cos\psi\rangle \to \cos\phi$ as $\delta = \Delta\lambda/\lambda_0 \to 0$.

The tuning curve calculated from (3) and (19) to (21) is shown in Figure 4 for $\beta = 12.5°$ and $\delta = 4\%$. The measured values were obtained by correctly setting H_0 using the above procedure, and then varying the current in the coil, using the same coil as for Figure 3. The agreement in the region of interest (the tuning point $\Gamma = \Gamma_0$) is very satisfactory. The disagreement at high currents has not been considered in detail, but probably results from coil non-uniformity, an imperfectly triangular spectrum, and effects of the type considered by Van Laar et al. [10].

Figure 5 shows equivalent data for $\delta = 40\%$. It is emphasized that measured parameters only are used in the theoretical calculation, and no parameter fitting is involved. The same coil was used throughout. It will be noted that the correctly tuned coil currents are in the inverse ratio of the mean wavelengths used, as follows from (15).

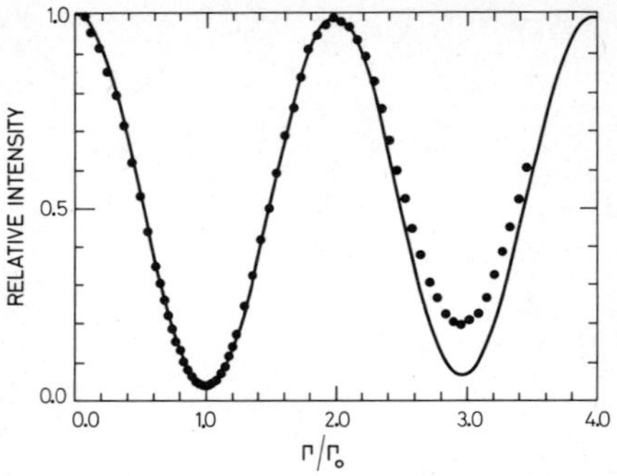

Fig. 4. Intensity *vs.* coil current expressed as Γ/Γ_0 for π-turn coil tuning. ——— Calculated using the spectrum of Figure 3. ● Measured at $\lambda_0 = 4.3$ Å, $\Delta\lambda/\lambda_0 = 4\%$. ($P_1P_2 = 0.93$, $\beta = 12.5°$, coil current $J_0 = 1.62$ ampères at $\Gamma = \Gamma_0$)

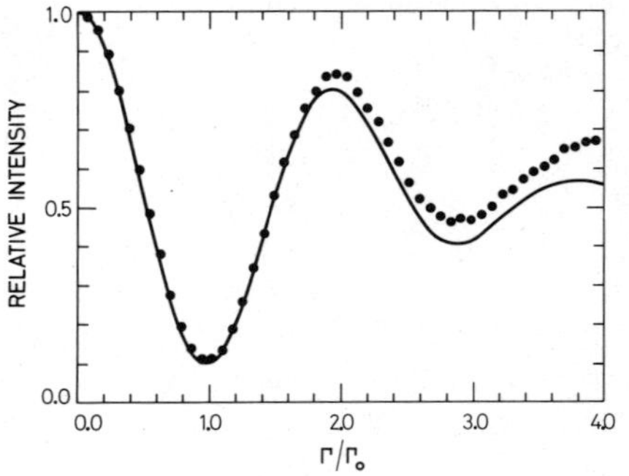

Fig. 5. Measured and calculated π-turn tuning curves corresponding to Figure 4 with $\lambda_0 = 5.5$ Å, $\Delta\lambda/\lambda_0 = 40\%$. ($P_1P_2 = 0.92$, $\beta = 12.5°$, $J_0 = 1.27$ ampères)

Similar agreement between theory and experiment is obtained at other values of β in the range in which measurements were made ($5° \leqq \beta \leqq 30°$).

2.4. The $\pi/2$ Spin-Turn

When $\beta = 0$, (5)–(7) reduce to

$$\cos\theta = \Gamma/(1+\Gamma^2)^{1/2} \tag{22}$$

$$\sin\theta = 1/(1+\Gamma^2)^{1/2} \tag{23}$$

$$H_N = H_0(1+\Gamma^2)^{1/2} \tag{24}$$

Projection of z onto x, and vice versa, is accomplished by using π radians precession about $x=y$ or $x=-y$. Thus $\theta=\pi/4$ or $3\pi/4$ and hence $\Gamma_0=\pm 1$ for correct setting.
If the coil is then to be perfectly tuned at wavelength λ_0, we have from (11)

$$\lambda_0 = \lambda_r = 67\cdot 824/\sqrt{2}\,H_0 d_0 \tag{25}$$

and hence

$$H_c = H_0 = 47\cdot 959/\lambda_0 d_0 \tag{26}$$

Since in the spin-echo configuration we may refer the π-turn operators $\hat{\mathbf{F}}$ to axes with the x-direction defined by the first $\pi/2$ coil, we need only consider $\pi/2$ turns where $\mathbf{H}_c$ is parallel or antiparallel to x. The corresponding operators may then be written down from (12) using (22) and (23); ψ is given by (19) where now

$$\phi = (1+\Gamma^2)^{1/2}\,\pi/\sqrt{2} \tag{27}$$

In particular, for the correctly set cases ($\Gamma_0=\pm 1$) the operators are given by

$$\hat{\mathbf{C}}_{\pm x} = \frac{1}{2}\begin{pmatrix} 1+\cos\psi & -\sqrt{2}\sin\psi & \pm(1-\cos\psi) \\ \sqrt{2}\sin\psi & 2\cos\psi & \mp\sqrt{2}\sin\psi \\ \pm(1-\cos\psi) & \pm\sqrt{2}\sin\psi & 1+\cos\psi \end{pmatrix} \tag{28}$$

with $\psi=\lambda\pi/\lambda_0$. When $\lambda=\lambda_0$, the matrices (28) are described by the schematic transformations

$$\hat{\mathbf{C}}_{\pm x}\colon\ (x,y,z)\rightarrow(\pm z, -y, \pm x) \tag{28'}$$

We now consider the practical problem of tuning a $\pi/2$ coil, which requires a different procedure from that used in the case of a π-turn. For an arbitrary incident spectrum, (1), (2) and (28) give the z-component after a correctly set $\pi/2$ turn:

$$P_z(\pi/2) = (1+\langle\cos\psi\rangle)/2 \tag{29}$$

For the same spectrum, (16) and (17) show that the same coil used as a π-turn yields a z-component

$$P_z(\pi) = \langle\cos\psi\rangle. \tag{30}$$

Equations (3), (29) and (30) then give the relative analysed intensity after a correctly set $\pi/2$ turn:

$$I_1/I_0 = (R+1)/2R \tag{31}$$

where R is the flipping ratio measured using the same coil as a π-turn.

The setting procedure for a $\pi/2$ turn coil is therefore clear, and may be accomplished without any knowledge of P_1P_2 or the incident spectrum:
(i) Turn the coil to any angle $\beta>0$ and set it for a π-turn as outlined in Section 2.3. Measure the flipping ratio, R, and the tuning current J_0.
(ii) Turn the coil back to the $\pi/2$ geometry ($\beta=0$) and set the coil current to $J_0/\sqrt{2}$ (this follows from (15) and (26)).
(iii) Adjust the guide field H_0 until the relative transmitted intensity is given by (31).

Correct tuning may be confirmed by measuring the intensity as a function of coil current and comparing it with the tuning curve calculated in a similar fashion to that for a π-turn, using an assumed triangular spectrum and (3), (19), (21) and (27). Figures 6 and 7 show such

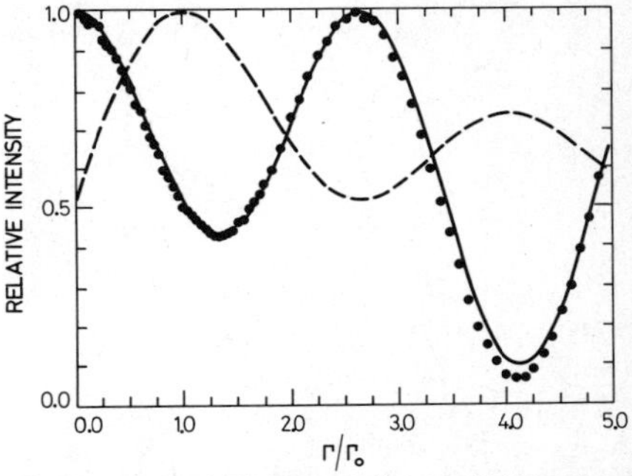

Fig. 6. Intensity *vs.* Γ/Γ_0 for $\pi/2$ turn coil. ——— Calculated assuming spectrum of Figure 3. --- Corresponding calculated x-component (see text). ● Measured at $\lambda_0=4.3$ Å, $\Delta\lambda/\lambda_0=4\%$. ($P_1P_2=0.93$, $J_0=1.15$ ampères)

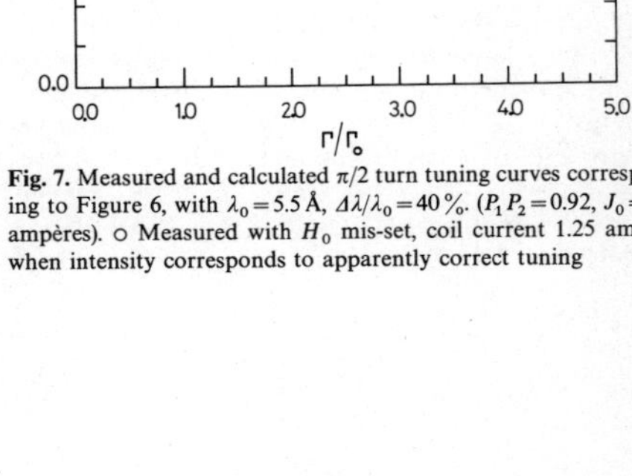

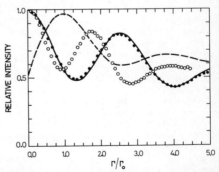

Fig. 7. Measured and calculated $\pi/2$ turn tuning curves corresponding to Figure 6, with $\lambda_0=5.5$ Å, $\Delta\lambda/\lambda_0=40\%$. ($P_1P_2=0.92$, $J_0=0.89$ ampères). ○ Measured with H_0 mis-set, coil current 1.25 ampères when intensity corresponds to apparently correct tuning

comparisons for $\delta=4\%$ and $\delta=40\%$, corresponding to Figures 4 and 5, respectively.

Replacing P_z by P_x in (3) yields the intensity which would be transmitted by an analyser sensitive to the x-component after the coil. This is also shown for reference in the figures. It will be noted that it is maximised at the tuning point, as is required for a projection of z onto x.

At first sight the more direct optimisation technique which was used for setting the π-turn may appear applicable to a $\pi/2$ coil as well. We shall demonstrate the necessity for the prescribed indirect procedure by considering the simple case of a $\hat{\mathbf{C}}_x$ operation on a monochromatic incident beam with $\langle\mathbf{P}_i\rangle=(0,0,P_1)$. In this case it follows from (28′) that $\langle\mathbf{P}_f\rangle=(P_1,0,0)$ and the analysed relative intensity is $(1+P_1P_2)^{-1}$, corresponding to $P_z=0$.

The latter is therefore a necessary condition for a correct $\pi/2$ turn. That it is not a sufficient condition is seen by examining the action of a $\pi/2$ coil set with arbitrary values of H_0 and Γ. Application of (12) in this case shows that $P_z=0$ whenever $\cos\psi=-\tan^2\theta$. From (22) and (23) this is satisfied when

$$\cos\psi=-1/\Gamma^2 \tag{32}$$

with ψ given by (10) and (11). For any $\Gamma\geqq 1$, therefore, it is always possible to find a value of H_0 which satisfies (32), and hence there is a continuum of settings which yield $P_z=0$ after the coil.

Correct setting also requires $P_y=0$, however, since the coil must project z onto x. From (12), (22), (23) and (32) the other components of the final polarization are $P_x=1/\Gamma$ and $P_y=-(\Gamma^2-1)^{1/2}/\Gamma$, and the only solution for correct setting is $\Gamma=1$ (or $\Gamma=-1$ in the case of $\hat{\mathbf{C}}_{-x}$).

The argument is easily extended to the polychromatic case. It is thus not feasible to proceed in the manner used for tuning a π-turn, since there is an infinity of (Γ, H_0) pairs which yield an intensity satisfying (31), but which give a final polarization with mixed x and y components.

An example is shown in Figure 7, where a measured tuning curve is presented for a $\pi/2$ coil with H_0 deliberately mis-set from the correct value (obtained by the method described above). It is still possible to find a value of Γ such that the intensity corresponds to apparent tuning, but where the x-component has an arbitrary value. The same comments apply, of course, to the (infinite) set of points at which the correctly tuned intensity is obtained with $\Gamma>\Gamma_0$, even when H_0 is correctly set.

2.5. *The Two-Coil Flipper*

As an example of the operator technique, we shall briefly examine a commonly used geometry involving sequential spin-turns: the two-coil flipper [1]. A π-turn is obtained by the successive application of two $\pi/2$ turns, and the operator for the overall system may be written down as the product of the individual operators. For correctly set coils the result is

$$\hat{\mathbf{C}}_{\pm x}\hat{\mathbf{C}}_{\mp x}=\frac{1}{2}\begin{pmatrix} 2\cos\psi-\sin^2\psi & -\sqrt{2}\sin\psi(1+\cos\psi) & \pm\sin^2\psi \\ \sqrt{2}\sin\psi(1+\cos\psi) & 2\cos^2\psi & \sqrt{2}\sin\psi(1-\cos\psi) \\ \pm\sin^2\psi & \sqrt{2}\sin\psi(1-\cos\psi) & 2\cos\psi+\sin^2\psi \end{pmatrix} \tag{33}$$

When $\lambda=\lambda_0$, the individual operators commute and the resulting transformation is

$$\hat{\mathbf{C}}_{\pm x}\hat{\mathbf{C}}_{\mp x}\colon\ (x,y,z)\to(-x,y,-z). \tag{33′}$$

The configuration is therefore equivalent, in the monochromatic case, to $\hat{\mathbf{F}}_{\pm y}$ (17′). For the polychromatic case, however, it is evident from the 33-element of (33) that the efficiency is less than for a single π-turn coil.

3. The Spin-Echo Operator

To calculate the operator $\hat{\mathbf{T}}$ corresponding to the sequence of spin-turns which make up the spin-echo spectrometer configuration, we simply take operator products. Since Larmor precession regions are involved, we also need the operator corresponding to free precession about z. This is

$$\hat{\mathbf{L}}_{\pm z}=\begin{pmatrix} \cos\Omega & \mp\sin\Omega & 0 \\ \pm\sin\Omega & \cos\Omega & 0 \\ 0 & 0 & 1 \end{pmatrix} \tag{34}$$

Here, $\Omega=2\pi N_0\lambda/\lambda_0$ where N_0 is the number of precessions at the centre wavelength λ_0. For a precession region with uniform field H (oersted) over length l(cm), $N_0=7.37\times10^{-3}Hl\lambda_0$. Generally, we need only consider $\hat{\mathbf{L}}_z$, and the subscript $\pm z$ will be omitted.

If the spectrometer is drawn schematically with the beam running from right to left, the operators are multiplied in the order they are written down. For example, in the arrangement of Figure 1b,

$$\hat{\mathbf{T}}=\hat{\mathbf{C}}_{-x}\hat{\mathbf{L}}_2\hat{\mathbf{F}}_y\hat{\mathbf{L}}_1\hat{\mathbf{C}}_x \tag{35}$$

It is easily verified, using (17′), (28′) and (34) that for a monochromatic beam and no energy transfer, $\hat{\mathbf{T}}$ is just the unit matrix if $\Omega_2=\Omega_1$. Energy transfer at the sample is included in the calculation by shifting λ an appropriate amount at the sample, and then integrating over all such shifts correctly weighted by the scattering law.

The operator $\hat{\mathbf{T}}$ may be evaluated analytically, but the

Table 1. The sign of spin-echo polarization as a function of spectrometer configuration. The notation is described in the text

C	F	C	P_z
x	$\pm x$	x	+
$-x$	$\pm y$	x	+
x	$\pm y$	x	−
$-x$	$\pm x$	x	−

general white-beam calculation is most easily performed by numerical matrix multiplication followed by numerical integration over the incident spectrum. This is particularly true for more complex geometries such as that shown in Figure 1c, where the π-turn has been replaced by separated $\pi/2$ turns to stop precession of the original z-component in the sample region. This is used, for example, when path differences in the sample may be important.

The P_z generated in this manner may be compared with experiment using (3), and some examples will be given in a later paper.

As we shall see below, it is useful when setting up the spectrometer to be able to choose the sign of the final polarization. This is obtained by calculating P_z for the focussed configuration with elastic scattering. We are free to choose $\Omega_1 = \Omega_2 = 2k\pi$ in this case (k integer), so that $\hat{\mathbf{L}}$ is the identity matrix. The effect of any successive operations such as (16′), (17′) or (28′) is then easily determined. Configurations of the type shown in Figure 1b may be denoted simply by the relative orientations of the spin-turn coils, and Table 1 shows the sign of the final polarization, or echo, for some commonly used geometries.

Those coil arrangements which give a final negative polarization are equivalent in their nett action to a π-turn. For the polychromatic case, however, numerical calculation of P_z yields the interesting result that the spin-echo configuration is always more efficient than a single π coil. This has been confirmed by experiment.

3.1. Spin-Flip Scattering

A case of particular practical interest occurs when the sample scattering is spin-incoherent, giving rise to spin-flip processes on scattering. In the case of an isotopically pure sample, spin incoherence will give rise to 2/3 spin-flip on scattering. If several isotopes are present, the calculation of the relative amount of spin-flip becomes more complicated. Lovesey [15] has treated the monatomic case, and Copley and Lovesey [16] have extended the calculation to several chemical species. For a general incoherent scattering sample, we will denote the proportion of spin-flip by α, $0 \leqq \alpha \leqq 2/3$. We now show that the measurement of spin-echo polarization may be effected without a knowledge of α [17]. Spin-flip may be integrated into the calculation of the preceding section by including a spin-flip operator at the sample position for that proportion α of neutrons which are spin-flipped. Since nuclear spin-flip reverses all three components of the polarization vector [15], this operator is just $-\hat{\mathbf{1}}$, however, and hence may be introduced anywhere in the spin-turn sequence for the purposes of calculation. This means that we may absorb the effect of spin-flip into the incident beam polarization, so that the echo polarization purely reflects the sample dynamics.

The fact that nuclear spin-flip commutes with all other spin-turn operations means that it cannot be used to replace the π-turn to produce spin-echo focussing. The essential difference is evident from (16′) or (17′): reversal of only two spin components will reverse the direction of precession [3], while reversal of all three components merely corresponds to a jump of π-radians in the Larmor precession.

An incoherent scatterer may therefore be treated as a non-spin-flip sample coupled with a flipper of efficiency α, and this flipper may be introduced into the sequence immediately after the polarizer. The revised form of (3) is

$$I/I_0 = [1 + (1-2\alpha) P_1 P_2 P_z]/[1 + (1-2\alpha) P_1 P_2] \tag{36}$$

If $\hat{\mathbf{T}}$ in Figure 1a is a π-turn of known efficiency f, α may be measured from the flipping ratio. For measurement of P_z in a spin-echo configuration, however, we can obtain a more interesting result.

Let I_0 be the intensity scattered from the sample in the absence of any spin-turn operators, I_1 the intensity in the spin-echo configuration, and I_2 the intensity with only a π-turn coil of efficiency f in operation. Then application of (36) to the π-turn gives the product $(1-2\alpha) P_1 P_2$; substituting back into (36) in the spin-echo case gives for the spin-echo polarization

$$P_z = 1 - 2f(I_0 - I_1)/(I_0 - I_2) \tag{37}$$

The polarization may therefore be obtained free from the effects of spin-flip, without specifically measuring either $P_1 P_2$ or α. In the particular case when $f=1$, (37) becomes $P_z = 2(I_1 - \langle I_1 \rangle)/(I_0 - I_2)$, where $\langle I_1 \rangle$ is the average level of the spin-echo signal, which may be shown, using (31), to be given by $(I_0 + I_2)/2$.

It will be noted from (36) that $\alpha > 0$ diminishes the modulation due to P_z, and for $\alpha > 0.5$ the modulation changes sign. In the latter case, positive modulation may be obtained, if desired, by choosing a spin-echo configuration where the final polarization would normally be negative, as described in the previous section.

4. Summary

We have described a rotation matrix technique for analyzing the general action of neutron spin-turn coils. The response of π and $\pi/2$ spin-turn coils to a polychromatic beam has been calculated, and a general practical technique for tuning these coils has been presented and compared with experiment.

The matrix representation of a spin-turn coil provides a simple way of evaluating the effect of a sequence of spin-turns, and in particular the sequence resulting in spin-echo focussing. A general means of calculating the response of a spin-echo spectrometer under white beam conditions has been presented, in terms of a relative transmitted intensity which may be compared directly with experiment.

Finally, the effect of spin-flip scattering from spin-incoherent samples has been considered, and it has been shown that dynamic effects may be decoupled from spin-flip effects in the analysis of the spin-echo intensity.

References

1. Mezei, F.: Z. Physik **255**, 146 (1972)
2. Hayter, J.B.: ORNL Report CONF-760601-P2, 1074 (1976)
3. Hayter, J.B.: Polarized neutrons. In: Neutron diffraction (ed. H. Dachs) Berlin-Heidelberg-New York: Springer 1978
4. Rekveldt, M.Th.: Z. Physik **259**, 391 (1973)
5. Mezei, F.: Physica **86–88**B, 1049 (1977)
6. Okorokov, A.I., Runov, V.V., Volkov, V.I., Gukasov, A.G.: Preprint FTI-106, Leningrad Inst. Nucl. Phys. (1974)
7. Jones, T.J.L., Williams, W.G.: Rutherford Laboratory Internal Report RL-77-079/A (1977). See also references cited therein
8. Badurek, G., Westphal, G.P.: Nucl. Instr. and Meth. **128**, 315 (1975)
9. Badurek, G., Westphal, G.P., Ziegler, P.: Nucl. Instr. and Meth. **120**, 351 (1974)
10. Van Laar, B., Maniawski, F., Mijnarends, P.E.: Nucl. Instr. and Meth. **133**, 241 (1976)
11. Kendrick, H., Werner, S.A., Arrott, A.: Nucl. Instr. and Meth. **68**, 50 (1969)
12. Pynn, R.: Submitted to J. Phys. E (1978)
13. Maier, B.: Neutron Beam Facilities at the H.F.R., Grenoble, Institut Laue-Langevin publication (1977)
14. Greene, G., Ramsey, N.F., Mampe, W., Pendlebury, J.M., Smith, K., Dress, W.B., Miller, P.D., Perrin, P.: Phys. Lett. **71**B, 297 (1977)
15. Lovesey, S.W.: Dynamics of solids and liquids by neutron scattering (eds. S.W. Lovesey, T. Springer), pp. 54–58. Berlin-Heidelberg-New York: Springer 1977
16. Copley, J.R.D., Lovesey, S.W.: Individual and collective Motion in Non-Molecular Liquids. Bristol, England. Proceedings of the 3rd International Conference on Liquid Metals (1976)
17. Hayter, J.B.: Neutron Spin-Echo Data Analysis; Grenoble, ILL Internal Scientific Report 78HA50 (1978)

John B. Hayter
Institut Max von Laue-
Paul Langevin
B.P. 156X
F-38042 Grenoble Cédex
France

Z. Physik B 35, 199–205 (1979)

Zeitschrift für Physik B

Neutron Spin-Echo Integral Transform Spectroscopy

John B. Hayter
Institut Laue-Langevin, Grenoble, France

J. Penfold
Rutherford Laboratory, Chilton, Didcot, Oxon., England

Received May 9, 1979

The matrix analysis of neutron spin-echo is extended to the general case of an asymmetric spectrometer configuration. It is shown that an integral transform of the incident or scattered beam spectrum may be measured by scanning the asymmetry, defined as the difference between the number of precessions in the two arms. The problem of inverting the integral transform is considered, and we show how the fast Fourier transform technique may be used as the basis for an efficient numerical solution. Experimental results confirming the calculation are presented. Since only a few precessions of asymmetry are needed to measure the spectrum of a broad incident beam, the technique provides a simple means of spectral analysis suitable for use on medium flux reactors.

1. Introduction

Various neutron spin-echo configurations are available for the measurement of small inelastic linewidths [1–3]. These configurations all make use of the focussing property, which requires the numbers of precessions in the two arms of the spectrometer to be equal. In the case of quasielastic scattering, focussing is achieved simply by making the magnetic paths through the spectrometer symmetric about the central π-turn coil. The cosine transform of the lineshape is then measured by scanning the total number of precessions, maintaining the symmetry between the two arms. In a practical spectrometer, with a broad incident wavelength spread, a rather more complicated integral transform of the lineshape is obtained, but the spectrometer response is always exactly calculable provided a model of the scattering law is available. Details of the calculation have been given in a previous publication [2], which we shall refer to below as Paper I.

In the present paper we demonstrate the utility of a different, asymmetric scan in which the scanning parameter is the difference between the numbers of precessions in the two arms. Such a scan measures an integral transform of the wavelength spectrum of the beam. In the ideal case, this integral transform is the Fourier cosine transform. We shall calculate the spectrometer response in the real case and compare it with the ideal case and with experiment. We shall also show that in the real case the beam spectrum may still be recovered from the measured transform with reasonable accuracy *via* a straightforward fast Fourier transformation.

If the total number of precessions is kept low, the spectrometer will be insensitive to the small energy transfers encountered in quasi-elastic scattering. Under these conditions the asymmetric scan technique may also be used to measure a scattered wavelength spectrum. Since, in quasielastic scattering, the momentum transfer Q is simply related to the wavelength λ at any scattering angle $2\theta(Q=4\pi\sin\theta/\lambda)$, the measurement of incident and scattered wavelength spectra provides a fast estimate of $S(Q)$ in, for example, small-angle scattering. The measurement is also a useful way of obtaining the true spectrum for spectrometer response calculations in cases where the incident spectrum has been hardened by sample absorption, or softened by small-angle scattering.

Since the spectrometer response to an asymmetric scan is approximately the Fourier transform of the incident spectrum, it is damped fairly quickly if the incident spectrum is broad. An asymmetric scan therefore provides a means of quickly determining

0340-224X/79/0035/0199/$01.40

the symmetric setting of the spectrometer. We remark in passing that this is the best practical way of setting up the spectrometer for high resolution symmetric scans.

2. Calculation of Spectrometer Response

To calculate the general response, we shall use the matrix method given in Paper I, and we shall adhere to the notation used in that paper. We take the specific case of a $\pi/2-\pi-\pi/2$ spin-turn configuration, so that the spectrometer is described by operator products $\hat{\mathbf{T}}$ of the form $\hat{\mathbf{C}}\hat{\mathbf{L}}_2\hat{\mathbf{F}}\hat{\mathbf{L}}_1\hat{\mathbf{C}}$ (a particular example is given by (35) and Fig. 1b of Paper I). Here $\hat{\mathbf{L}}_1$ and $\hat{\mathbf{L}}_2$ are the Larmor precession operators in the first and second arms of the spectrometer, and $\hat{\mathbf{C}}$ and $\hat{\mathbf{F}}$ the operators corresponding to $\pi/2$ and π spin-turns, respectively.
We recall first the general expression for spectrometer intensity [2]. For a normalised incident beam spectrum $I(\lambda)$, the transmitted intensity I_1 is given by

$$I_1/I_0=(1+\mathbf{P}_1\mathbf{P}_2\langle\mathbf{P}_z\rangle)/(1+\mathbf{P}_1\mathbf{P}_2) \quad (1)$$

where I_0 is the intensity with no spin-turn coils operating, P_1P_2 is the polariser-analyser efficiency product, and the mean final polarisation $\langle P_z\rangle$ is given by

$$\langle\mathbf{P}_z\rangle=\int_0^\infty I(\lambda)\,T_{33}[\mathbf{S}(Q,\omega),\lambda]\,d\lambda \quad (2)$$

where $S(Q,\omega)$ is the sample scattering law. The general method of calculating the matrix element T_{33} is given in Paper I. Basically it consists of evaluating the operator product $\hat{\mathbf{T}}$ for each possible energy transfer $\hbar\omega$, then integrating the result over all values of ω correctly weighted by the scattering law, to obtain $T_{33}[S(Q,\omega)\,\lambda]$ at each incident wavelength λ. Since ω is introduced into the calculation of $\hat{\mathbf{T}}$ in terms of the wavelength shift it produces at that point in the matrix product which corresponds to the sample position, such a calculation is in principle exact, the final accuracy depending on the numerical integration method employed.
Under focussing conditions the Larmor precession operators $\hat{\mathbf{L}}_1$ and $\hat{\mathbf{L}}_2$ are equal. Our purpose now is to explore the general case with $\hat{\mathbf{L}}_1\neq\hat{\mathbf{L}}_2$. In the interest of clarity, we shall first place the calculation in context by considering an ideal case which, with a few reasonable approximations, may be solved analytically.

2.1. Ideal Spectrometer Response to Diffusive Scattering

We define an ideal spectrometer as one in which the spin-turn coils operate independently of wavelength. Such coils are then described for all λ by (16′), (17′) and (28′) of Paper I. In the absence of any energy transfer on scattering we obtain $T_{33}=\pm\cos(\Omega_2-\Omega_1)$ for all configurations of type $\hat{\mathbf{C}}\hat{\mathbf{L}}_2\hat{\mathbf{F}}\hat{\mathbf{L}}_1\hat{\mathbf{C}}$, where $\Omega_i=2\pi N_i$ is the total precession angle in arm i $(i=1,2)$ of the spectrometer, N_i being the number of precessions. (For the choice of sign, see Table 1 of Paper I.) At mean incident wavelength λ_0, let N_0 and $N_0+\Delta N_0$ be the numbers of precessions in the first and second arms, respectively, ΔN_0 thus being a measure of the asymmetry. At a general incident wavelength λ, let $\Delta\lambda$ be the wavelength change on scattering with an energy transfer $\hbar\omega$. (To first order, $\Delta\lambda\sim m\lambda^3\omega/2\pi h$ for quasielastic scattering.) We then obtain, to first order in asymmetry and wavelength shift,

$$\cos(\Omega_2-\Omega_1)=\cos[2\pi(N_0\Delta\lambda+\lambda\Delta N_0)/\lambda_0]. \quad (3)$$

The cosine is expanded in terms of the individual arguments, $\Delta\lambda$ written in terms of ω and the result substituted in (2). For quasielastic scattering, $S(Q,\omega)$ may be taken as an even function of ω (since the detailed balance factor is effectively unity). The sine transform is then zero, and the result is

$$\langle\mathbf{P}_z\rangle=\int_0^\infty I(\lambda)\cos(2\pi\lambda\Delta N_0/\lambda_0)\,d\lambda$$
$$\int_{-\infty}^{\infty}\mathbf{S}(Q,\omega)\cos[\omega\,t(\lambda)]\,d\omega \quad (4)$$

where the time variable $t(\lambda)=N_0 m\lambda^3/h\lambda_0$, m being the neutron mass and h Planck's constant. If N_0 is produced by a field integral of J oe-cm, $N_0=7.37\times10^{-3}J\lambda_0$ and $t(\lambda)=1.86\times10^{-16}J\lambda^3$ s for λ in Å.
We now take the specific example of scattering from a system undergoing self-diffusion, for which the scattering law is Lorentzian:

$$\mathbf{S}(Q,\omega)=DQ^2/\pi[(DQ^2)^2+\omega^2] \quad (5)$$

with D the diffusion coefficient. The integration over ω yields

$$\langle\mathbf{P}_z\rangle=\int_0^\infty I(\lambda)\cos(2\pi\lambda\Delta N_0/\lambda_0)\exp[-DQ^2\,t(\lambda)]\,d\lambda. \quad (6)$$

At fixed scattering angle 2θ the exponent may be written $DQ^2\,t(\lambda)=D\,\beta\,N_0\,\lambda$ where $\beta=(4\pi\sin\theta)^2m/h\,\lambda_0$. Thus in this particular case, the strong λ^3 dependence

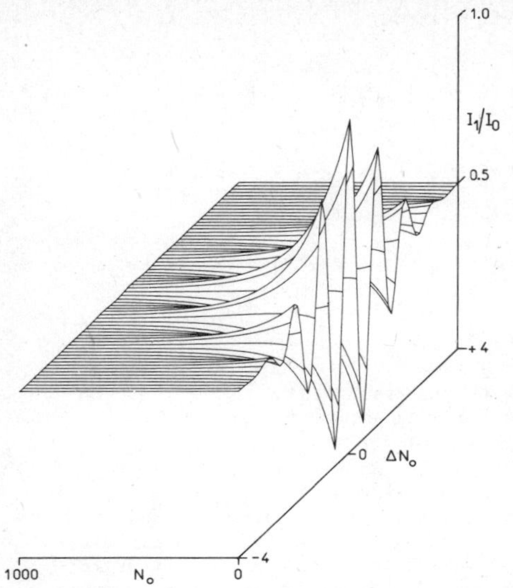

Fig. 1. Calculated spin-echo intensity from a diffusive scatterer ($D = 2.0 \times 10^{-5}\,\mathrm{cm}^2\,\mathrm{s}^{-1}$) in an ideal spectrometer with $P_1 P_2 = 1$, $\lambda_0 = 6\,\text{Å}$, $\Delta\lambda_0/\lambda_0 = 0.33$, scattering angle 10° ($Q = 0.18\,\text{Å}^{-1}$). N_0 Number of precessions in the first arm; ΔN_0 Difference between the numbers of precessions in the two arms

of $t(\lambda)$ has been reduced to the much weaker linear dependence on λ in the final result because of the Q^2 term.

For a *symmetric* spectrometer setting ($\Delta N_0 = 0$) the final polarisation in this particular case is therefore the Laplace transform of the incident spectrum, the Laplace variable being $D\beta N_0$. The orthogonal *asymmetric* scan along $N_0 = 0$, on the other hand, gives the Fourier transform of $I(\lambda)$, with Fourier variable $2\pi \Delta N_0/\lambda_0$.

In the general case, we may evaluate (6) analytically if $I(\lambda)$ is taken as a normalised Gaussian with standard deviation $\sigma = \Delta\lambda_0/2.3548$, $\Delta\lambda_0$ being the FWHM of the incident spectrum. Provided λ_0 and $\Delta\lambda_0$ are such that $I(\lambda)$ has fallen to zero at very short wavelength, the integration limits may be formally extended to $(-\infty, \infty)$, yielding

$$\langle \mathbf{P}_z \rangle = \exp[-D\beta N_0 \lambda_0 + (D\beta N_0 \sigma)^2/2 - 2(\pi\sigma \Delta N_0/\lambda_0)^2] \cdot \cos[2\pi \Delta N_0(1 - D\beta N_0 \sigma^2/\lambda_0)]. \quad (7)$$

This function is shown in Fig. 1, where (1) has been used to convert $\langle P_z \rangle$ to intensity.

Figure 1 illustrates all the essential features of a quasielastic neutron spin-echo experiment:

(i) Once the spectrometer has been set to within a few precessions of the symmetric position by dead reckoning, an asymmetric scan at fixed, small N_0 allows the precise determination of the exact centre of symmetry. We shall show below how to recover $I(\lambda)$ from this scan.

(ii) The scan to examine $S(Q,\omega)$ is then made by varying N_0 keeping $\Delta N_0 = 0$. (An example of such a measurement is given in [4].)

(iii) The mean level at large values of either N_0 or ΔN_0 is constant. In the ideal case its value corresponds to $\langle P_z \rangle = 0$, or complete randomisation of the spins. The general case will be discussed below.

(iv) Only a few precessions of asymmetry are needed to measure the incident spectrum. Such a measurement may therefore be undertaken with a simple, low-field spectrometer. Many more precessions are needed for a high resolution symmetric scan. In the present figure, where the Lorentzian FWHM has been taken as 8.5 μeV, the signal is fully damped only after about 500 precessions. (A practical spectrometer such as IN11 [5] may achieve better than 5,000 precessions.) Only the initial slope of the transform need be measured in the case of a known Lorentzian lineshape, of course; this would require about 25 precessions in the present case.

2.2. General Response of a Real Spectrometer

The intensity detected in a real spectrometer at a general setting (N_1, N_2) is given by (1) and (2). The calculation is described in Paper I and we shall not consider the general case any further here. Instead, we shall concentrate on the practical use of the asymmetric scan to measure the beam spectrum $I(\lambda)$. That is, we shall henceforth take $S(Q,\omega) = \delta(\omega)$. The results obtained will therefore be valid at any value of N_0 for measurements on the direct beam. From the previous discussion and Fig. 1 it is also clear that the results will be a good description of the intensity of the scattered beam, provided in this case that N_0 is kept small relative to the value at which the energy transfers have caused significant depolarisation.

The operator $\hat{\mathbf{T}}$ may be evaluated by direct numerical matrix multiplication, and the response to any $I(\lambda)$ calculated from (2). It is instructive, however, to calculate the matrix product analytically for the common configurations which we are considering. We assume correctly set coils, so that the relevant angle appearing in the operators of Paper I is $\psi = \lambda\pi/\lambda_0$. We then obtain for the eight configurations of the type $\hat{\mathbf{C}}\hat{\mathbf{L}}_2\hat{\mathbf{F}}\hat{\mathbf{L}}_1\hat{\mathbf{C}}$ the general result

$$T_{33} = R(\Omega_1, \psi)\sin\Omega_2 + S(\Omega_1, \psi) \cdot \cos\Omega_2 + Q(\Omega_1, \psi) \quad (8)$$

where R, S and Q all have the form $A_1(\psi)\sin\Omega_1 + A_2(\psi)\cos\Omega_1 + A_3(\psi)$, the A_i being given in the Appendix. In all cases, if $\lambda=\lambda_0$ we obtain $R=\pm\sin\Omega_1$, $S=\pm\cos\Omega_1$ and $Q=0$, so that $T_{33}=\pm\cos(2\pi\Delta N_0)$, as expected, and (2) becomes just the Fourier cosine transform of $I(\lambda)$.

Remembering the λ-dependence of Ω_1, Ω_2 and ψ, the final polarisation may therefore be written

$$\langle \mathbf{P}_z \rangle = \int_0^\infty I(\lambda)\, R(N_0,\lambda) \sin[2\pi(N_0+\Delta N_0)\lambda/\lambda_0]\, d\lambda$$

$$+ \int_0^\infty I(\lambda)\, \mathbf{S}(N_0,\lambda) \cos[2\pi(N_0+\Delta N_0)\lambda/\lambda_0]\, d\lambda$$

$$+ \int_0^\infty I(\lambda)\, Q(N_0,\lambda)\, d\lambda \tag{9}$$

where R, S and Q now explicitly contain the dependence on N_0 and λ.

The first two integrals of (9) tend to zero as $\Delta N_0 \to \infty$, since $I(\lambda)\to 0$ as $\lambda\to\infty$ and R, S and Q are bounded functions of N_0 and λ [6]. The third integral, however, remains small but finite for a finite incident wavelength spread, becoming zero only for $I(\lambda)=\delta(\lambda-\lambda_0)$. In general, therefore, the mean level to which the transform tends at large values of the scan parameter ΔN_0 does not quite correspond to complete depolarisation of the beam. We shall refer to this mean level, which is given by the third integral of (9), as $\langle P_z\rangle_\infty$.

3. Experimental

Measurements were performed on the direct beam of the IN11 spin-echo spectrometer at the Institut Laue-Langevin [5]. The spectrometer configuration is shown in Fig. 2. Incident wavelengths may be selected in the range $4\,\text{Å} \leqq \lambda_0 \leqq 8\,\text{Å}$, with wavelength spreads $\Delta\lambda_0/\lambda_0 \leqq 50\%$. The guide fields $\mathbf{H}_1$ and $\mathbf{H}_2$ are produced by 133 cm long solenoids co-axial with the z (beam) direction. The field integrals are variable up to 80 kOe-cm (corresponding to about 3,500 precessions at $\lambda_0=6\,\text{Å}$). The $\pi/2$ spin-turn coils are oriented in the x-direction, and the π-turn coil oriented either along x or y. (Orientations of the opposite sign are achieved by reversing the current in the relevant coil.) The operators $\hat{\mathbf{C}}$, $\hat{\mathbf{L}}$ and $\hat{\mathbf{F}}$ are as defined in Paper I. All spin-turn coils consisted of a single layer of 1 mm enamelled aluminium wire, machine wound with a 1.1 mm pitch on a flat aluminium plate 6 mm thick, with surface dimensions of $150\times100\,\text{mm}^2$. (A hole of $50\times50\,\text{mm}^2$ was cut in the centre of each plate before winding to allow passage of the beam.) Each coil was mounted with its large surface perpendicular to the beam, and the π-turn coil was then tilted 5°. The coils were separately tuned by the methods of Paper I, and then all switched on together to produce the spin-echo configuration. The identical solenoids $\mathbf{H}_1$ and $\mathbf{H}_2$ were energised in series in a symmetric spectrometer geometry, so that varying their current produced a symmetric scan of N_0 with $\Delta N_0=0$. At any value of N_0, asymmetric scans could be made independently by passing current through an additional 30-turn coil close-wound around the second solenoid. This coil could be energised bi-directionally, so that the number of precessions in the second arm could be increased or decreased at will. It is easily shown that the axial field integral produced by such a coil on $(-\infty,\infty)$ is 1.257 oe-cm per ampère-turn, independent of the coil radius. The cut-off introduced by the finite separation of the π and $\pi/2$ coils is small provided their separation corresponds to at least several diameters of the scan coil. In the present case, a scan current of ±0.6 ampère produced $\Delta N_0=\pm1$ precession at $\lambda_0=6\,\text{Å}$.

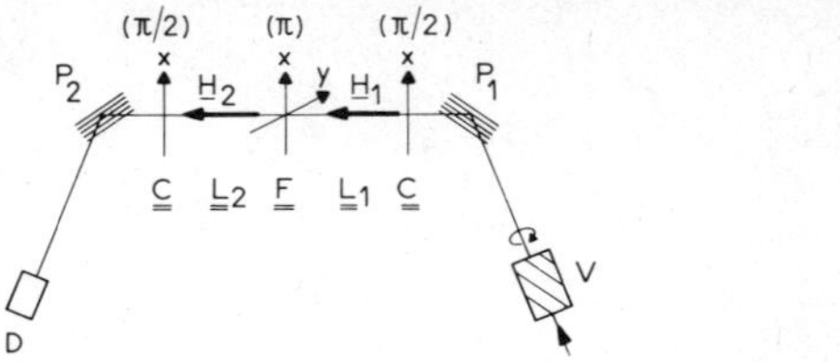

Fig. 2. Experimental arrangement. P_1 Polariser; P_2 Analyser; $\mathbf{H}_1$, $\mathbf{H}_2$ Precession fields; V Helical velocity selector; D Detector. Other notation is defined in the text

Asymmetric scans were made at main solenoid currents of 10, 20 and 40 ampères, corresponding to N_0 values of 147, 294 and 588 precessions at $\lambda_0=6\,\text{Å}$. The velocity selector transmitted a wavelength spread $\Delta\lambda_0/\lambda_0=0.42$. Scans were performed for all eight spin-echo configurations, and gave predicted results. Typical data for a complementary pair of scans are shown in Fig. 3, together with the theoretical intensities calculated from (1) and (9) assuming $I(\lambda)$ to be Gaussian. The phase-shift between theory and experiment in the wings of the data is due to the slightly non-Gaussian nature of the (known) real selector spectrum, and demonstrates the sensitivity of the method.

It is worth noting that in the focussed configuration ($\Delta N_0=0$), complementary spin-echo measurements provide the best measurement of $\langle P_z\rangle$. (Configurations are considered complementary if they are related by reversal of the current in one $\pi/2$ coil.) If I_1 is the intensity in the spin-echo configuration, and $\bar{I}_1$

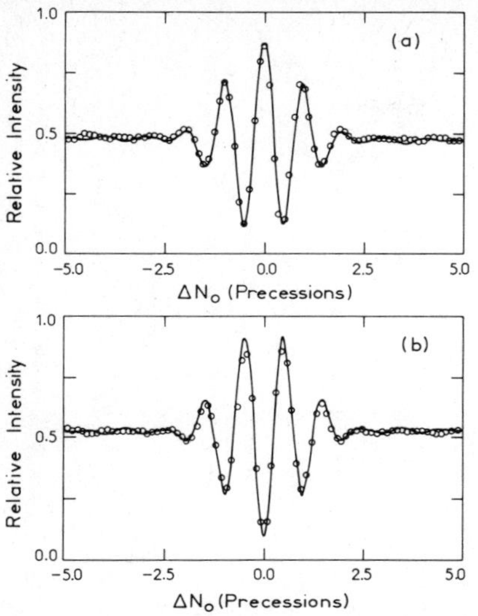

Fig. 3. Spin-echo intensity as a function of ΔN_0 for $N_0=588$, $\lambda_0=6$ Å, $\Delta\lambda_0/\lambda_0=0.42$, $P_1P_2=0.9$. ○ Measured. —— Calculated assuming $I(\lambda)$ Gaussian. (a) Configuration $\hat{\mathbf{C}}_{-x}\hat{\mathbf{F}}_y\hat{\mathbf{C}}_x$. (b) Complementary configuration $\hat{\mathbf{C}}_x\hat{\mathbf{F}}_y\hat{\mathbf{C}}_x$

the complementary intensity, (37) of Paper I applied in the two cases yields

$$\langle \mathbf{P}_z \rangle = f(I_1 - \bar{I}_1)/(I_0 - I_2) \tag{10}$$

where I_0 is the intensity with all coils off ("spin-up"), and I_2 is the intensity with only the π-coil of efficiency f operating ("spin-down"). As shown in Paper I, (10) yields a polarisation free from any effects of spin-flip in the sample. By normalising the measured polarisation (10) against that for a standard elastic scatterer, the true spin-echo polarisation is obtained free from machine inhomogeneity and sample spin-flip effects, without needing to measure either P_1P_2 or f. If the scattering is purely non-spin-flip, (1) applied to the various intensities appearing in (10) further yields

$$\langle \mathbf{P}_z \rangle = (I_1 - \bar{I}_1)/\mathbf{P}_1\mathbf{P}_2(I_1 + \bar{I}_1) \tag{10'}$$

where the same remarks about normalisation apply. In the latter case, fewer measurements are needed.

4. The Reverse Integral Transform

The general problem in using the asymmetric scan to measure an unknown spectrum is to invert (9) to obtain $I(\lambda)$ given $\langle P_z \rangle$. Since the polarisation will be measured at a finite number of ΔN_0 values, we shall be faced with using a discrete transform method, and by far the most efficient such method is the matrix factoring technique known as the fast Fourier transform [7]. We shall now show that under certain slight restrictions, application of the fast Fourier transform technique allows direct inversion of (9).

We first note that the use of a discrete transform implies certain properties of the data, most notably that of periodicity. If the measured set of data points constitutes a "frame", such frames are assumed to repeat indefinitely in either direction; that is, if we have measured M points of data, points with indices n and $n+kM$ (k integral) are assumed equal. This in turn will imply the same periodicity in the resulting $I(\lambda)$. Further, in the resulting $I(\lambda)$, points in the transform with indices greater than $M/2$ are redundant results corresponding to $I(-\lambda)$. These points are an artefact of the transform method, negative wavelength having no physical significance. It is not our purpose to enter into further detail here, but the reader unfamiliar with the properties of discrete transforms is strongly urged to consult an appropriate reference (for example, [7]).

The implied periodicity in the data means that a cyclic rearrangement will enter the transform as a phase-factor, and therefore have no effect upon the transform modulus. This has the following useful implication: if we measure M points of data from $-\Delta N_0(\min)$ to $+\Delta N_0(\max)$, using a fixed scan interval of $1/m$ precessions, we may label the indices of the data points from 0 to $M-1$, so that the scan parameter ΔN_0 appears to run from 0 to $(M-1)/m$, without affecting the transform modulus. That is, we can disregard the absolute value of the starting point of the scan in representing the data.

The scan is therefore made by starting at a high enough negative asymmetry for the polarisation to have been damped to the mean level $\langle P_z \rangle_\infty$, and then stepping ΔN_0 in units of $1/m$ precession until $\langle P_z \rangle_\infty$ is reached on the other side of the symmetric position. (Typically, we choose $m=8$ points per precession.) The measured points are labelled $n=0, 1, \ldots, M-1$. The mean level is now subtracted from the data. The resulting set of M points, which we shall call $p_n = \langle P_z(n/m) \rangle - \langle P_z \rangle_\infty$, are then given by the first two integrals of (9) with appropriate ΔN_0 values. (Note that $p_0 = p_{M-1} = 0$.)

We now apply the discrete transform

$$D_k = \frac{1}{M} \sum_{n=0}^{M-1} p_n \exp(-2\pi i n k/M) \tag{11}$$

which may be evaluated very quickly and accurately

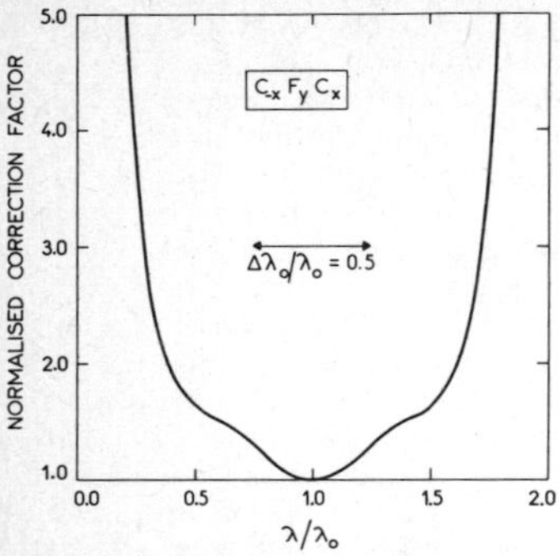

Fig. 4. Inverse transform correction factor calculated for $\hat{\mathbf{C}}_{-x}\hat{\mathbf{F}}_y\hat{\mathbf{C}}_x$ with $N_0=512$, $M=256$, $m=16$

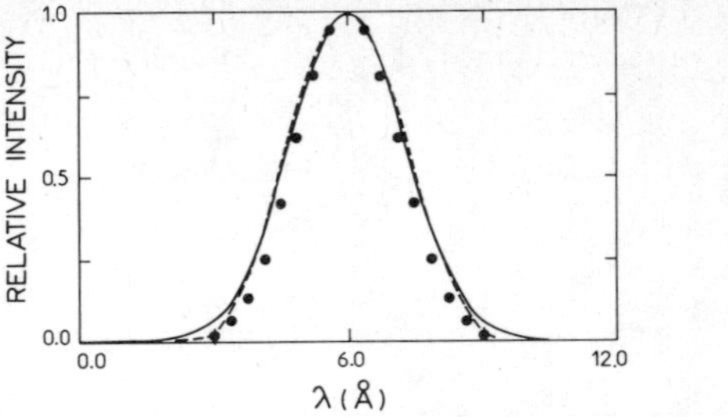

Fig. 5. Comparison of uncorrected and corrected inverse transforms. Data to be transformed was calculated by applying (9) to a Gaussian spectrum with $\Delta\lambda_0/\lambda_0=0.5$, $N_0=588$. —— Original spectrum. ● Inverse transform using (11). --- Inverse transform using (15) with N_0 taken as 512

by the fast Fourier transform algorithm provided M is an exact power of 2. In our case, this condition may be satisfied for an arbitrary number of data points by simply adding an appropriate number of zeros to the dataset. (Typically we choose $M=256$ or 512.)

To apply (11) to (9) first re-write the integrals as sums (remembering that the third integral has been subtracted from the data) and note that the sampling is such that the k'th point in each sum corresponds to

$$\lambda_k = k\, m\, \lambda_0/M. \tag{12}$$

N_0 must be chosen so as to remove the double periodicity which would arise from the forms of R and S. This means restricting N_0 to the values

$$N_0 = jM/m \qquad (j=0,1,2,\ldots). \tag{13}$$

Applying (11) to the summation form of (9) and using the periodic properties of the complex exponential then yields

$$D_k = I(\lambda_k)\,\mathbf{S}(N_0,\lambda_k) + I(\lambda_{-k})\,\mathbf{S}(N_0,\lambda_{-k}) - i[I(\lambda_k)\,R(N_0,\lambda_k) - I(\lambda_{-k})\,R(N_0,\lambda_{-k})] \tag{14}$$

where λ_k is given by (12). As noted previously, we have generated terms involving $\lambda_{-k}=-\lambda_k$; from periodicity, $\lambda_{-k}=\lambda_{M-k}$ and these are just the redundant terms corresponding to a reflection of the result about $n=M/2$. It is easily seen that the real part of D is even ($\mathrm{Re}\,D_k=\mathrm{Re}\,D_{M-k}$) and the imaginary part is odd, from which we find $I(\lambda_{-k})=I(\lambda_k)$. Taking the modulus of (14) and rearranging yields the desired result

$$I(\lambda_k) = |D_k|/\{[R(N_0,\lambda_k) - R(N_0,\lambda_{M-k})]^2 + [\mathbf{S}(N_0,\lambda_k) + \mathbf{S}(N_0,\lambda_{M-k})]^2\}^{\frac{1}{2}} \tag{15}$$

where λ_k is given by (12) and N_0 is restricted to the values (13).

The form of the correction function $(R^2_{\mathrm{odd}} + S^2_{\mathrm{even}})^{-1/2}$ is shown in Fig. 4 for a typical configuration. Since we know from physical considerations that the continuous integral (9) cannot be sensitive to the precise value of N_0, we may further conclude that the restriction (13) is a result of using a discrete transform. We therefore propose using (15) on data measured at any value of N_0, merely using the nearest value of N_0 which satisfies (13) in the calculation of the correction factor. Figure 5 shows the result of such an approach, and it is seen that reasonable accuracy is obtained. Further, reference to Fig. 4 shows that direct use of the spectrum $|D_k|$ resulting from a fast Fourier transform gives better than 5% accuracy for $\Delta\lambda_0/\lambda_0<0.2$, without any correction at all.

Finally, we note that $|D_k|$ has zeroes at points corresponding to $\lambda=0, 2\lambda_0, 4\lambda_0, \ldots$ and the correction function has corresponding poles. This is expected on physical grounds, since the action of the spin-turn coils on even multiples of λ_0 corresponds to $\pi-2\pi-\pi$, and hence there can be no resulting echo. From Fig. 4, the practical limit of the inversion technique is $\Delta\lambda_0/\lambda_0<0.5$ for the total spectrum, or $\Delta\lambda_0/\lambda_0<1$ for measurement of the FWHM ($\Delta\lambda_0$).

5. Summary

The technique of asymmetric scan neutron spin-echo provides a simple means of wavelength analysis in white-beam neutron spectroscopy. Since a few precessions of asymmetry are sufficient, only a straightforward low-field spin-echo configuration is required. The wavelength resolution is of order one per cent, and is independent of the magnitude of the guide field. Numerical inversion of the measured transform is straightforward and may be performed in a few seconds on small computers of the size normally employed for spectrometer control.

The method is also suitable for analysing scattered wavelength spectra, for example in small-angle scattering. It has no particular flux requirements, and should therefore prove useful in small reactor installations.

Appendix

To find explicit forms of the functions R, S and Q, set

$$4R(\Omega_1,\psi)=R_1(\psi)\sin\Omega_1+R_2(\psi)\cos\Omega_1+R_3(\psi)$$
$$4S(\Omega_1,\psi)=S_1(\psi)\sin\Omega_1+S_2(\psi)\cos\Omega_1+S_3(\psi)$$
$$4Q(\Omega_1,\psi)=Q_1(\psi)\sin\Omega_1+Q_2(\psi)\cos\Omega_1+Q_3(\psi),$$

where $\psi=\lambda\pi/\lambda_0$ and $\Omega_1=2\pi N_0$. Let

$$r_1=1-4\cos\psi+\cos^2\psi+2\cos^3\psi$$
$$r_2=\sqrt{2}\sin^3\psi$$
$$r_3=\sqrt{2}\sin^2\psi(1+\cos\psi)$$
$$s_1=2-\cos\psi-\cos^3\psi$$
$$s_2=\sqrt{2}\sin^3\psi$$
$$s_3=\sin^3\psi$$
$$q_1=r_3$$
$$q_2=s_3$$
$$q_3=\cos\psi(1+\cos\psi)^2$$

Then the R_i, S_i and Q_i are given in Table 1.

Table 1. Terms appearing in the spectrometer response function for configurations of the type $\hat{\mathbf{C}}\hat{\mathbf{L}}_2\hat{\mathbf{F}}\hat{\mathbf{L}}_1\hat{\mathbf{C}}$

C:	x	x	−x	−x	−x	−x	x	x
F:	x	−x	x	−x	y	−y	y	−y
C:	x	x	x	x	x	x	x	x
Echo sign	+	+	−	−	+	+	−	−
R_1	s_1	s_1	$-s_1$	$-s_1$	r_1	r_1	$-r_1$	$-r_1$
R_2	s_2	s_2	$-s_2$	$-s_2$	$-r_2$	$-r_2$	r_2	r_2
R_3	s_3	$-s_3$	$-s_3$	s_3	r_3	$-r_3$	$-r_3$	r_3
S_1	r_2	r_2	$-r_2$	$-r_2$	$-s_2$	$-s_2$	s_2	s_2
S_2	r_1	r_1	$-r_1$	$-r_1$	s_1	s_1	$-s_1$	$-s_1$
S_3	$-r_3$	r_3	r_3	$-r_3$	s_3	$-s_3$	$-s_3$	s_3
Q_1	q_2	$-q_2$	q_2	$-q_2$	q_1	$-q_1$	q_1	$-q_1$
Q_2	$-q_1$	q_1	$-q_1$	q_1	q_2	$-q_2$	q_2	$-q_2$
Q_3	q_3	q_3	q_3	q_3	q_3	q_3	q_3	q_3

References

1. Mezei, F.: Z. Physik **255**, 146 (1972)
2. Hayter, J.B.: Z. Physik B**31**, 117 (1978)
3. Pynn, R.: J. Phys. E**11**, 1133 (1978)
4. Richter, D., Hayter, J.B., Mezei, F., Ewen, B.: Phys. Rev. Letts. **41**, 1484 (1978)
5. Hayter, J.B., Mezei, F.: In preparation
6. Titchmarsh, E.C.: Introduction to the Theory of Fourier Integrals, 2nd ed., Oxford: University Press 1948
7. Brigham, E.O.: The Fast Fourier Transform, New Jersey: Prentice-Hall 1974

John B. Hayter
Institut Max von Laue
Paul Langevin
156 X Centre de Tri
F-38042 Grenoble Cedex
France

J. Penfold
Rutherford Laboratory
Chilton, Didcot, Oxon.
England

IAEA-SM-219/56

NEUTRON SPIN ECHO AND POLARIZED NEUTRONS

F. MEZEI
Institut Laue-Langevin,
Grenoble
and
Central Research Institute for Physics,
Budapest,
Hungary

Abstract

NEUTRON SPIN ECHO AND POLARIZED NEUTRONS.

The Larmor precession of polarized neutrons provides an intrinsic measure of a selected component of the momentum of a given neutron, and a way of storing this information on the same neutron via a neutron spin memory effect. These phenomena are utilized in the neutron spin echo approach to neutron scattering, which consists of replacing the classical measurement of ingoing and outgoing neutron momenta by a single measurement of a quantity related directly to the change of momentum of a given neutron in the scattering process, in much the same way as selected parts of the scattering function depend on the momentum change. This technique breaks the strong relation between resolution and neutron intensity which the other methods are submitted to. Consequently, the neutron spin echo approach offers a much improved resolution capability and better neutron economy even for moderate resolutions for the major part of inelastic scattering and for certain classes of elastic scattering experiments. Its application could substantially improve the utilization efficiency of continuous beam reactors, or eventually, help to improve resolution on pulsed sources. For illustration, sample experimental results are presented, which were obtained with the instrument built and operating at the Institut Laue-Langevin (ILL).

1. Larmor Precessions in Polarized Neutron Beams

The behaviour of the spins of free neutrons can be described in an essentially classical mechanical way[1]. The spin of a given neutron is represented by a classical unit vector $\vec{S}$ defined e.g. by the polar angles θ and ϕ. The polarization of a given ensemble of neutrons is then defined by the ensemble average

$$\vec{P} = \langle \vec{S} \rangle$$

The components of $\vec{P}$ are the expectation values, i.e. first moments of the distributions of the components of the $\vec{S}$ vectors in the ensemble and the only quantum mechanical restriction to this classical picture is that higher moments of these distributions cannot be measured, so they do not make sense.

The equation of motion for $\vec{S}$ is simply given by

$$\frac{d\vec{S}}{dt} = \gamma[\vec{B}(t) \times \vec{S}(t)] \qquad (1)$$

where $\vec{B}(t)$ is the magnetic field seen by the given pointlike neutron following the trajectory $\vec{r}(t)$, i.e. for static fields

$$\vec{B}(t) = \vec{B}(\vec{r}(t))$$

Therefore the equation of motion for P can only be given if the $\vec{B}(t)$ is the same for all neutrons in the ensemble.

This picture breaks down in all cases where the neutron wave-functions play a role, i.e. the neutrons cannot be considered as pointlike. Examples for this are the scattering processes. On the other hand, neutron beams can be treated in the above manner. It has to be kept in mind, however, that the precise definition of the ensemble of neutrons characterized by $\vec{P}$ is essential, e.g. we can speak of the polarization of a neutron beam at a given point in space.

If $\vec{B}$ and $\vec{S}$ are not parallel, eq.(1) will describe the well known Larmor precession, which in turn will be a measure of $\vec{B}(t)$. Since for a static field the time dependence in $\vec{B}(t)$ only comes from the neutron velocity, we have here an intrinsic measure of this latter quantity. A very essential feature of this Larmor precession velocity determination is that it is made for all individual neutrons in the beam separately and the result of this "measurement", the Larmor precession angle, is stored on the neutron (as the direction of the spin) and can either be read out by 3-dimensional polarization analysis or can be used to compare with what will happen to the same neutron later. Thus we have a handy neutron spin memory effect.

It is of particular interest to use the Larmor precession technique to determine a selected component of the neutron velocity, as shown in Fig. 1. The neutron spin is parallel to the magnetic (guide) field before the 90° turn coil, which initiates the precession. The total angle of Larmor precession between the 90° coil and the reference surface F is given as

$$2\pi N = \frac{A}{\vec{n}\,\vec{k}}$$

where $\vec{n}$ is the unit vector perpendicular to the magnet faces, $\vec{k}$ is the momentum of the neutron, and the constant A is proportional to the line integral of the field across the magnet. For example, for a $\lambda = 4$ Å neutron and a magnet of 20 cm width and of 500 Oe field, $2\pi N \simeq 2000$ rad (for $\vec{n} \parallel \vec{k}$) which illustrates the resolution power of the Larmor precession velocity determination method.

2. The Spin Echo Approach to Neutron Scattering

The usual way neutron scattering experiments are performed consists of two basic steps: preparation of the incoming beam and analysis of the outgoing beam. (This applies to all elastic and inelastic experiments with the exception of single crystal diffractometry). The preparation of the beam means the selection of neutrons with momenta within a given range, and it represents a first measurement in the quantum mechanical sense (projection to a subspace). The analysis is a second similar process. Both of these processes involve selecting out a small fraction of the neutrons present. (In time-of-flight methods the chopping basically makes part of the analysis and not the beam preparation).

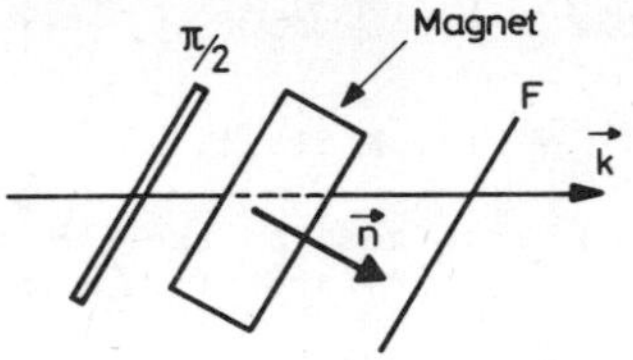

FIG. 1. The Larmor precession measurement of a given component of the neutron velocity.

However, the relevant parameters of the scattering function S are not the ingoing and outgoing neutron momenta $\vec{k}_o$ and $\vec{k}_1$, respectively, but the combinations $\vec{\kappa} = \vec{k}_1 - \vec{k}_o$ and $\omega = \hbar^2(k_1^2 - k_o^2)/2m$. This is why the usual approach to neutron scattering is a poorly adapted one, with the main drawbacks:

a) The resolution in κ and ω depends on the scatter of $\vec{k}_o$ and $\vec{k}_1$, i.e. both have to be severely restricted. High resolution is difficult to achieve;

b) These resolution restrictions affect the neutron intensity very sensitively, particularly since they apply twice, both for $\vec{k}_o$ and $\vec{k}_1$. Neutron intensity is inversely proportional to a high power of the resolution.

A more adequate method of measuring $S(\vec{\kappa},\omega)$ has to be related to the change of the state of a given neutron in the scattering process, the same way as it is in $S(\vec{\kappa},\omega)$. In other words, the two quantum mechanical measurements, that of $\vec{k}_o$ and that of $\vec{k}_1$ have to be replaced by a single one related to the difference between $\vec{k}_1$ and $\vec{k}_o$. Such a method a priori offers the advantages that

a) In measuring directly a quantity related to the difference between $\vec{k}_1$ and $\vec{k}_o$, better resolution can be achieved;

b) The resolution and neutron intensity become, in principle, independent.

For the practical realization of such a difference method, we need a neutron memory effect, i.e. a way to have stored on each scattered neutron the information on its state before the scattering. The only available coding parameter for information storage on individual neutrons is the spin, and, as pointed out before, the neutron spin memory effect is easily controllable. The resulting experimental method has a strong formal, but no conceptual resemblance to the NMR spin echo technique. This is why in the first place, with a superficial understanding of the phenomenon, the name Neutron Spin Echo was chosen [2].

It has to be emphasized that the Neutron Spin Echo (NSE) is not just another method in neutron (inelastic) scattering, but a conceptually different approach to all the others. It could be defined in the following way: Direct, single measurement of the change of the neutron state in the scattering process using the spin memory effect, as opposed to the classical way of measuring incoming and outgoing neutron momenta in two, separate steps. Consequently there is no common measure in the resolution and neutron intensity relations for the two cases. Whereas for all of the classical methods the Liouville space considerations give roughly identical resolution-intensity conditions (depending on monochromatizations, collimations etc), for the NSE, these restrictions do not apply at all. The crucial, and only real problem for intensity in the NSE is the luminosity and efficiency of the neutron polarizer and polarization analyser. This problem, constituting for a long time the main obstacle to the proliferation of polarization analysis studies, is now solved by the development of neutron supermirrors [3].

It is obvious that the NSE approach will be superior in neutron economy to any of the classical methods in any type of experiment where the phase space limitations due to resolution requirements become essential. A consideration of the technical details suggests that this is the case generally if either the required momentum or energy resolution is better than about 2 - 5% of the incoming value. In addition, the practical limits of the resolution for the NSE are better than for the classical methods, namely 10^{-3} to 10^{-5}, depending on the case, and this is achieved with a neutron intensity constant over the whole resolution range.

2.1. Formal Theory of NSE

There is a theoretical possibility of coding more than one component of $\vec{k}_o$ into the neutron spin direction. However, the way of doing this is very impractical, and in addition, the gain in information is not too useful anyway. Instead, we will only consider the case when only a given component of $\vec{k}_o$ is taken, in the way described above, and compared to another given component of $\vec{k}_1$ for each scattered neutron. The essential thing is the appropriate choice of these components, which makes in most cases a more general approach unnecessary.

Thus, in general, the quantity we can measure, the Larmor precession angle of neutron spins is given by the equation

$$2\pi N = \frac{A}{\vec{n}_o\vec{k}_o} + \frac{B}{\vec{n}_1\vec{k}_1}$$

where N is the total number of Larmor precession which is the algebraic sum of precessions before the scattering (first term) and after the scattering (second term). The unit vectors $\vec{n}_o$ and $\vec{n}_1$ defining the components of $\vec{k}_o$ and $\vec{k}_1$ which are considered, are freely chosen together with the parameters A and B (which can be positive and negative), within practical limits, of course.

It is easy to see that a given value of N corresponds to a 5-dimensional surface (which we shall call a NSE surface) of the 6-dimensional variable space $(\vec{k}_o, \vec{k}_1)$. Under certain special conditions the N = const. equation describes a 3-dimensional surface also in the 4-dimensional space of the relevant variables $(\vec{\kappa},\omega)$. For simplicity, let us assume for the moment that this is the case. So in a NSE scan (below we will come back to what this means in practice) one essentially probes the scattering function $S(\kappa,\omega)$ by moving the NSE surface around in the $(\vec{\kappa},\omega)$ space, which surface now replaces the classical resolution ellipsoid. This situation is schematically illustrated in 2 dimensions in Fig. 2, where the scattering function $S(\kappa,\omega)$ is represented by a phonon dispersion surface. It is obvious that a single scan, in which the NSE surface "moves" parallel to itself, does not give a point by point map of $S(\kappa,\omega)$, which in fact is rarely needed either. This limited, one-dimensional NSE information can be very efficiently used, as illustrated in Fig. 2 by the shaded area. Within this area the dispersion surface and the NSE surface are parallel, so the NSE scan can give precise measurement of the width, position and slope of the dispersion relation. In practice such a measurement is a combination of a very rough resolution classical, selection-type procedure (e.g. triple axis) that singles out the shaded area, and a fine NSE scan within this area. It remains to be shown that the slope of the NSE surface can be tuned to any given direction in the $(\vec{\kappa},\omega)$ space. (In principle, scanning the same area with different NSE surfaces of, say, perpendicular slopes gives an equivalent to a point by point map).

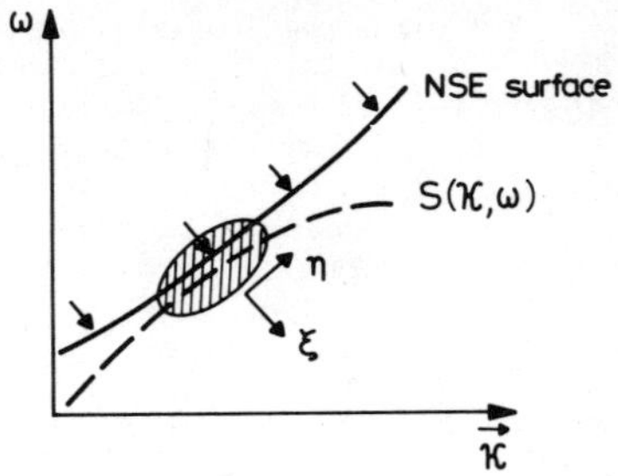

FIG.2. The principle of the NSE focussing.

Using a series expansion around the centre of gravity values $\vec{k}_o^o$, $\vec{k}_1^o$, $\vec{\kappa}^o = \vec{k}_1^o - \vec{k}_o^o$, $\omega_o = \hbar^2((k_1^o)^2 - (k_o^o)^2)/2m$, the NSE surface and the dispersion relation are given by the equations (2) and (3), respectively,

$$dN = -\frac{A}{(\vec{n}_o\vec{k}_o^o)^2}\,\vec{n}_o d\vec{k}_o - \frac{B}{(\vec{n}_1\vec{k}_1^o)^2}\,\vec{n}_1 d\vec{k}_1 = 0 \tag{2}$$

$$\vec{\alpha}\, d\vec{\kappa} + d\omega = 0 \tag{3}$$

where $d\vec{k}_o = \vec{k}_o - \vec{k}_o^o$, etc.

The equivalence of (2) and (3) requires that

$$\begin{aligned} \vec{n}_o &= -\frac{\vec{\alpha} + \hbar^2\vec{k}_o/m}{|\vec{\alpha} + \hbar^2\vec{k}_o/m|} \\ \vec{n}_1 &= +\frac{\vec{\alpha} + \hbar^2\vec{k}_1/m}{|\vec{\alpha} + \hbar^2\vec{k}_1/m|} \\ \frac{A}{B} &= -\frac{|\vec{\alpha} + \hbar^2\vec{k}_o/m|}{|\vec{\alpha} + \hbar^2\vec{k}_1/m|} \cdot \frac{(\vec{n}_o\vec{k}_o^o)^2}{(\vec{n}_1\vec{k}_1^o)^2} \end{aligned} \tag{4}$$

Obviously the set of equations (4) automatically assures that to first order the NSE surface fulfils the conditions to be represented also in the $(\vec{\kappa},\omega)$ space, which we assumed above for Fig.2.

We conclude that the first order focussing, i.e. the tuning of the vector slope of the NSE surface to a dispersion surface is generally possible. Consequently, the resolution of the corresponding NSE scan will only be limited by second order effects, i.e. by eventual differences in the curvatures of these two surfaces. It turns out that for typical acoustic dispersion relations these second order effects will be equivalent to about 10^{-5} - 10^{-6} eV resolution broadening while for quasi-elastic scattering and dispersionless excitations (flat part of optical branches, molecular transition etc), i.e. when $\vec{\alpha} = 0$, they vanish, and the resolution is only limited by the details of magnetic precession field configuration, to about 10^{-7} - 10^{-8} eV (see below).

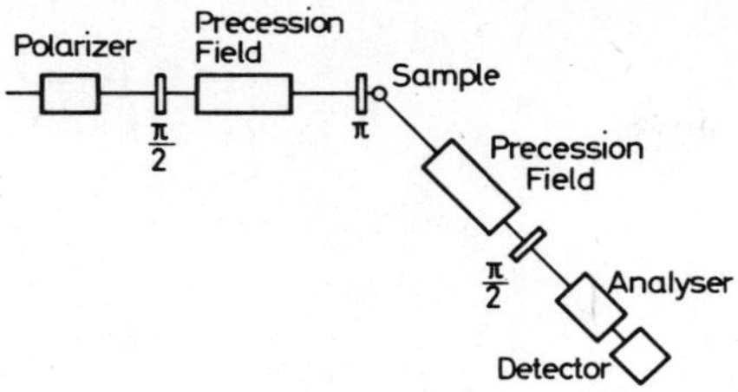

FIG.3. The general lay-out of the NSE Spectrometer IN11 of the ILL.

2.2. The Fourier Aspects of the NSE Measurements

In the original paper[3] introducing the NSE, it was shown that inherently the NSE means the measurement of a Fourier transform of the scattering function. Here we will only point out the basic features of this.

Obviously, the quantity that is physically measured is not N itself, but the vector polarization of a scattered beam at a given surface, which defines the end limit of the Larmor precession region. This way a directly measured component of this polarization will be given e.g. by

$$P_X = \langle \cos(2\pi N) \rangle$$

where the average is taken over all the neutrons in the analysed beam. Since N varies in the direction perpendicular to the NSE surface, which direction is taken as the ξ axis of a new coordinate system (see Fig. 2), the distribution of N in the beam is proportional to the cross section of $S(\kappa,\omega)$ taken in the ξ direction. Hence

$$P_X(t) = \int S_\eta(\xi) \cos(2\pi t\xi) d\xi$$

where $$S_\eta(\xi) = \int S(\xi,\eta) d\eta$$

and the integrals are taken over the shaded selection area (properly weighted with the pertinent transmission function). The proportionality constant t is obviously proportional to A (or B) which is the only experimental parameter left free by eqs.(4), which determine only A/B.

In first approximation, this proportionality constant t can be taken as independent of η . So by scanning A and B so that A/B stays constant, the Fourier transform of $S_\eta(\xi)$ can be determined. As mentioned above, there is a way of obtaining the full two-dimensional Fourier transform of $S(\xi,\eta)$. It consists, in principle, of independent scanning of A and B. However, if $S(\xi,\eta)$ has a very different behaviour in ξ and η directions, the scan has to be asymmetric too. Practically this means scanning say A in the first place, and then B around the value calculated from the last equation of the set (4).

Finally there is a remark to be made about Fourier transform measurements. The relatively discouraging experiences with mechanical Fourier choppers are due to two facts. Namely, that the Fourier signal was contaminated by higher order harmonics, (or other frequencies), and that the whole scattering function with a number of different structures (elastic and different inelastic scattering processes) were to be sorted out simultaneously. None of these difficulties emerge here. The Larmor precessions

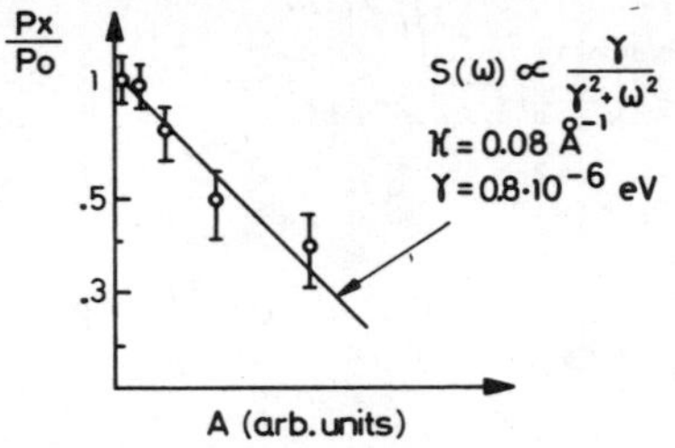

FIG.4. NSE spectrum of the quasi-elastic diffuse scattering of water.

give a perfectly pure cosine or sine modulation, and, as it was made clear above, we only use the NSE-Fourier method inside a pre-selected $(\vec{\kappa},\omega)$ domain (the shaded area in Fig. 2.).

3. Experimental Examples

A Neutron Spin Echo spectrometer has recently been completed at the Institut Laue-Langevin in Grenoble. It was designed for the very high resolution study of quasi-elastic scattering (Fig. 3). In this case, as it follows from Eqs.(4) for $\vec{\alpha} = 0$ and $|\vec{k}_o| = |\vec{k}_1|$, $\vec{n}_o \| \vec{k}_o$, $\vec{n}_1 \| \vec{k}_1$ and $A/B = -1$. Since these solutions are independent of k_o and k_1, eqs.(2) and (3) can be integrated, and the focussing conditions are seen to be fulfilled exactly, not only to first order. So the resolution will be limited by (second order) effects concerning the neutron passage through the precession fields. These are basically the cross-sectional inhomogeneities of these fields and the neutron path length differences in the precession region due to finite divergence. For the 200 cm long solenoidal precession fields actually these effects give, for a 3 cm beam and sample diameter 8×10^{-5} and 2×10^{-5}, respectively, resolution broadening relative to the incoming energy. (The first of these effects will be reduced by shimming the solenoids). The spectrometer operates from 4 - 8 Å incoming wavelength. The limiting resolution, taking into account also neutron statistics for a 10% (coherent) diffuse scattering sample in a 1 - 2 days NSE scan, is actually 2×10^{-8} eV, and should be improved by a factor of 5 within the next year. The maximum polarized neutron flux at the sample is 7×10^6 n/cm^{2}sec, which uniquely high value at long wavelengths is due to the use of a supermirror polarizer (though a temporary one), and the poor monochromatization required by the NSE (±20%). This flux will be improved by a factor of about 3 - 4 in the next 6 months by improving the incoming beam and polarizer geometry. The instrument is interactively on line controlled by a PDP 11/20 computer, which is capable of performing basic data reduction, including the calculation of Fourier transforms of up to 1k data point spectra immediately. The construction costs of the apparatus were much inferior ($ 200.000) to those of other types of inelastic spectrometers. This spectrometer was developed in collaboration with J.B. Hayter and P.A. Dagleish.

Fig. 4 shows the NSE spectrum of the diffuse, low angle scattering of water. For spin incoherent scattering the polarization of the scattered beam is - 1/3 of that of the incoming one, which reduces the data collection rate by about a factor of 10 as compared with coherent or

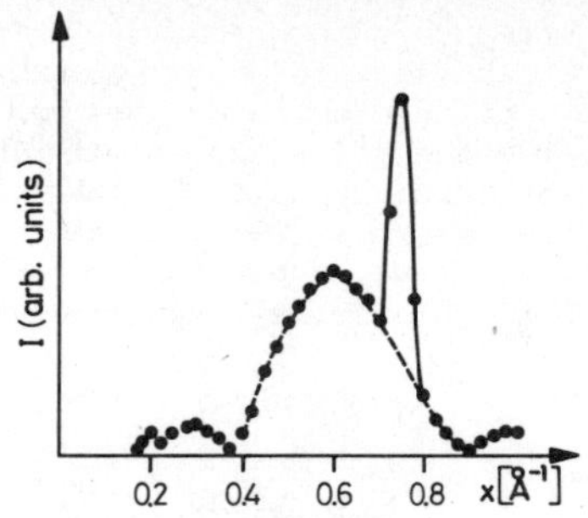

FIG.5. An elastic scattering spectrum measured by a two-dimensional NSE-Fourier scan.

isotopic incoherent scattering. The data presented here was collected at $2\theta = 4^\circ$ scattering angle in about 1 hour per point. The observed behaviour of $P_X(A)$ is consistent in this logarithmic plot with the straight line representing the diffusion constant of water extrapolated from measurements at higher κ values, assuming that the Lorentzian width γ follows the classical $\gamma = D\kappa^2$ relation (D is the diffusion constant). Note that the measured cosine Fourier transform of a Lorentzian line is the exponential function, which is easier to evaluate than the direct Lorentzian spectrum. It is also important that the instrumental resolution broadening only affects the value of P_o, i.e. the polarization of the NSE signal for elastic scattering, which is used to normalize P_X. In other words, the instrumental resolution, which broadens the direct spectra via a convolution, turns up in the NSE spectra only as a simple multiplicative normalization factor, just due to the Fourier transform performed naturally by the NSE. This is a very important general advantage, which in effect improves the information collection rate.

Finally, Fig. 5 shows an example of the above-mentioned two-dimensional NSE scan, with $\xi = \omega$ and $\eta = \kappa$ in this case. The small angle scattering spectrum of an Al sample holder is shown here, calculated as the Fourier transform of the NSE spectrum obtained by scanning B for a fix A, which corresponds to the measurement of κ in this case. The focussing NSE scan for $\xi = \omega$ (A/B = const.=-1) was used here only to establish that the scattering was elastic, and to filter out the inelastic air scattering. The broad peak (dashed line) approximately reproduces the incoming spectrum, and it comes from the diffuse part of the scattering. (To obtain the slow κ dependence of this, the data has to be normalized to the exact incoming spectrum, actually measured by the same NSE-Fourier method). The sharp peak is due to a double Bragg reflection effect. This spectrum gives a nice illustration of the good two-dimensional, both ω and κ resolution (10^{-4} and 2×10^{-2}, respectievly), achieved with a very poorly monochromated beam. This is at the same time an example of how to use the NSE method in elastic scattering.

4. Conclusion

The Neutron Spin Echo approach to neutron scattering, by breaking the close, unfavourable relation between resolution and neutron intensity, offers substantial improvements both in neutron economy and resolution as compared to the other techniques based on the usual concepts. These predictions have been experimentally verified in a few cases, and the systematic work in this field is under way now at the Institut Laue-Langevin in Grenoble. The systematic use of the NSE principle would substantially enhance (by an estimated factor of 3 - 10) the efficiency of the use of available neutron flux at continuous reactors, which has to be taken into consideration in the comparisons with pulsed sources. On the other hand, the use of the NSE can be interesting to improve resolution on pulsed sources too, by using the pulsed time-of-flight as the rough pre-selection for the finer NSE analysis.

REFERENCES

[1] For an introduction to 3-dimensional neutron polarization and for references on experimental methods see MEZEI, F., Physica 86-88B, (1977) 1049. The technique used for Spin Echo is described in Ref.[2].

[2] MEZEI, F., Z. Phys. 255 (1972) 146.

[3] MEZEI, F., Comm. on Physics 1 (1976) 81;
MEZEI, F. and DAGLEISH, P.A., Comm. on Physics 2 (1977) 41.

DISCUSSION

W. SCHMATZ: Are the analogies between the spin echo method and the TOF technique based on statistical choppers? In particular, does the spin echo method need sharp signals even against a negligible background?

F. MEZEI: The analogy is that both methods measure an integral transform of the direct spectrum. Consequently, the optimal data collection rate is achieved by an appropriate mixture of the direct and the transform methods, depending on the structure of the measured spectra. For statistical choppers this means the use of less than 50% duty cycle sequences. For spin echo, the equivalent optimization involves using classical preselection in combination (shaded area in Fig.2).

B.N. BROCKHOUSE: Has your very promising instrument already been used for actual experiments, or just for the demonstration experiments you mentioned in your talk?

F. MEZEI: The instrument has just been completed. The only real experiment performed so far involved looking for a dynamic width for the familiar central peak in $SrTiO_3$ antiferrodistorsive phase transitions. The central peak was found to be elastic to within 2×10^{-8} eV experimental error. More experiments are to be performed during the next few months.

D.K. ROSS: Can you distinguish between incoherent and coherent scattering?

F. MEZEI: Spin incoherent scattering changes the polarization to $-1/3$ as compared with coherent and isotopic incoherent scattering. In particular, this is a powerful way of distinguishing between proton incoherent scattering and coherent scattering in hydrogenous samples.

J.W. WHITE *(Chairman)*: Would you comment on the use of the spin echo technique to get energy resolutions in the range 10 to 100 μeV without the intensity penalties which normally accompany crystal techniques?

F. MEZEI: This is one of the main points I wanted to make with the illustration of a phonon line in Fig.2. The advantage of spin echo in such a case is that the monochromatization on the incoming and outgoing sides can be very relaxed as compared with the resolution of 10 to 100 μeV, e.g. 13 ± 2 meV.

J. Phys. E: Sci. Instrum., Vol. 11, 1978. Printed in Great Britain

Neutron spin-echo and three-axis spectrometers

R Pynn
Institut Laue–Langevin, 156X, 38042 Grenoble Cedex, France

Received 8 March 1978

Abstract A detailed calculation is presented of the resolution properties of a three-axis neutron spectrometer equipped with a neutron spin-echo device. The calculation shows that the Fourier transform of the spectral density of an elementary excitation may be measured with a resolution which is substantially better than that available with a conventional three-axis spectrometer. In contrast to conventional spectrometers the spin-echo technique allows focusing of both longitudinal and transverse phonon modes.

1 Introduction

A little over a quarter of a century ago, Hahn (1950) discovered and explained a number of transient responses of nuclear spin systems and aptly named these effects spin echoes. In the simplest demonstration of the echoes (Pake 1956) a nuclear spin system, aligned in the z direction by a steady magnetic field, is subjected to a pulse of RF power which turns the spins into the y direction. The spins start to precess about the direction of the steady field and if the latter is inhomogeneous, different spins precess at different rates. At some time τ after the first RF pulse, a second such pulse is applied which reverses the x component of all nuclear spins but does not change the sense or rate of precession. At a time τ after the second RF pulse the spins will again be in phase along the y axis and may induce a signal in a suitably placed pick-up coil.

An analogy to the phenomena described by Hahn was presented by Mezei (1972), who introduced into the vocabulary of neutron scattering the term neutron spin echo. To understand this term one may consider a non-monochromatic beam of neutrons initially polarised in the z direction. By the application of a suitable spin-turn field the spins may be brought into the y direction before traversing a guide field directed in the z direction. The spins will precess about this direction at the Larmor frequency. At any point along the guide field there will be a distribution of precession angles which is determined by the distribution of neutron velocities. In the language of Hahn the spins will 'fan out' with respect to the spin of a neutron with the mean velocity. Rather than fanning out as a function of time as in NMR, the spins fan out as a function of position along the neutron beam. At some point in the beam the x component of the spins may be reversed and, as the spins continue precessing, the 'unfanning' process described by Hahn occurs. If the guide field is uniform all spins will again be in phase (in the y direction) at a point which is half as far from the x-reversal point as from the start of the precessing region. Evidently this situation is entirely analogous to the spin-echo phenomena described by Hahn with the spatial dimension playing the role of the time variable of the NMR case. The situation is unchanged by the presence of a scattering sample placed in the neutron beam, provided that the scattering is elastic and involves no spin flip. Any change of neutron velocity at the scattering process will cause a spatial spread of the echo points and thus a loss of echo signal at a given point. This effect is the basis of a spectrometer (called IN11) with very high energy resolution which has been constructed at the Institut Laue–Langevin by J B Hayter and F Mezei.

The same principle, used in conjunction with a three-axis spectrometer, can be used to make high-resolution measurements of the spectral densities of the elementary excitations of a crystal†. In the present paper the resolution properties of such a spectrometer are calculated in detail. To introduce the subject, §2 essentially repeats the calculation (Mezei 1972) of the resolution properties of an IN11-type spectrometer. In §3 a combination of the spin-echo part of IN11 and a three-axis spectrometer is considered and shown to be inadequate for the study of elementary excitations of finite group velocity. However, the calculation demonstrates clearly the requirements for the design of a focusing three-axis spectrometer, and in §4 a suggestion made by Mezei (1977) for the realisation of such a spectrometer is considered in detail. Finally, in §5, examples are presented to demonstrate the potential utility of the method.

2 The focusing spin-echo spectrometer

To summarise the concept of spin echo and to define some necessary notation we begin by repeating the calculation (Mezei 1972) of the resolution of a focusing spin-echo spectrometer similar to the IN11 machine at the Institut Laue–Langevin. A diagram of this instrument is shown in figure 1. A helical velocity selector is used to prepare a beam of neutrons with up to about $\pm 20\%$ velocity spread. This beam then passes through the various components of the instrument as follows:

(i) The polariser. This is either a Soller guide (Hayter *et al* 1978) or a super-mirror (Mezei 1976, Gukasov *et al* 1977) arrangement which ideally transmits only neutrons of a single spin state. For definiteness we may assume the spin to be perpendicular to the scattering plane, i.e. to be (0, 0, 1).

(ii) The $\pi/2$ spin-turn coil. This coil (Mezei 1972) generates a field which, when added to the fringe field of the (z-direction) guide field, yields a resultant which is at an equal angle to the z and x directions. The neutron spin precesses about the resultant field and the dimensions of the coil are chosen such that the neutron leaves the $\pi/2$ coil with its spin changed from (0, 0, 1) to (1, 0, 0). Since the angle of precession of the neutron within the spin-turn coil depends on the time the neutron spends in the field of the coil, the latter can be tuned to give the desired change of spin state only for one neutron velocity. However, for velocity spreads around $\pm 10\%$ the decrease in the principal spin component is less than 5% (Hayter 1978).

(iii) Incident guide field. This field, which is perpendicular to the plane of figure 1, causes the neutron spin to precess

† To the best of the author's knowledge the combination of a neutron spin-echo device with non-tilted guide fields and a three-axis spectrometer was made in 1974 by F Mezei.

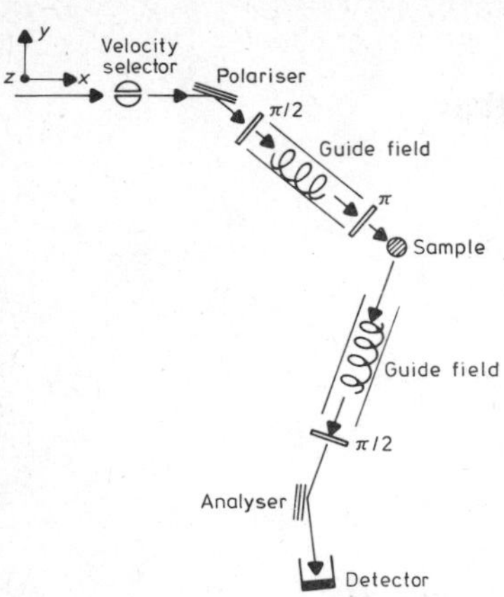

Figure 1 Diagram of the IN11 spectrometer. At any point in the instrument the neutron spin is defined with respect to Cartesian axes which have the x axis parallel to the neutron flight path and the z axis directed upwards out of the plane of the figure.

in the x–y scattering plane such that, after a time t in the field, the spin vector is ($\cos \omega t$, $\sin \omega t$, 0) where ω is the Larmor precession frequency given by

$$\omega = 2\mu_n B/\hbar = 1{\cdot}84\ B \text{ rad s}^{-1}. \tag{1}$$

Here μ_n is the neutron magnetic moment and B (T) is the guide field induction. The value of ωt at the end of the incident guide field, denoted by $2\pi N_i$, is related to the line integral of the field by

$$\begin{aligned} 2\pi N_i &= 2\, L\bar{B}\mu_n m/\hbar^2 \boldsymbol{k}_i . \boldsymbol{n}_I \\ 2\pi N_i &= 0{\cdot}292\, L\bar{B}/\boldsymbol{k}_i . \boldsymbol{n}_I \equiv A/\boldsymbol{k}_i . \boldsymbol{n}_I \end{aligned} \tag{2}$$

where $L\bar{B}$ is the field line integral (product of mean path length and field) (T μm), $\boldsymbol{k}_i$ is the neutron wavevector ($2\pi/\lambda$) (units of $10\ \text{nm}^{-1}$) and $\boldsymbol{n}_I$ is a unit vector parallel to the mean neutron flight path.

(iv) π spin-turn coil. Similar in construction to the $\pi/2$ spin-turn coil, the π coil changes the neutron spin vector from ($\cos 2\pi N_i$, $\sin 2\pi N_i$, 0) to ($\cos 2\pi N_i$, $-\sin 2\pi N_i$, 0).

(v) Final guide field. This field imparts N_f precessions to the neutron spin and thus changes the latter to ($\cos 2\pi(N_i - N_f)$, $-\sin 2\pi(N_i - N_f)$, 0) with N_f given by

$$2\pi N_f = B/\boldsymbol{k}_f . \boldsymbol{n}_F. \tag{3}$$

Here $\boldsymbol{k}_f$ is the wavevector of a scattered neutron and $\boldsymbol{n}_F$ is a unit vector in the mean direction of the scattered beam.

(vi) $\pi/2$ spin-turn coil. Like its counterpart in the incident beam this coil rotates the spin to give a spin vector (0, $\sin 2\pi(N_i - N_f)$, $\cos 2\pi(N_i - N_f)$).

(vii) Polarisation analyser. Constructed in a similar way to the polariser, this unit reflects only spin-up neutrons (i.e. neutron spins with classical value (0, 0, 1)). Thus the detector signal is proportional to $[1 + \cos 2\pi(N_i - N_f)]/2$ for neutrons incident with wavevector $\boldsymbol{k}_i$ and scattered with wavevector $\boldsymbol{k}_f$.

As intimated earlier, the above account of the spin history of a neutron as it traverses the spectrometer is an idealised view. In practice the efficiency of the components is less than 100% and depends upon the neutron wavelength, a fact which must be accounted for in a calculation of the true detector counting rate (Hayter 1978, Hayter and Penfold 1978). However, for the purposes of this section, which is intended to be pedagogic rather than exact, we calculate the counting rate assuming wavelength-independent properties for all spin-turn coils, etc. In this case the total detector signal is given by $(1+P)/2$ and the polarisation P is proportional to the integral of $\cos 2\pi(N_i - N_f)$ over the distribution of initial and final wavevectors. Thus

$$P \approx \int \mathrm{d}\boldsymbol{k}_i \int \mathrm{d}\boldsymbol{k}_f p(\boldsymbol{k}_i, \boldsymbol{k}_f) S(\boldsymbol{Q}, \omega) \cos 2\pi(N_i - N_f) \times \delta(\hbar\omega - E_i + E_f)\delta(\boldsymbol{Q} + \boldsymbol{k}_i - \boldsymbol{k}_f) \tag{4}$$

where $p(\boldsymbol{k}_i, \boldsymbol{k}_f)$ is the probability that a neutron with initial wavevector $\boldsymbol{k}_i$ and final wavevector $\boldsymbol{k}_f$ will be transmitted by the instrument if the spin-turn coils are removed. $\hbar\boldsymbol{Q}$ and $\hbar\omega$ are the changes of neutron momentum and energy at the scattering process which obey the conservation relations

$$\hbar\omega = \frac{\hbar^2}{2m}(\boldsymbol{k}_i{}^2 - \boldsymbol{k}_f{}^2) = E_i - E_f \tag{5a}$$

and

$$\boldsymbol{Q} = \boldsymbol{k}_f - \boldsymbol{k}_i. \tag{5b}$$

$S(\boldsymbol{Q}, \omega)$ is the sample fluctuation spectrum which is proportional to the probability that a neutron is coherently (i.e. without spin flip) scattered from an initial wavevector $\boldsymbol{k}_i$ to a final wavevector $\boldsymbol{k}_f$.

To evaluate the phase angle $2\pi(N_i - N_f)$ we first define

$$\boldsymbol{k}_i = \boldsymbol{k}_I + \boldsymbol{u}_i + \boldsymbol{v}_i + \boldsymbol{z}_i \tag{6}$$

where $\boldsymbol{k}_I$ denotes the most probable value of the incident wavevector and $\boldsymbol{u}_i$, $\boldsymbol{v}_i$, $\boldsymbol{z}_i$ are three mutually orthogonal vectors with $\boldsymbol{u}_i$ parallel to $\boldsymbol{k}_I$ and $\boldsymbol{z}_i$ perpendicular to the (horizontal) scattering plane. The definition (6) holds for scattered wavevectors if the subscript replacements $i \to f$ and $I \to F$ are made.

To simplify the calculation somewhat, let us assume that the fluctuation spectrum consists of a well defined constant-frequency excitation. Then the energy conservation law equation (5a) gives

$$\omega - \omega_0 = \delta\omega = \frac{\hbar}{2m}(\boldsymbol{k}_i{}^2 - \boldsymbol{k}_f{}^2 - \boldsymbol{k}_I{}^2 + \boldsymbol{k}_F{}^2) \tag{7}$$

and $\delta\omega$ is measured with respect to the maximum, ω_0, of the fluctuation spectrum, independent of $\boldsymbol{Q}$. Substituting equation (6) in equation (7) yields

$$u_f{}^2 + 2k_F u_f + (v_f{}^2 + z_f{}^2 - v_i{}^2 - z_i{}^2) - 2k_I u_i - u_i{}^2 + 2m\delta\omega/\hbar = 0, \tag{8}$$

which may easily be solved for u_f. The latter may then be substituted in the relation

$$2\pi(N_i - N_f) = \frac{A}{k_I + u_i} - \frac{B}{k_F + u_f} \tag{9}$$

which follows immediately from the definitions (2) and (3). To lowest order the result is

$$N_i - N_f = N_I - N_F - k_I\left(\frac{N_I}{k_I{}^2} - \frac{N_F}{k_F{}^2}\right)u_i - \frac{N_F \hbar\delta\omega}{2E_F}. \tag{10}$$

From this equation it is clear that if we choose

$$N_I/k_I{}^2 = N_F/k_F{}^2 \tag{11}$$

the Larmor phase factor $2\pi(N_i - N_f)$ is a function only of $\delta\omega$ and not, in first order, of the initial and final neutron wavevectors. Hence, in this approximation, the polarisation

Neutron spin-echo and three-axis spectrometers

P is simply given by the Fourier transform in frequency of the fluctuation spectrum. Thus

$$P \approx \int F(\delta\omega)\cos 2\pi\left[N_F\left(\frac{k_I^2}{k_F^2}-1\right)-\frac{N_F\hbar\delta\omega}{2E_F}\right]\mathrm{d}(\delta\omega) \qquad (4a)$$

where the function $F(\delta\omega)$ describes the lineshape of the (assumed) Einstein excitation. The integral of equation (4*a*) is to be performed over the ω-acceptance of the instrument, determined from $p(\boldsymbol{k}_i, \boldsymbol{k}_f)$. In principle, this acceptance function should be included in equation (4*a*) but it may be omitted if $F(\delta\omega)$ is sharp enough.

From the above calculation it would appear that the spin-echo technique has infinitely good resolution. In fact, finite resolution arises in two distinct ways. Firstly, technical problems are encountered in setting up a spectrometer which conforms strictly to the specifications assumed here. The spin-turn coils, for example, are not equally efficient for all neutron wavelengths, an effect which can result in non-negligible corrections for the large wavelength spreads customarily used on IN11. Further, a well defined and acceptable guide-field configuration can only be achieved with a restricted degree of accuracy. Neither of these constraints ought, however, to be important for the neutron spin-echo three-axis spectrometer described in the next section, because the intended resolution is significantly less than that available with IN11. Thus, here, attention is restricted to the second main limitation of resolution, namely that the Larmor phase factor, while coupled directly to $\delta\omega$, also depends in second order on the wavevector spread of the incident neutrons. In fact, if condition (11) is satisfied, one finds that in place of equation (10) one should write

$$N_i - N_f = (N_I - N_F) + \frac{3}{2}k_I^2\left(\frac{N_I}{k_I^4}-\frac{N_F}{k_F^4}\right)u_i^2 - \frac{N_F\hbar\delta\omega}{2E_F} + \frac{3}{2}\frac{k_I}{k_F}\frac{u_i}{k_F}\frac{\hbar\delta\omega N_F}{E_F} - \frac{N_F}{2k_F^2}(v_f^2+z_f^2-v_i^2-z_i^2) + \dots \qquad (10a)$$

For reasonable collimations (i.e. FWHM $\approx 0{\cdot}5°$) the final term of equation (10*a*) is less than about $10^{-5}N_F$ and can be accounted for if it is (say) about 10^{-1}. This fixes a reasonable value of N_F to about 10^4 and the minimum width of an excitation which can be measured (not just detected) to between $10^{-4}E_F$ and $10^{-5}E_F$ or, with $E_F = 2$ meV, to between 0·02 and 0·2 μeV. Equation (10*a*) also displays another important result, namely the privileged situation of quasi-elastic scattering. With $k_I = k_F$ and $N_I = N_F$ the u_i^2 term in equation (10*a*) vanishes and the Larmor phase factor depends on the spread of incident wavelength only via the $u_i\delta\omega$ term. Since this term is less than or approximately equal to u_i/k_I, beams with large wavelength spread (about $\pm 10\%$) can be used without detriment to the resolution. Evidently this is a claim which no classical method, limited by the phase-space density of neutrons, can make.

3 Resolution of a three-axis spectrometer with neutron spin echo (NSE)

The calculation for the IN11 spectrometer presented in §2, while instructive, is deficient in several ways. Most importantly it is restricted to the situation in which $S(\boldsymbol{Q}, \omega)$ is described by a constant-frequency excitation. Further, the role of the energy and wavevector acceptance of the instrument was not discussed. To rectify this we now consider the detailed form of the resolution function for a three-axis spectrometer with a simple spin-echo attachment. The instrumental configuration envisaged is that depicted in figure 2.

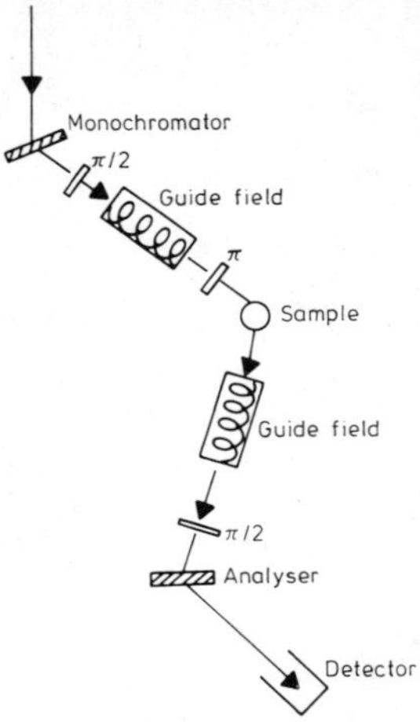

Figure 2 Diagram of a three-axis spectrometer with a simple NSE attachment. The monochromator and analyser are polarising crystals.

The calculation of this section, although of limited practical applicability in itself, is the natural bridge between the simple considerations of §2 and the discussion of focusing spin-echo given in §4. Here the method of calculation is introduced in detail and the criteria for successful application of the spin-echo method are developed. In addition it is shown that the spin-echo system considered in §2 cannot be used for the measurement of any excitations (e.g. phonons) with finite group velocity. The modifications to the spin-echo technique which are required to achieve focusing are evident from the calculation of this section and are developed in detail in §4.

For an understanding of the properties of the instrument sketched in figure 2, we may proceed along lines similar to those developed by Cooper and Nathans (1967) for a description of the resolution of a conventional three-axis spectrometer. As in §2, the quantity measured is a polarisation defined by (cf equation (4))

$$P = \frac{1}{n_0}\int \mathrm{d}^3k_i \int \mathrm{d}^3k_f S(\boldsymbol{Q}, \omega)p(\boldsymbol{k}_i, \boldsymbol{k}_f)\cos 2\pi(N_i - N_f). \qquad (4b)$$

The normalisation factor n_0 is identical to the integral of equation (4*b*) with the cosine term omitted. Thus n_0 is the integral calculated by Cooper and Nathans in their study of the resolution of a classical three-axis spectrometer.

The scattering cross-section $S(\boldsymbol{Q}, \omega)$ depends only on the energy and wavevector transfers ω and $\boldsymbol{Q}$ defined by equations (5*a*) and (5*b*). Since $\boldsymbol{Q}$ and ω represent only four independent variables, two of the integrations of equation (4*b*) may be carried out independently of the form of $S(\boldsymbol{Q}, \omega)$. Thus one may write

$$P = \frac{1}{n_0}\int \mathrm{d}^3Q \int \mathrm{d}\omega\, S(\boldsymbol{Q}, \omega)R(\boldsymbol{Q}-\boldsymbol{Q}_0, \omega-\omega_0) \qquad (12)$$

where the resolution function $R(\boldsymbol{Q}-\boldsymbol{Q}_0, \omega-\omega_0)$ is defined as

$$R(\boldsymbol{Q}-\boldsymbol{Q}_0, \omega-\omega_0) = \int_{\text{constant } \boldsymbol{Q},\, \omega} \mathrm{d}^3k_i \int_{\text{constant } \boldsymbol{Q},\, \omega} \mathrm{d}^3k_f \times p(\boldsymbol{k}_i, \boldsymbol{k}_f)\cos 2\pi(N_i - N_f). \qquad (13)$$

Q_0 and ω_0 are the most probable wavevector and energy transfer, obtained by replacing $\boldsymbol{k}_i$ and $\boldsymbol{k}_f$ by $\boldsymbol{k}_I$ and $\boldsymbol{k}_F$ in equations (5*a*) and (5*b*). Q_0 and ω_0 are included explicitly in equation (13) as a reminder that the resolution depends both on ω and Q *and* on the spectrometer configuration.

The standard (Cooper–Nathans) method of evaluating an integral like that of equation (13) is to write $p(\boldsymbol{k}_i, \boldsymbol{k}_f)$ and $(N_i - N_f)$ in terms of $(Q - Q_0)$, $(\omega - \omega_0)$ and two components of (say) $\boldsymbol{k}_f$ and to perform the integral over the two redundant components. A convenient choice for the latter are y_f and z_f, the components of $(\boldsymbol{k}_f - \boldsymbol{k}_F)$ perpendicular to Q_0 in and perpendicular to the (horizontal) scattering plane (cf figure 3). Assuming, as is traditional, that the spectrometer transmission functions are Gaussian, $p(\boldsymbol{k}_i, \boldsymbol{k}_f)$ may be written as

$$p(\boldsymbol{k}_i, \boldsymbol{k}_f) = \exp[-\tfrac{1}{2}(W_H + W_V)] \tag{14}$$

where the horizontal part W_H is given by

$$W_H = V_0^2 y_f^2 + 2 \sum_l V_0 . V_l X_l y_f + \sum_l \left(X_l V_l + 2 \sum_k X_k V_k \right) . V_l X_l \tag{15}$$

where $l = 1, 2, 4$.

In equation (15) the vector X represents the difference between the actual and nominal trajectories; the first three components of X are defined by

$$X = Q - Q_0 \tag{16}$$

while the fourth component is

$$X_4 = \omega - \omega_0. \tag{17}$$

The vectors V_l in equation (15) are defined in terms of the scattering geometry, collimations and monochromator and analyser mosaics. Suitable expressions for the V_l are given elsewhere (Werner and Pynn 1971, Pynn 1975). The detailed form of the expression for W_V (which is quadratic in X_3 and z_f) is immaterial for the purposes of the present calculation, because the Larmor phase factor is independent of vertical beam divergence in first order (cf equation (18)). Thus the integral over z_f in equation (13) is identical with that in the definition of the normalisation factor n_0 and the Cooper–Nathans result can be used directly in both cases.

To complete the evaluation of P we expand the Larmor phase factor $2\pi(N_i - N_f)$ in terms of the differences between neutron wavevectors and their mean values, as in §2. Recalling the definition equation (6) the required expansion is

$$(N_i - N_f) \approx N_I \left(1 - \frac{u_i}{k_I}\right) - N_F \left(1 - \frac{u_f}{k_F}\right) \tag{18}$$

to first order in u, v and z. The quantities u_i and u_f may be expressed in terms of X and y_f by coordinate transformation and by using a linearised version of the energy conservation law equation (5*a*) (see figure 3 and appendix). Thus the definitions of u_i and u_f given by equations (A.1) and (A.2) are consistent with equation (18) but may not be used if the expansion of the Larmor phase factor is carried to higher order.

From equations (18), (A.1) and (A.2) it is clear that the Larmor phase factor depends linearly on one of the integration variables, y_f of equation (13). Physically this means that neutrons scattered with a particular Q and ω have different values of $(N_i - N_f)$. Evidently if the spread of values of the latter quantity is sufficiently large, the average over the cosine term of equation (13) will be vanishingly small and the scattered beam will be completely depolarised. More quantitatively, we may combine equations (13), (14), (18), (A.1) and (A.2) to obtain a resolution function of the form

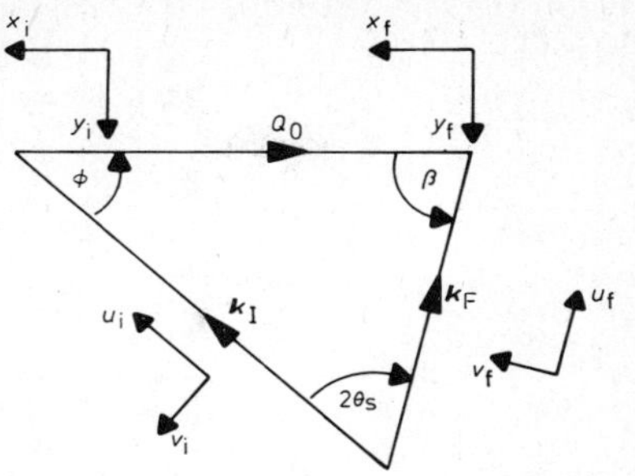

Figure 3 Scattering triangle together with the definitions of coordinate systems used in the text.

$$R(Q - Q_0; \omega - \omega_0) = \exp\left(-\frac{1}{2} \sum_{kl} M_{kl} X_k X_l\right) \times \exp\left[-2 \frac{\pi^2}{V_0^2} k_I^2 \sin^2 \phi \left(\frac{N_I}{k_I^2} - \frac{N_F}{k_F^2}\right)^2\right] \times \cos[2\pi(N_I - N_F + \alpha X_1 + \beta X_2 + \delta X_4)] \tag{19}$$

where

$$\alpha = \alpha' \left(\frac{N_I}{k_I^2} - \frac{N_F}{k_F^2}\right), \tag{20a}$$

$$\beta = \beta' \left(\frac{N_I}{k_I^2} - \frac{N_F}{k_F^2}\right), \tag{20b}$$

$$\delta = \delta' \left(\frac{N_I}{k_I^2} - \frac{N_F}{k_F^2}\right) - \frac{m N_F}{\hbar k_F^2}, \tag{20c}$$

$$n_0 = \int d^3Q \int d\omega\, S(Q, \omega) \exp\left(-\frac{1}{2} \sum_{kl} M_{kl} X_k X_l\right) \tag{21}$$

and $1/V_0$ is the standard deviation of y_f at the spectrometer set point. M_{kl} represents the elements of the Cooper–Nathans resolution matrix and α', β', δ' are quantities, defined in the appendix, which are expressible in terms of the geometry and transmission properties of the spectrometer. The second exponential in equation (19) reflects the dependence of $(N_i - N_f)$ upon y_f. Evidently this exponential will be unity (i.e. maximum) if

$$\frac{N_I}{k_I^2} - \frac{N_F}{k_F^2} = 0, \tag{22}$$

a condition (cf equation (11)) which was also fulfilled in our calculation of the properties of the IN11 spectrometer. We refer to the fulfilling of equation (22) as 'obtaining the echo' since if the condition is not satisfied, the measured polarisation P disappears rapidly as N_I or N_F is increased.

The measured polarisation is given, according to equation (12), by convoluting the resolution function (equation (19)) with $S(Q, \omega)$. If $S(Q, \omega)$ represents a well defined excitation (e.g. a phonon) we may write

$$S(Q, \omega) = S(Q)\delta(\omega - \omega_Q) \tag{23}$$

where the delta function expresses the fact that ω is a particular function, ω_Q, of Q on the dispersion surface of the excitation, and $S(Q)$ accounts for the variation of scattering cross-section over the dispersion surface; for phonons, for example, the variation of phonon eigenvectors with $\boldsymbol{Q}$ is included in $S(\boldsymbol{Q})$. Evidently equation (23) is adequate only for excitations of infinite lifetime. Let us examine the effects of combining the δ function of equation (23) with the cosine term of equation (19). Writing

$$\omega_Q = \omega_0 + a'X_1 + b'X_2 + c' \tag{24}$$

the argument of the cosine becomes

$$2\pi[N_I - N_F + (\alpha + \delta a')X_1 + (\beta + \delta b')X_2 + \delta c'] \tag{25}$$

which, since equation (22) is assumed to be satisfied, is independent of X_1 and X_2 if $a' = b' = 0$, i.e. if the excitation represented by equation (24) is of constant frequency. In this case equations (12) and (19) combine to give the result obtained in equation (4*a*) of §2.

The case $a' = b' = 0$ corresponds to a situation in which lines of constant Larmor phase, referred to as spin-echo lines, are parallel to the dispersion surface defined by equation (24). This is displayed in figure 4 which shows both the

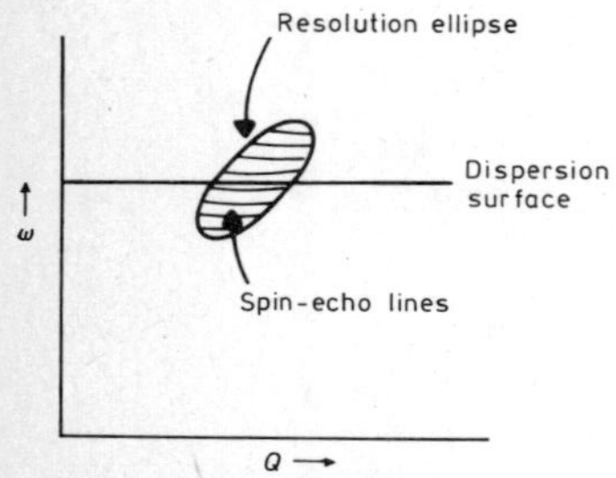

Figure 4 Representation of spin-echo focusing for a dispersionless excitation. The spin-echo lines have zero slope at the set point (ω_0, Q_0) of the spectrometer and may have non-zero curvature.

(flat) dispersion curve (in only one Q dimension), the Cooper–Nathans resolution ellipse (the first exponential in equation (19)) and the spin-echo lines. We refer to figure 4 as the focused case since it is only when the spin–echo lines and the dispersion surface are parallel that the NSE technique directly measures the spectral shape ($F(\delta\omega)$) of equation (4*a*)) of the dispersion surface. In order to focus an excitation of finite group velocity it is necessary to incline the spin-echo lines. This requires that both α and β (equations (20*a*) and (20*b*)) should be related to the principal gradients of the dispersion surface. Evidently this implies a finite value for $(N_I/k_I{}^2 - N_F/k_F{}^2)$ which cannot be achieved unless the condition (22) is relaxed and the echo partially lost. Even if this is done one cannot, in general, match both gradients (a' and b') by manipulating only N_I and N_F. Focusing can only be achieved by careful choice of the parameters α', β' and δ' in equations (20); this corresponds to choosing all the spectrometer parameters (collimations, crystal mosaics, etc) in a specific manner – a hopeless task.

Thus we are forced to the conclusion that the spectrometer depicted in figure 2 cannot be used to measure linewidths of phonons with Q-dependent frequencies. Nevertheless the calculation presented is useful because it demonstrates how the limitations may be overcome and a focusing spectrometer designed.

4 The focusing spin-echo three-axis spectrometer

In the previous section the problem of the design of a focusing NSE spectrometer was reduced to the satisfaction of two criteria. Firstly, the spin echo has to be obtained. Physically this means that all neutrons scattered with a given $\boldsymbol{Q}$ and ω should have (at least in lowest order) the same value of the Larmor phase factor independent of their initial and final wavevectors. The second condition involves matching the gradient of the spin-echo surface to that of a dispersion surface. Since measurements are generally made with the scattering plane as a mirror plane of the sample only two components of the gradient need to be matched to achieve focusing. Thus, in total, three variables are required to match the required conditions.

It is clear that the problem of obtaining the spin echoes, as it is stated above, cannot be solved if the total number of precessions depends only on the components of $\boldsymbol{k}_i$ and $\boldsymbol{k}_f$ parallel to the mean flight paths. A given ω and $\boldsymbol{Q}$ can be achieved by many pairs of $\boldsymbol{k}_i$ and $\boldsymbol{k}_f$ and both the moduli and orientations of these vectors vary from pair to pair, whence, unless the net number of precessions depends in some way on both the magnitude and orientation of both $\boldsymbol{k}_i$ and $\boldsymbol{k}_f$, a given $\boldsymbol{Q}$ and ω will in general be associated with a range of values of $(N_i - N_f)$ and the echo will be lost.

A way in which the line integral of the guide field may be made to depend on the orientation of $\boldsymbol{k}$ has been suggested by Mezei (1977) and is shown in figure 5. The rectangle

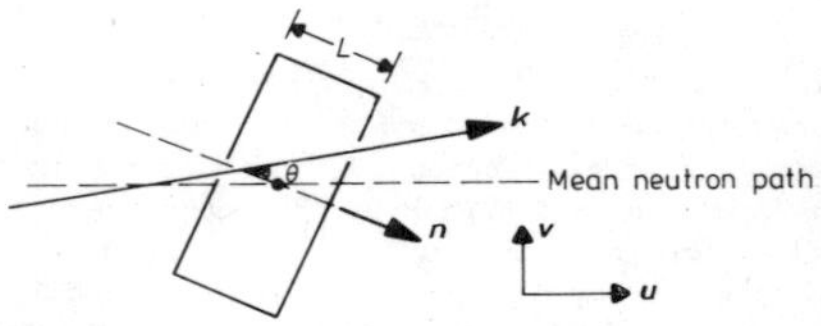

Figure 5 Diagram of the guide field for a focusing NSE three-axis spectrometer.

denotes the horizontal area over which the (vertical) guide field is applied. In the notation of equation (2) the number of precessions generated by this guide field is given by

$$2\pi N = \frac{A}{k\cos\theta} = \frac{A}{\boldsymbol{k}.\boldsymbol{n}} \tag{26}$$

where $\boldsymbol{n}$ is a unit vector perpendicular to the end face of the magnetic field volume. In terms of the components $\boldsymbol{u}$, $\boldsymbol{v}$, $\boldsymbol{w}$ introduced in equation (6) (see also figure 3) one may rewrite equation (26) as

$$2\pi N = \frac{A}{n_1 k_0}\left(1 + \frac{u}{k_0} + \frac{n_2 v}{n_1 k_0}\right)^{-1} \tag{27}$$

where k_0 represents either k_I or k_F and (n_1, n_2) are the components of $\boldsymbol{n}$ parallel to ($\boldsymbol{u}$, $\boldsymbol{v}$). In view of equation (27), a spin-echo spectrometer built with guide fields like those of figure 5 has three adjustable parameters (N_I/N_F and n_2/n_1 for initial and final guide fields) which can be used to fulfil the criteria for obtaining and focusing the echo.

Under normal operating conditions for a three-axis spectrometer the standard deviations of the wavevector spreads (u/k_0) or the collimations (v/k_0) are likely to be about 10^{-2}. Hence only with $N_I \lesssim 10^2$ may we completely ignore second-order terms in the expansion of equation (27). With $N_I \approx 10^3$, which is reasonable for a first-generation machine of this type, second-order terms are not completely negligible since they correspond to a variation of the Larmor phase factor of about $\pm 30°$. However, in order to formulate the focusing conditions we consider a first-order calculation and return to second-order corrections in a subsequent section.

Expanding equation (27) to lowest order and substituting for the *u*s and *v*s from equations (A.1)–(A.4), the Larmor phase factor is given by

$$N_{\mathrm{i}}-N_{\mathrm{f}}=N_{\mathrm{I}}-N_{\mathrm{F}}-\frac{N_{\mathrm{I}}}{k_{\mathrm{I}}}\Bigg(ay_{\mathrm{f}}+bv_1X_1+(bu_2-a)X_2+bu_4X_4 +\frac{n_{\mathrm{I2}}}{n_{\mathrm{I1}}}[by_{\mathrm{f}}-av_1X_1-(au_2+b)X_2-au_4X_4]\Bigg) +\frac{N_{\mathrm{F}}}{k_{\mathrm{F}}}\Bigg(Ay_{\mathrm{f}}+Bu_1X_1+Bu_2X_2+Bu_4X_4 +\frac{n_{\mathrm{F2}}}{n_{\mathrm{F1}}}(By_{\mathrm{f}}-Au_1X_1-Au_2X_2-Au_4X_4)\Bigg). \quad (28)$$

The first-order condition for obtaining the spin echo is satisfied by demanding that $(N_{\mathrm{i}}-N_{\mathrm{f}})$ be independent of y_{f}. Gathering terms, this gives

$$\frac{N_{\mathrm{I}}}{k_{\mathrm{I}}}\left(a+\frac{n_{\mathrm{I2}}b}{n_{\mathrm{I1}}}\right)=\frac{N_{\mathrm{F}}}{k_{\mathrm{F}}}\left(A+\frac{n_{\mathrm{F2}}}{n_{\mathrm{F1}}}B\right) \quad (29)$$

which reduces to the flat-branch result, equation (22) if $n_{\mathrm{I2}}=n_{\mathrm{F2}}=0$, i.e. if the end faces of both guide fields are perpendicular to the neutron flight paths. To obtain the focusing condition we substitute the dispersion relation equation (24) in equation (28) and demand that the result be independent of X_1 and X_2. In this case the spin-echo lines will (to lowest order) be parallel to the dispersion surface provided that the latter has no gradient perpendicular to the scattering plane. The focusing conditions may, after a little simple algebra, be expressed as

$$\frac{n_{\mathrm{I2}}}{n_{\mathrm{I1}}}=\frac{b'\cos\phi-a'\epsilon_{\mathrm{s}}\sin\phi}{a'\cos\phi+b'\epsilon_{\mathrm{s}}\sin\phi+\hbar k_{\mathrm{I}}/m} \quad (30a)$$

and

$$\frac{n_{\mathrm{F2}}}{n_{\mathrm{F1}}}=\frac{-b'\cos\beta-a'\epsilon_{\mathrm{s}}\sin\beta}{-a'\cos\beta+b'\epsilon_{\mathrm{s}}\sin\beta+\hbar k_{\mathrm{F}}/m} \quad (30b)$$

where the angles ϕ and β are defined in figure 3 and ϵ_{s} is $+1$ (-1) for scattering to the left (right) at the sample. Equations (30) give $n_{\mathrm{I2}}=n_{\mathrm{F2}}=0$ for the dispersionless branch case $(a'=b'=0)$ and, in a backscattering configuration, for any dispersion surface which has $b'=0$ (zero gradient perpendicular to $\boldsymbol{Q}_0$). Otherwise, either $n_{\mathrm{I2}}=0$ or $n_{\mathrm{F2}}=0$ may be achieved separately by judicious choice of instrument geometry. Substituting equations (30) into the condition (29) for the echo one finds that the latter becomes

$$\frac{N_{\mathrm{I}}}{N_{\mathrm{F}}}=\frac{k_{\mathrm{I}}(a'\cos\phi+b'\epsilon_{\mathrm{s}}\sin\phi+\hbar k_{\mathrm{I}}/m)}{k_{\mathrm{F}}(-a'\cos\beta+b'\epsilon_{\mathrm{s}}\sin\beta+\hbar k_{\mathrm{F}}/m)}. \quad (31)$$

Equations (30*a*), (30*b*) and (31), which are identical to results given by Mezei (1977), provide a prescription for setting up a spin-echo three-axis spectrometer focused on an excitation of finite group velocity. With the spectrometer set in this manner and in the linear approximation which we have been using, the measured polarisation is given simply by

$$P=\frac{1}{n_0}\int \mathrm{d}\omega'\int \mathrm{d}^3Q S(\boldsymbol{Q})F(\omega')\exp\left(-\frac{1}{2}\sum_{kl}M_{kl}X_kX_l\right)\times\cos 2\pi(N_{\mathrm{I}}-N_{\mathrm{F}}+\delta\omega'+\delta c') \quad (32)$$

where

$$\delta=\frac{-N_{\mathrm{I}}}{k_{\mathrm{I}}}\frac{1}{(a'\cos\phi+b'\epsilon_{\mathrm{s}}\sin\phi+\hbar k_{\mathrm{I}}/m)}$$

and

$$X_4=\omega'+\omega_Q-\omega_0.$$

ω_Q is given by equation (24) and the M_{kl} are, as in equation (19), the elements of the Cooper–Nathans resolution matrix. $F(\omega')$ is the spectral weight (lineshape) for the phonon and, in writing equation (32) we have assumed that

$$S(\boldsymbol{Q},\omega)=S(\boldsymbol{Q})F(\omega-\omega_Q) \quad (33)$$

so that $F(\omega')$ is independent of $\boldsymbol{Q}$ within the transmission function of the instrument. Just as in the case of the IN11 spectrometer, the quantity measured is essentially the Fourier transform of the phonon lineshape modified by the transmission function of the spectrometer. For the flat-branch case $(a'=b'=0)$ equation (32) is, apart from the explicit inclusion of the transmission function, identical to the result given by equation (4*a*).

5 Examples of spin-echo focusing

At this point we gather together some of the formulae given in the previous section and use them to evaluate the spin-echo signals which would be expected in two specific (and reasonably typical) cases. To simplify the analysis we shall consider planar phonon dispersion surfaces on which the scattering amplitude $S(\boldsymbol{Q})$ is constant within the spectrometer transmission function. This is, of course, the standard case considered by Stedman (1968), Cooper and Nathans (1967) and others. These authors have evaluated the $\boldsymbol{Q}$ integral of equation (32) in order to determine the width σ of a neutron group obtained during a conventional constant-Q scan. In terms of σ, the polarisation given by equation (32) is (for $c'=0$)

$$P=\frac{\int \mathrm{d}\omega\, F(\omega)\exp-(\omega^2/2\sigma^2)\cos 2\pi(N_{\mathrm{I}}-N_{\mathrm{F}}+\delta\omega)}{\int \mathrm{d}\omega\, F(\omega)\exp-(\omega^2/2\sigma^2)}. \quad (34)$$

This equation is most simply evaluated if the phonon lineshape is assumed to be Gaussian with standard deviation γ. In this case one finds that

$$P=\exp-\left(\frac{2\pi^2N_{\mathrm{I}}^2}{k_{\mathrm{I}}^2(1/\sigma^2+1/\gamma^2)(a'\cos\phi+\epsilon_{\mathrm{s}}b'\sin\phi+\hbar k_{\mathrm{I}}/m)^2}\right)\cos 2\pi(N_{\mathrm{I}}-N_{\mathrm{F}}) \quad (35)$$

which is a suitably generalised form of the flat-branch result equation (4*a*).

Equation (35) implies generally that the polarisation is more sensitive to γ when the width of a constant-Q scan, σ, is *large*. If σ is small the instrumental transmission function suppresses the wings of $F(\omega)$ and information is lost. In practice this observation means that one may want to *avoid* the classical, three-axis focused configuration (Cooper and Nathans 1967) in which the major axis of the transmission function coincides with the dispersion surface.

5.1 *Transverse acoustic phonon*

As a typical transverse acoustic phonon we choose the [110] T_2 phonons of Nb measured by Shapiro *et al* (1975). The conditions for the measurement are summarised in figure 6. Focusing (in the conventional sense discussed by Cooper and Nathans (1967)) is almost perfect in this case and, for a planar dispersion surface one finds a (constant-Q) phonon peak of width (FWHM) $0 \cdot 12$ meV.†

To determine the guide-field configuration required for a spin-echo measurement we use equations (30) and (31). For any transverse acoustic branch which is focused (spectrometer in W configuration) the product $\epsilon_s b'$ is negative so that $|n_{I2}/n_{I1}|$ and $|n_{F2}/n_{F1}|$ are generally large for this configuration. In fact for the measurement summarised in figure 6 the totally impracticable solutions $n_{I2}/n_{I1} = -3 \cdot 21$ and $n_{F2}/n_{F1} = 6 \cdot 53$ are found. Thus if we choose to maximise the detector signal by using the focused (W) configuration, the tilt angles for the guide fields become unrealistically large.

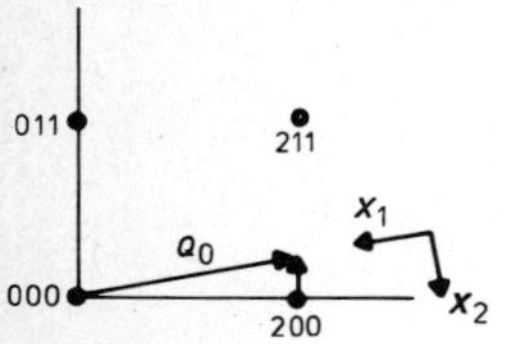

Figure 6 Summary of measurement conditions for the [110] T_2 phonon of Nb measured by Shapiro *et al* (1975). $Q_0 = (2{\cdot}0, 0{\cdot}07, 0{\cdot}07)$; $E_I = 14{\cdot}6$ meV, $E_F = 12{\cdot}2$ meV; collimation 20′–20′–20′–20′; $\epsilon_m = -1$, $\epsilon_s = 1$, $\epsilon_A = -1$; $a' = 0$, $b' = -1{\cdot}26$ meV nm.

In the case of the Nb phonon of figure 6, the solution is to use the U configuration ($\epsilon_m = \epsilon_s = \epsilon_A = -1$) which reduces the peak phonon intensity by a factor of about 2, increases the FWHM of a constant-Q scan to $0 \cdot 46$ meV and gives

$$\frac{n_{I2}}{n_{I1}} = -0 \cdot 51, \qquad \frac{n_{F2}}{n_{F1}} = 0 \cdot 49, \qquad \frac{N_I}{N_F} = 1 \cdot 106.$$

In this case the tilt angles (between $\boldsymbol{n}$ and the mean beam path) are less than 30° for both guide fields. A further advantage of the U configuration is that δ (equation (32)) is smaller than for the W configuration and thus the polarisation given by equation (35) does not decay as rapidly with increasing N_I. The resolution is therefore improved. In addition, the sensitivity to γ is enhanced by the increase of σ.

Rather than try to measure the oscillating polarisation given by equation (35), it will probably turn out in practice to be more reasonable to set up the conditions (30) and (31) and then to fine-tune N_F (say) to maximise the cosine term of equation (35). The result is then a measure of the exponential decay term of this equation. As a 'bench-mark' we note that for the Nb phonon of figure 6, if $\gamma \approx 50\ \mu$eV as found by Shapiro *et al* (1975), the exponential term of equation (35) is 1/e for $N_I \approx 250$ or for an integrated initial guide field of about $1 \cdot 3 \times 10^4$ T μm.

5.2 *Longitudinal acoustic phonon*

Once again we choose a niobium phonon as a typical case: the [100] L-branch measured at $Q_0 = (2 \cdot 1, 0, 0)$. We then have

$$E_I = 14 \cdot 6 \text{ meV}, \quad E_F = 8 \text{ meV}, \quad b' = 0, \quad a' = -3 \cdot 5 \text{ meV nm},$$

which, with collimations of $(20')^4$ gives, in the W configuration, a constant-Q scan with an FWHM of $0 \cdot 6$ meV. This large width reflects the impossibility of achieving classical focusing for low-energy longitudinal modes. Usually this implies that the spectral widths of such modes are difficult to measure.

From equations (30) and (31) the guide-field configuration for this case is found to be

$$\frac{n_{I2}}{n_{I1}} = -0 \cdot 729, \qquad \frac{n_{F2}}{n_{F1}} = 0 \cdot 554, \qquad \frac{N_I}{N_F} = -0 \cdot 79,$$

which implies reasonable tilt angles for the guide magnets. The negative value of the ratio N_I/N_F implies simply that the π-turn coil (figure 2) needs to be omitted from the spectrometer. As with the transverse phonon discussed above, the exponential term of equation (35) decays to 1/e for $N_I \approx 250$ if the phonon halfwidth γ is about 50 μeV. Clearly this resolution, which is very easy to obtain (for comparison, IN11 may achieve N_I of the order of 10^4), is substantially better than one may hope to achieve by classical three-axis spectrometry.

6 Higher-order corrections

Higher-order corrections to the calculation of §4 arise from several distinct sources. Firstly, neutrons which are scattered with a given Q and ω transfer may have different values of the Larmor phase factor. This will be manifested as a higher-order dependence of equation (27) upon y_f. Secondly, either the spin-echo surfaces or the phonon dispersion surface may have a finite curvature within the transmission function of the spectrometer. Both of these effects appear as terms of the form $X_i X_j$ in equation (28). In practical cases the effect of curvature of the dispersion surface is likely to be the most serious. Any constant-Q peak which is asymmetric or which displays effects similar to those discussed by Werner and Pynn (1971) will be a poor candidate for a spin-echo measurement. Even the [110] T_2 branch of Nb discussed above has significant curvature and, with 80′ (FWHM) vertical collimation the phonon frequency varies by about $0 \cdot 02$ meV within the (half-height) transmission function. This sets a lower limit to a measured spectral width which can only be decreased by tightening the vertical collimation. The improvement cannot be more than a factor of 2 in this case, however, because there is also a significant curvature of the dispersion surface in the [100] direction! For transverse branches, especially those with a large gradient, a further limit to resolution will be introduced by the crystal mosaic of the sample which can be included as in the conventional three-axis resolution calculation (Werner and Pynn 1971).

7 Concluding remarks

We have considered in this paper the performance characteristics of a three-axis spectrometer equipped with a neutron spin-echo device. The calculations show, in detail, how such an instrument can be focused on any phonon branch and the Fourier transform of the phonon spectral density measured. Resolution which is substantially better than that available in a conventional measurement may be achieved, especially for longitudinal phonon modes. However, setting up such a spectrometer is expected to be a non-trivial task; it is not envisaged that NSE could take the place of conventional three-axis spectroscopy but rather that it may be an important supplementary technique in specific cases.

Acknowledgments

I have benefited from ideas, facts and advice which have been offered by J B Hayter, J Joffrin, F Mezei and J Penfold.

† The inclusion of the curvature of the dispersion surface, vertical collimation, etc increases the width of a constant-Q scan of the Nb [110] T_2 phonon to 0·2 meV for the W configuration and to 0·5 meV for the U configuration.

R Pynn

Appendix

The deviations of the neutron wavevectors from their average values may be written to lowest order in $(\boldsymbol{Q}-\boldsymbol{Q}_0)$ and $(\omega-\omega_0)$ as

$$u_{\rm i}=ay_{\rm f}+bv_1X_1+(bu_2-a)X_2+bu_4X_4 \quad (A.1)$$

$$u_{\rm f}=Ay_{\rm f}+Bu_1X_1+Bu_2X_2+Bu_4X_4 \quad (A.2)$$

$$v_{\rm i}=by_{\rm f}-av_1X_1-(b+au_2)X_2-au_4X_4 \quad (A.3)$$

$$v_{\rm f}=By_{\rm f}-Au_1X_1-Au_2X_2-Au_4X_4 \quad (A.4)$$

where

$$a=\epsilon_{\rm s}\sin\phi \quad (A.5)$$

$$b=\cos\phi \quad (A.6)$$

$$A=\epsilon_{\rm s}\sin\beta \quad (A.7)$$

$$B=-\cos\beta \quad (A.8)$$

$$u_1=\frac{m\omega_0}{\hbar Q_0^2}+\frac{1}{2} \quad (A.9)$$

$$v_1=u_1-1 \quad (A.10)$$

$$u_2=\epsilon_{\rm s}\left[\left(\frac{k_{\rm I}}{Q_0}\right)^2-v_1^2\right]^{1/2} \quad (A.11)$$

$$u_4=m/\hbar Q_0 \quad (A.12)$$

with ϕ and β as defined in figure 3 and $\epsilon_{\rm s}=+1$ (-1) for anticlockwise (clockwise) scattering at the sample. Equations (A.1)–(A.4) are obtained from a linearised version of equation (8) by simple coordinate transformation (figure 3). For a higher-order calculation (cf §6) all terms of equation (8) must be retained.

The constants α', β', δ' which appear in equations (20*a*, *b*, *c*) may, after substantial algebraic effort, be written as

$$\alpha'=k_{\rm I}[\cos\phi(\tfrac{1}{2}-m\omega_0/\hbar Q_0^2)+\boldsymbol{V}_0\cdot\boldsymbol{V}_1\epsilon_{\rm s}\sin\phi/|\boldsymbol{V}_0|^2] \quad (A.13)$$

$$\beta'=k_{\rm I}[\tfrac{1}{2}-m\omega_0/\hbar Q_0^2+\boldsymbol{V}_0\cdot\boldsymbol{V}_2/|\boldsymbol{V}_0|^2]\epsilon_{\rm s}\sin\phi \quad (A.14)$$

$$\delta'=-k_{\rm I}[m\cos\phi/\hbar Q_0-\boldsymbol{V}_0\cdot\boldsymbol{V}_4\epsilon_{\rm s}\sin\phi/|\boldsymbol{V}_0|^2] \quad (A.15)$$

where the $\boldsymbol{V}_l$ are defined by Werner and Pynn (1971) and Pynn (1975).

References

Cooper M J and Nathans R 1977 The resolution function in neutron diffractometry
Acta Crystallogr. **23** 357

Gukasov A G, Ruban V A and Bedrizov M N 1977 Interference magnification of the region of specular reflection of neutrons by multilayer quasimosaic structures
Sov. Phys. – Tech. Phys. **3** 52

Hahn E L 1950 Spin echoes
Phys. Rev. **80** 580

Hayter J B 1978 Matrix analysis of neutron spin echo
Z. Phys. in the press

Hayter J B and Penfold J 1978 Fourier spectroscopy with neutron spin echo
unpublished

Hayter J B, Penfold J and Williams W G 1978 Compact polarising Soller guides for cold neutrons
J. Phys. E: Sci. Instrum. **11** 454

Mezei F 1972 Neutron spin echo: A new concept in polarised thermal neutron techniques
Z. Phys. **255** 146

Mezei F 1977 Neutron spin echo and polarised neutrons
Proc. Int. Symp. on Neutron Scattering, Vienna 1977

Pake G E 1956 Nuclear magnetic resonance
Solid St. Phys. **2** 1

Pynn R 1975 Lorentz factor for triple-axis spectrometers
Acta Crystallogr. B **31** 2555

Shapiro S M, Shirane G and Axe J D 1975 Measurement of the electron–phonon interaction in Nb by inelastic neutron scattering
Phys. Rev. B **12** 4899

Stedman R 1968 Energy resolution and focusing in inelastic scattering experiments
Rev. Sci. Instrum. **39** 878

Werner S A and Pynn R 1971 Resolution effects in the measurement of phonons in sodium metal
J. Appl. Phys. **42** 4736

Dynamical Scaling in Polymer Solutions Investigated by the Neutron Spin-Echo Technique

D. Richter
Institut für Festkörperforschung der Kernforschungsanlage Jülich, Jülich, Federal Republic of Germany

and

J. B. Hayter and F. Mezei
Institut Laue-Langevin, 38042 Grenoble Cédex, France

and

B. Ewen
Institut für Physikalische Chemie der Universität Mainz, Mainz, Federal Republic of Germany
(Received 11 August 1978)

Applying the recently developed technique of neutron spin-echo spectroscopy, we found it possible to verify unambiguously the scaling predictions of the Zimm model: $\Delta\omega \sim q^3$ for the dynamics of a single chain. At higher polymer concentrations, experimental indications for the predicted crossover to many-chain behavior, which is accompanied by a change in exponent from 3 to 2, were observed. Line-shape analysis rules out a simple exponential decay, indicating the importance of memory effects.

The thermodynamic behavior of polymer solutions shows a close similarity to critical phenomena near second-order phase transitions.[1] Certain static scaling predictions have recently found remarkable experimental proof.[2] As in the general development of the theory of phase transitions, the static scaling laws have been extended to dynamical scaling.[3,4] An experimental proof of this theory can in principle be achieved by quasielastic scattering experiments which probe the relaxation spectrum of the polymer. Neutron-scattering experiments, in particular, reach the required region of momentum and energy transfer q and ω. However, until very recently there was no instrument available which provided high energy resolution at small enough momentum transfers. As a result, apart from some attempts to determine the exponent for the q dependence of quasielastic linewidths in dilute solutions,[5,6] little progress has so far been made. In the present work we have applied the newly developed neutron spin-echo technique,[7,8] and for the first time it has been possible (i) to reach the q region of interest in the dilute case, (ii) to perform a line-shape analysis in the time regime, and (iii) to investigate the influence of concentration. In particular, the scaling prediction of the Zimm model in dilute solutions was confirmed unambiguously; the decay of correlations was found to deviate considerably from a simple exponential law and the crossover from single- to many-chain behavior was observed.

The theoretical description of polymer relaxation in solution on the scale of segmental diffusion (Flory radius $R_F > r >$ segment length l) has so far been based on hydrodynamic equations.[9,10] In these theories, deviations from equilibrium are counterbalanced by entropic restoring forces, and the hydrodynamic interaction between the monomers has to be included (Zimm model[11]) in order to obtain physically realistic predictions. With respect to scaling, the excluded-volume interaction has also been considered.[3,4] The influence of concentration was treated by introducing a correlation length ξ which is roughly the distance between entanglement points of different polymers.[3,4] ξ scales with the concentration as $c^{-3/4}$ (Ref. 2) in good solvents. For spatial dimensions $r < \xi$ one expects to observe essentially the behavior of a single chain, while for large r the many-chain behavior should be dominant. For the single-chain regime the half-width of the relaxation spectrum is predicted to scale with $(T/\eta)q^3$, independent of excluded-volume effects, where T is the temperature and η the viscosity of the solvent.[4,12] In the many-chain regime, the influence of entanglements becomes important. For short times, this gives rise to a gel-like rigidity modulus $E \propto 1/\xi^3$.[2,3] The half-width of the relaxation spectrum is predicted to scale with q^2.[3,4] The proportionality factor may be taken as the cooperative diffusion coefficient D_c of the gel. From the condition of smooth crossover between single- and many-chain regimes $D_c \sim c^{3/4}$ follows.

Scaling theories do not give proportionality factors, nor do they provide information about the shape of the scattering law. For this purpose direct calculations on model systems are necessary. The first direct attack on this problem was undertaken by Dubois-Violette and de Gennes,[9] who treated the Zimm model under θ conditions. For the half width at half maximum they found $\Delta\omega$ (μeV) $= 0.28(T/\eta)q^3$ K/cP Å^3. The calculated line shape exhibits considerable deviations from an exponential decay (see Fig. 1). Employing Kirkwood's generalized diffusion equation together with Mori's projection-operator technique, Akcasu and Gurol[10] have calculated the initial slope $\Omega(q,t)$ in the t regime which (apart from a numerical factor) is identical with the half width $\Delta\omega$.[13] Because this approach uses only equilibrium properties of the polymer, it has the further advantage of yielding results for good solvents, where a 30% increase in $\Delta\omega$ is predicted compared with that for a θ solvent at the same temperature and viscosity.[13] The transition regions towards smaller and larger q may also be treated. The theory has the disadvantage of giving no information on line shape as long as memory effects are not included. Calculations with this method have shown that earlier experiments have been performed mainly in the transition region between segment diffusion and "monomer diffusion."[6]

The experiments were performed at the neutron spin-echo spectrometer IN11 at the Institut Laue-Langevin, Grenoble. The neutron spin-echo method has been discussed elsewhere.[7,8] Performing a neutron spin-echo experiment, the final neutron polarization $P(q,H)$ is measured as a function of the applied magnetic field H. In contrast to conventional neutron-scattering techniques—for an ideally defined q—$P(q,H)$ is proportional to the real part of the intermediate scattering law $S(q,t)$, where t is a linear function of H.

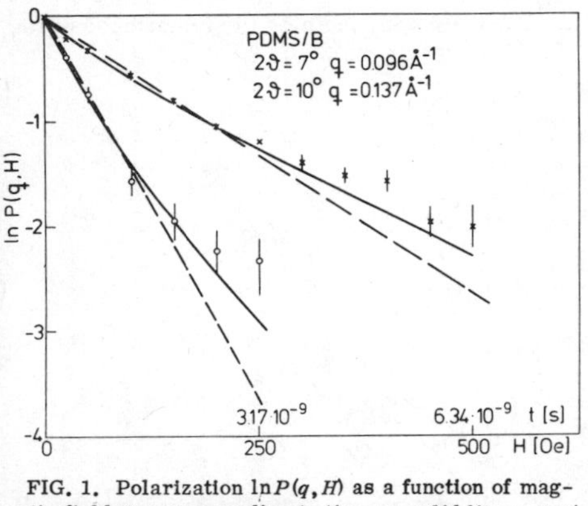

FIG. 1. Polarization $\ln P(q,H)$ as a function of magnetic field H corresponding to time t: solid lines, scattering law as calculated by Dubois-Violette and de Gennes (Ref. 9); dashed lines, exponential decay function.

The scattering experiments were performed using two different protonated polymers: atactic polymethylmethacrylate (PMMA) (mol. weight 7×10^6) and polydimethylsiloxane (PDMS) (mol. weight 3×10^4) dissolved in deuterated benzene. The polymer concentrations were 0.05, 0.15, and 0.30 g/cm^3 in the case of PDMS and 0.05 g/cm^3 for PMMA. The measurements were carried out at 72°C at a mean incident wavelength of 8 Å, with scattering angles 2θ between 2° and 10°, corresponding to q values between 0.027 and 0.137 Å^{-1}. The width of the incident-wavelength distribution was ± 17%; this width is effectively unrelated to instrumental resolution in the spin-echo method. The instrumental resolution was determined using a 9:1 mixture of deuterated and protonated polyethylene, which gives rise to strong elastic coherent scattering in the relevant angular region. The solvent scattering was measured in separate runs. The instrumental background was negligibly small.

The quasielastic intensity $I(q,t)$ scattered by the polymer solutions contains three contributions so that

$$I(q,t) = cK^2 S_{\rm pol}^{\rm coh}(q,t) + c\,\frac{\sigma_{\rm inc}^{\rm pol}}{4\pi} S_{\rm pol}^{\rm inc}(q,t) + (1-c)\frac{\sigma_{\rm tot}^{\rm sol}}{4\pi} S_{\rm sol}(q,t), \qquad (1)$$

where K is the scattering-length contrast between polymer and solvent, the $S(q,t)$ are the intermediate scattering laws and σ the corresponding cross sections. For our samples the coherent cross section of the polymer was always predominant. The data were corrected for the solvent scattering and instrumental resolution, the latter being a simple scale factor in the spin-echo technique. For the incoherent scattering of the polymer, no correction was made. This is justifiable as a very good approximation, since (i) the incoherent scattering was always only a few percent of coherent scattering, (ii) due to the spin-flip scattering the incoherent part enters only with a factor[8] $\frac{1}{3}$, and (iii) the coherent and incoherent contributions are a result of the same dynamical process—the scattering laws differ only slightly.[9]

The corrected data were fitted with the scattering law $S_{\rm DV}(q,t)$ calculated by Dubois-Violette and de Gennes,[9] and also by an exponential decay function. The fit parameter in both cases was $T/\eta z$, where z allows for deviations in the prefactor (nominally 0.28). Two typical results are shown in Fig. 1 together with the experimental spectra. In addition, a fit of all spectra obtained from one sample was performed with common variation of the parameter. If the q^3 power law is valid, the $T/\eta z$ values for one sample should coincide at all q values. In Fig. 2, $\Delta\omega/q^3 \propto T/\eta z$ is plotted versus q for PMMA and PDMS at $c = 0.05$ g/cm^3. The solid lines give the values obtained by the common fit. There is excellent agreement between the predicted q^3 power law and the experimental results. A fit of the exponent x by common variation of parameters for all spectra of one sample yields $x = 3.02 \pm 0.03$ for PDMS and $x = 2.93 \pm 0.03$ for PMMA. The proportionality factors $T/\eta z$ for $x = 3$ are 515 ± 12 and 878 ± 14 K/cP for PMMA and PDMS, respectively. For $z = 1$, the expected value[9] is 960 K/cP.

The results at higher concentration are shown in Fig. 3. Here $(T/\eta z)q$ (corresponding to $\Delta\omega/q^2$) is plotted versus q. While the results at 0.05 g/cm^3 follow a straight line indicating the q^3 law, the data obtained at 0.30 g/cm^3 PDMS show up strong deviations from this behavior at small q. Indications for increased linewidth at small q are also found at 0.15 g/cm^3. After detecting this behavior using the above-mentioned fit procedure, the data points below 0.1 Å^{-1} for the 0.3 g/cm^3 and below 0.07 Å^{-1} for the 0.15-g/cm^3 samples were refined by fitting to an exponential decay function, which turned out to be a reasonable description in this region. The $1/e$ decay was taken as $\Gamma = D_c q^2$, yielding $D_c(0.15) = (3.06 \pm 0.15) \times 10^{-6}$ cm^2 s^{-1}, and $D_c(0.30) = (4.70 \pm 0.17) \times 10^{-6}$ cm^2 s^{-1}.

The experiments on dilute solutions and good solvents verify clearly the scaling prediction of the Zimm model. However, the prefactors $(T/\eta z)$ obtained from the fits require some discussion. The model of Dubois-Violette and de Gennes,[9] for example, derives $z = 1$ in the reference case of a θ solvent. For PDMS we have $z = 1.1$, whereas PMMA yields $z = 1.9$. Akcasu[12] predicts an increase of linewidth of about 30% due to the excluded-volume interaction passing over from a θ to a good solvent. In this case the deviations would be even larger. As a consequence, the proportionality factor between $\Delta\omega$ and q^3 has no universal nature, but it seems to differ from polymer to polymer (see also Ref. 6).

Concerning the line shape, the shape functions $S_{DV}(q,t)$ of Dubois-Violette and de Gennes shows clearly a better agreement with the experimental spectra than the exponential decay (Fig. 1). However, as already evident from the figure, at longer times still systematic deviations from this shape function appear in all data sets.

Both statements are supported by the evaluated normalized χ^2 values. For all but one experiment $S_{DV}(q,t)$ turned out to be a reasonable description of the data ($0.79 < \chi^2 < 1.74$; exponential decay function: PDMS at $2\theta = 10°$, $\chi^2 = 3.04$ vs $\chi^2 = 0.99$; or at $2\theta = 8°$, $\chi^2 = 4.6$ vs 1.11). One experiment was performed with better statistics (measuring time increased by a factor of 5). It showed significant deviations from $S_{DV}(q,t)$ ($\chi^2 = 4.0$; for exponential decay function, $\chi^2 = 11.5$).

The strong deviations from an exponential decay, which occurred in all data taken at the lowest concentration, indicate the importance of the

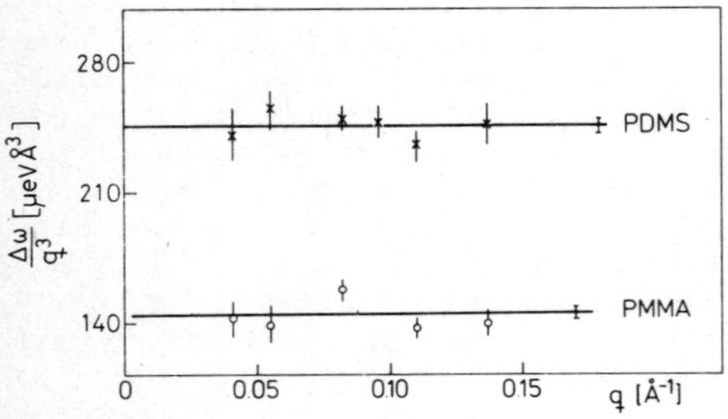

FIG. 2. Linewidth $\Delta\omega$ divided q^3 as a function of q: solid lines, result of a fit with common variation of parameters.

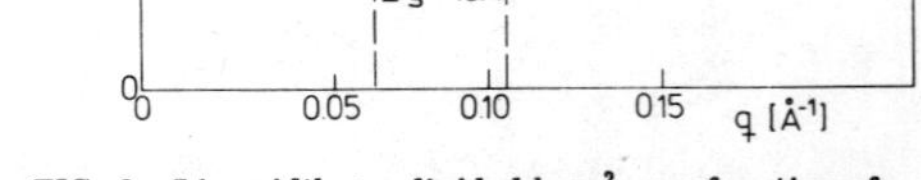

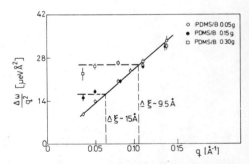

FIG. 3. Linewidth $\Delta\omega$ divided by q^2 as a function of q: solid line, result of a fit with common variation of parameters for the 0.05-g sample. The dashed lines are guides for the eye.

memory terms which have been neglected so far in Akcasu's approach. In view of the simplifications made in the calculations of Dubois-Violette and de Gennes, it is not surprising to also find some discrepancies at long times. While the preaveraging of the Oseen tensor has only a minor influence on the linewidth[12] and probably also on the line shape in the q region considered, the neglect of the excluded-volume interaction seems to be the most severe simplification. An experimental check of this statement will be carried out by future measurements under θ conditions.

At higher concentrations, Fig. 3 clearly shows strong deviations from the q^3 behavior in going to smaller q. This can be taken as an experimental indication for the crossover from a q^3 to a q^2 behavior, as predicted by de Gennes[4] for entangled systems. (This crossover is not to be confused with the expected q^2 behavior at very low q due to the diffusion of the polymer as a whole.) However, the small number of data points prevents an accurate determination of the exponent in the low-q range. For this purpose, more experiments will be necessary. Within the de Gennes theory the position of the crossover point q^* in momentum space allows a dynamical estimate of the screening length ξ, although in order to give a numerical value, one has to know the proportionality factor. The values given on Fig. 3 were obtained assuming $q^*\xi = 1$, and their order of magnitude agrees quite well with those found in static experiments on polystyrene in benzene.[2] Using $q^*\xi = \sqrt{6}$ as given by Akcasu[12] would increase ξ. To prove the predictions concerning the concentration dependence of ξ or D_c again more experiments will be necessary. For the two concentrations we have $D_c(0.3\ \mathrm{g/cm^3})/D_c(0.15\ \mathrm{g/cm^3}) = 1.54 \pm 0.09$ and for the monomer concentrations $(0.26/0.15)^{3/4} = 1.51$, which is in agreement with the proposed behavior.

We would like to thank Professor T. Springer for his suggestion of the problem, Dr. K. W. Kehr for many helpful discussions, and Dr. B. Lehnen for providing the PDMS sample.

[1] J. Des Cloiseaux, J. Phys. (Paris) 36, 281 (1975).
[2] M. Daoud *et al.*, Macromolecules 8, 894 (1975).
[3] P. G. de Gennes, Macromolecules 9, 587 (1976).
[4] P. G. de Gennes, Macromolecules 9, 594 (1976).
[5] A. Allen *et al.*, Chem. Phys. Lett. 38, 577 (1976).
[6] A. Z. Akcasu and J. S. Higgins, J. Polym. Sci. 15, 1745 (1977).
[7] F. Mezei, Z. Phys. 255, 146 (1972).
[8] J. B. Hayter, to be published.
[9] E. Dubois-Violette and P. G. de Gennes, Phys. (USA) 3, 181 (1967).
[10] Z. Akcasu and H. Gurol, J. Polym. Sci., Polym. Phys. Ed. 14, 1 (1976).
[11] B. Zimm, J. Chem. Phys. 24, 269 (1956).
[12] Z. A. Akcasu and M. Benmouna, to be published.
[13] M. Benmouna and Z. A. Akcasu, to be published.

NUCLEAR INSTRUMENTS AND METHODS 164 (1979) 153-156; © NORTH-HOLLAND PUBLISHING CO.

THE APPLICATION OF NEUTRON SPIN ECHO ON PULSED NEUTRON SOURCES

FERENC MEZEI

Institut Laue-Langevin, 156X, 38042 Grenoble, France and
Central Research Institut for Physics, 1525 Budapest, Hungary

Received 27 December 1978 and in revised form 17 April 1979

In the neutron-spin-echo technique a comparison is made between the before and after scattering times-of-flight for each neutron. On a pulsed source the most readily available quantity is the total time from neutron pulse to detection, i.e. the sum of the times-of-flight from source to sample and from sample to detector. These two items of information can be very naturally combined for high resolution inelastic spectrometry. The principle of such a spectrometer for μeV resolution is described. It is shown that the neutron-spin-echo method on a pulsed source makes it possible to utilize a large band of the wavelength spectrum at the same time, thus providing a substantial (up to 1000-fold) gain in neutron economy as compared to the classical methods of comparable resolution.

1. Introduction

The neutron spin echo[1]) (NSE) method has the unique advantage of providing good (energy) resolution without a corresponding monochromatization, i.e. without a related intensity loss. This is why it offers essential intensity gain in high resolution work, but also it has to be combined[2]) with a classical neutron scattering technique that provides an ordinary resolution in the other (momentum) dimensions of the $(\boldsymbol{\kappa}, \omega)$ scattering parameter space. Thus by this combination a very anisotropic resolution behaviour in $(\boldsymbol{\kappa}, \omega)$ is always obtained, e.g. 10^{-4} in energy and 10^{-2} in momentum.

Using a pulsed neutron source one has a natural classical instrument to be combined with the NSE: the time-of-flight diffractometer, i.e. an instrument where the total flight time of neutrons from the source to the detectors is measured in a multichannel analyser. It will be pointed out that the resulting combination provides a high NSE energy resolution together with an ordinary momentum resolution. It will be also shown that in some cases the use of an additional chopper can effectively filter disturbing background effects without excessive intensity losses.

2. Principle of a pulsed source NSE spectrometer

The scheme of the instrument is shown in fig. 1. The neutron spin polarizer and the analyser systems are made of supermirrors[4]) (SM). Technically a Soller type SM polarization analyser system can cover any detector area at a cost of the same order of magnitude as that of the detectors themselves. The $\pi/2$ and π spin flip coils are of the usual NSE type dc coils[1]) (typical currents 1–1.5 A).

The precession fields are produced by $l = 50$ cm long iron yoke electromagnets with about 50 cm

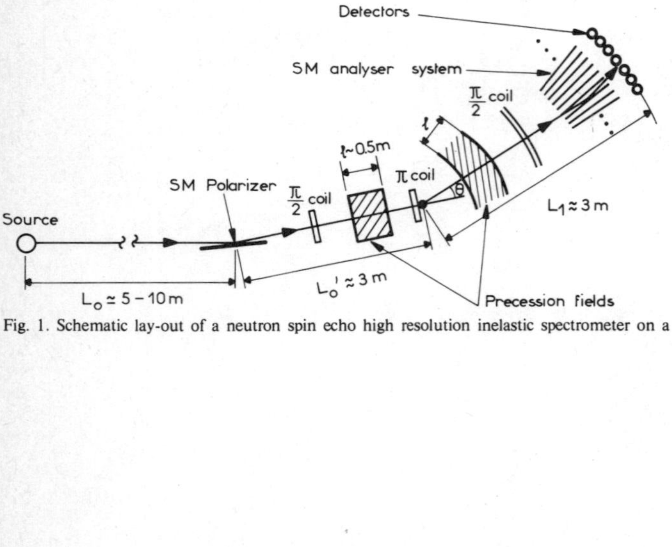

Fig. 1. Schematic lay-out of a neutron spin echo high resolution inelastic spectrometer on a pulsed neutron source.

gap and about 400 Oe peak field, which would provide better than 1 μeV resolution beyond 5 Å wavelength. Maximum sample size should be about 3×3 cm^2 and the vertical divergence of incoming and outgoing beam $\sim 3°$, corresponding to the homogeneity of the above precession field magnets.

Neutrons arriving at a given detector at a scattering angle θ carry two types of information: the total time-of-flight (TOF) with respect to the neutron source pulse and the spin polarization related to the NSE. For simplicity the case of quasielastic scattering will be considered in detail. (The generalization for the case of inelastic scattering on elementary excitations is straightforward on the lines of refs. 2 and 3.) The NSE polarization depends on the difference t of neutron times-of-flight through the equal magnetic precession fields before and after the scattering:

$$t = l/v_1 - l/v_0 ,$$

where v_0 and v_1 are the velocities of a neutron before and after scattering, respectively. On the other hand, the total time-of-flight T is the sum of two similar quantities (see fig. 1):

$$T = L_1/v_1 + (L_0 + L_0')/v_0 .$$

Obviously the knowledge of t and T provides full information on the neutron scattering event. In this scheme the NSE provides the unique feature of measuring directly for each neutron the difference quantity t.

In quantitative terms the NSE polarization is given as[1])

$$P_x = P_0 \int S(\kappa,\omega) \cos\left(\frac{\gamma_L H l}{v_1} - \frac{\gamma_L H l}{v_0}\right) d\omega ,$$

where

$$\omega = \tfrac{1}{2} m v_1^2 - \tfrac{1}{2} m v_0^2 .$$

($\gamma_L \simeq 3$ kHz/Oe is the Larmor constant, H the average precession field and P_0 the polarization efficiency of the NSE system). Since in high resolution quasielastic scattering one deals with $(v_1 - v_0) \ll v_0$, the momentum change κ and the total TOF T are given in an approximation good enough for the practical resolution of the TOF measurement ($\sim$0.5%) by

$$\kappa \simeq 2 \frac{m v_0}{\hbar} \sin\frac{\theta}{2} \quad \text{and} \quad T \simeq (L_0 + L_0' + L_1)/v_0 ,$$

respectively, i.e. a TOF value corresponds to a given quasielastic κ value. Furthermore, obviously

$$\frac{1}{v_1} - \frac{1}{v_0} \simeq \frac{v_0 - v_1}{v_0^2} \simeq \frac{\omega}{v_0^3 m} \simeq \left[\left(2 \sin\frac{\theta}{2}\right)^3 \frac{m^2}{\hbar^3 \kappa^3}\, \omega\right].$$

Finally the NSE polarization measured in a given time-of-flight channel of a given detector at angle θ becomes

$$P_x = P_0 \int S(\kappa,\omega) \cos\left[\left(2 \sin\frac{\theta}{2}\right)^3 \frac{\gamma_L H l m^2}{\hbar^3 \kappa^3}\, \omega\right] d\omega .$$

(The NSE focusing for quasielastic scattering is recognizable by the feature that the argument of the cosine becomes stationary with respect to κ at $\omega = 0$.)

To obtain full information on $S(\kappa,\omega) P_x$ must be measured for the same κ at several values of the Fourier parameter

$$\tau = \left(2 \sin\frac{\theta}{2}\right)^3 \frac{\gamma_L H l m^2}{\hbar^3 \kappa^3} ,$$

i.e. take the ω Fourier transform of $S(\kappa,\omega)$. A τ scan in NSE is normally achieved by changing the precession field H [1]). In this particular case, however, there is also an additional possibility for polycrystalline samples: looking at different θ's at the same κ. This means in practice that those TOF channels of detectors at different angles which correspond to the same κ (i.e. to different wavelengths) give different points on the NSE τ scan. Considering the band pass width of the polarizing SM systems the τ scan via θ will have a span of 1:8. Since the measured quantity in each TOF channel is polarization (obtained by relating subsequent counts in different spin configurations for each channel) and not intensity, the shape of the wavelength spectrum has no influence on the data. Of course, the usual H scan is also available and can be used both in some combination with the above θ scan, or alone (e.g. for single crystals).

To summarize, the three parameters for a given H magnetic field configuration, namely the total time-of-flight T, the scattering angle θ and the NSE polarization P_x measured for each TOF channel of each detector combine into a two dimensional array for measuring $S(\kappa,\omega)$: T and θ determines κ, and the different P_x values belonging to the same κ constitute a Fourier ω scan, in addition to the ordinary H scan[1]).

As mentioned above, these considerations can easily be generalized to the case of finite energy transfer experiments (e.g. study of the life-time of elementary excitations) as described in refs. 2 and 3.

3. Neutron spin echo with time-of-flight filtering

If a chopper phased to the neutron source pulses and placed just before the sample is added to the setup in fig. 1, an ordinary TOF spectrometer is obtained that works independently, but superposed on the NSE. The time T_0 between source pulse and chopper opening determines the monochromatization, which could vary reasonably between 2 and 20%, and the difference of the total TOF T and T_0, $T_1 = T - T_0$ gives the scattered beam velocity.

The superposition of the TOF and NSE inelastic spectrometry has the following advantage. Without the chopper in each T TOF channel of each detector neutrons are collected from a large chunk of the inelastic scattering spectrum $S(\boldsymbol{\kappa}, \omega)$, which are then sorted out by the NSE (fig. 2a). But the NSE acts as a magnifying lens focused at a given part of the scattering spectrum only. The rest of the scattering intensity will show up as an unpolarized background reducing the data collection efficiency for weak relative intensity processes. This is why it is beneficial in such a case to eliminate the other parts of the scattering spectrum. E.g., if we are interested in a weak inelastic scattering effect, the presence of a strong elastic intensity, say a Bragg peak, will reduce the counting statistics. In such cases, the additional TOF chopper can be used as a filter that makes a moderate or poor resolution preselection of the scattering spectra (fig. 2b). The NSE operates in the same way as above, only in some TOF channels there will be no interesting NSE information.

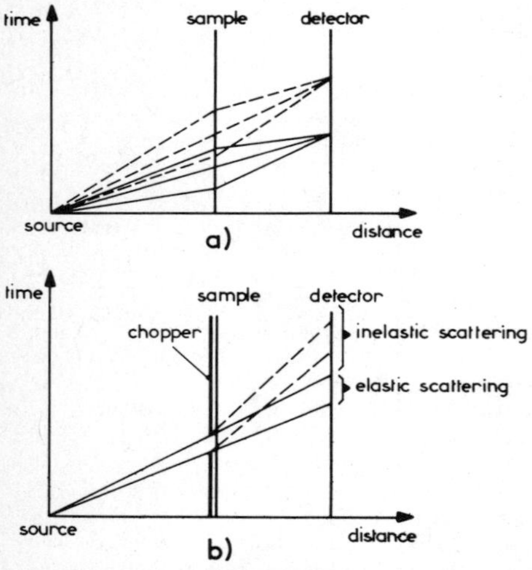

Fig. 2. (a) The time-of-flight diagram for pure NSE type of operation. (b) The principle of time-of-flight filtering for NSE-TOF type of operation.

The balance of the intensity loss due to monochromatization and the gain in statistics due to the background filtering decides which method, the pure NSE or the TOF-NSE, is better adapted to a given experiment. (Generally, if the instrumental background is high as compared to the total scattering intensity, the pure NSE is preferable. The whole situation is very similar to the relative merits of statistical and direct TOF spectrometry.)

In addition at ordinary resolutions (2–10%) the direct TOF is generally more favourable than the spin echo, if one removes the polarizer and analyser systems, gaining in this way about a factor of 10 in intensity. Thus the pure TOF operation with chopper can also be of interest.

4. Comparison of direct and NSE high resolution methods

The main interest of the NSE technique as compared to the direct monochromatization methods (TOF or backscattering) is that it provides a better neutron economy, since it permits using a large band of wavelengths simultaneously instead of a very narrow monochromatic band (10^{-4}–10^{-3} in $\delta\lambda/\lambda$). This intensity gain is partly offset by the finite transmission of the polarizer and analyser systems (about 40–50% for super-mirrors[4,5]) (SM) for wavelength above 3–4 Å) and the inherent 50% spin polarization loss. The bandwidth of the existing SM polarizer and analyser systems is about $\pm 40\%$ [5]) (e.g. from 4 to 8 Å or from 3 to 6 Å, fig. 3). Thus for example 1 μeV resolution at 6 Å incoming wavelength would require a monochromatization of $\delta\lambda/\lambda = 2\times10^{-4}$ for a classical type instrument as compared to the above bandwidth for NSE, i.e. the a priori net intensity gain for the NSE is 400 for the pure spin echo operation and 10–100 if 2–20% chopper filtering is used.

In addition the very long flight-path in an approx. 50 m long neutron guide tube, necessary for the high resolution monochromatization of the incoming beam in a classical approach, involves guide losses of about 50–70% at 6 Å and considerably more for longer wavelengths. Furthermore, the vertical divergence of the beam in a short guide (10 m) or in a direct beam tube could be increased

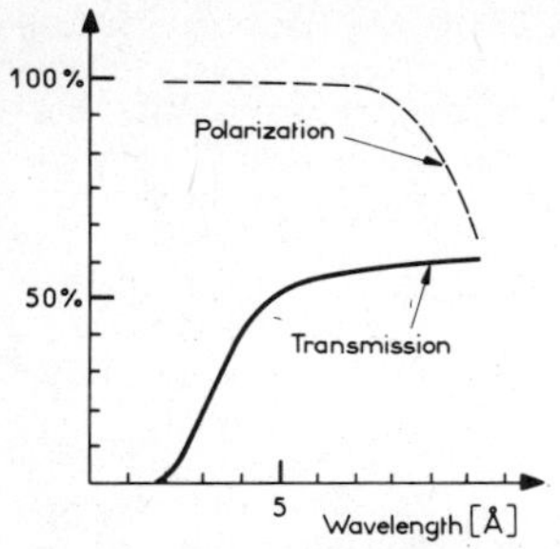

Fig. 3. The schematic wavelength dependence of the transmission and polarization efficiency of a Soller type supermirror analyser system (cf. ref. 5).

by a factor of 2.5–3 by the use of nonpolarising partially reflecting (~80%) SM top and bottom plates. The horizontal beam divergence for the SM polarizer is the same as that for a Ni guide.

The bandwidth of the NSE setup itself, i.e. that of the $\pi/2 - \pi - \pi/2$ flipper ensemble is about ±20% [6]). This means that if they are set to a given wavelength the NSE polarization would not drop more than 10% for ±20% different wavelengths. However, since different wavelength neutrons arrive at different times at each of these coils, they could be phased to the neutron source so that the full useful wavelength band of the polarizer and analyser system could be utilized with maximum efficiency.

5. Conclusion

It has been shown that the neutron spin echo technique could be readily adapted to a pulsed neutron source to obtain high energy resolution of about 1 μeV. The resulting, very natural combination of the total time-of-flight measurement and spin echo gives a good (.005 $Å^{-1}$ at $\lambda = 6$ Å or 0.5% at high scattering angles) momentum resolution at the same time. The NSE method offers a substantial gain in neutron intensity, partly because it uses a large incoming wavelength band simultaneously, and partly because of the use of a shorter source to sample distance. This intensity gain should be as much as 1000 e.g. for coherent quasielastic scattering studies (critical phenomena, molecular diffusion). For spin incoherent scattering there is a 9-fold data collection rate loss for the NSE since the scattered beam polarization is $-\frac{1}{3}$ as compared to the incoming one[7]). So the effective gain is more like 100 for problems like proton diffusion. The fact that the NSE measures a Fourier transformed spectrum is sometimes convenient[8]), sometimes not. The particularly disadvantageous case for a Fourier method is the study of a scattering spectrum with a great deal of structure, some of it weak as compared to the average intensity. This can be remedied by a TOF filtering technique if the different parts of the scattering spectra are sufficiently separated (more than about 100 μeV) at the expense of an intensity loss of a factor 4–40. Finally, it is of interest to emphasize that the NSE is the only known method which permits the μeV scale study of elementary excitation linewidths for non-zero slope dispersion relations, using the general focusing scheme[2]) which is also applicable in the present case.

References

[1]) F. Mezei, Z. Physik **255** (1972) 146.
[2]) F. Mezei, *Neutron inelastic scattering*, Conf. Proc. (IAEA, Vienna, 1978) vol. 1, p. 125.
[3]) F. Mezei, Proc. 3rd Int. School on *Neutron scattering*, Alushta, USSR (1978).
[4]) F. Mezei, Comm. Phys. **1** (1976) 81; F. Mezei and P. A. Dagleish, ibid. **2** (1977) 41.
[5]) F. Mezei, Diffuse neutron scattering Workshop, St. Pierre en Chartreuse, October 1978 (ILL report, to be published).
[6]) J. B. Hayter, Z. Physik **B31** (1978) 117.
[7]) See e.g. W. Marschall and S. W. Lovesey, *Theory of thermal neutron scattering* (Oxford, 1971).
[8]) For an example see D. Richter, J. B. Hayter, F. Mezei and B. Ewen, Phys. Rev. Lett. **41** (1978) 1484.

Journal of Magnetism and Magnetic Materials 14 (1979) 211–213

COMBINED THREE-DIMENSIONAL POLARIZATION ANALYSIS AND SPIN ECHO STUDY OF SPIN GLASS DYNAMICS

F. MEZEI and A.P. MURANI
Institut Laue–Langevin, 156X, 38042 Grenoble Cédex, France

We report direct measurements of the time-dependent spin correlation function $S(\kappa, t)$ for a Cu–Mn spin glass alloy over nearly 3 decades of time in a single scan in the range $10^{-12} < t < 10^{-9}$ s. This was achieved by using, for the first time, a combination of neutron spin echo and polarization analysis techniques. The results show most clearly the particular dynamics of spin glasses, namely a behaviour that can be described by a spectrum of relaxation times spreading over a wide range.

Neutron scattering techniques have been increasingly applied in recent years to study the dynamics of spin glass alloys [1]. With the conventional methods of measurement using unpolarized neutrons, the magnetic scattering law $S(\kappa, \omega)$ has been possible to determine because of its particular form – a broad quasi-elastic-like spectrum which is well separated from the phonon spectrum lying at high energies in classical spin glass system such as Cu–Mn. The elastic magnetic scattering effects are identified through their temperature dependence since non-magnetic elastic scattering is almost temperature independent in the temperature and κ-range of interest in the spin glass problem. Direct measurements of magnetic scattering using polarized neutrons have also been carried out earlier on some Cu–Mn spin glasses by Ahmed and Hicks [2]. In these measurements, however, no energy analysis was employed with the result that information about spin dynamics is lacking.

In the following we report the first measurements of the intermediate scattering law $S(\kappa, t)$ for a Cu–5 at% Mn spin glass where the magnetic scattering is directly observed by use of polarized neutrons. The measurements involve the neutron spin echo technique [3] and polarization analysis of the scattered beam in three dimensions with neutrons polarized sequentially along the three perpendicular axes of analysis.

In the neutron spin echo technique, a neutron polarized perpendicular to the magnetic field direction traverses a well-defined magnetic field region in which it carries out n Larmor precessions before impinging on the sample. The neutron is spin flipped at the sample and traverses another precession field with identical strength times length product. There it undergoes a further m precessions, which because of the spin flip appear to be in the opposite sense. Thus for elastic scattering from the sample $n - m = 0$ for all neutrons and the beam is phase focused at the analyzer i.e., its polarization is a maximum. This is equivalent to the spin echo condition in an nmr experiment. Any inelasticity in the scattering will then result in a value of m different from n and the measured polarization in the original direction x will be given by

$$\langle P_x \rangle = P \int S(\kappa, \omega) \cos\phi \, d\omega / \int S(\kappa, \omega) \, d\omega \,, \tag{1}$$

with

$$\phi = 2\pi(n - m) \simeq 2\pi n \frac{\omega}{2E_0} \tag{2}$$

and

$$S(\kappa, \omega) \simeq N \left[\frac{\omega}{1 - \exp(-\omega/kT)} \right] \times \frac{F^2(\kappa)}{\pi} \int e^{i\omega t} \langle S_\kappa(0) \, S_{-\kappa}(t) \rangle \, dt \,,$$

where P is the initial polarization, kT the thermal energy, κ and ω the neutron wave vector and energy transfer, respectively, and E_0 the incident energy. Hence,

$$\langle P_x \rangle = P \int S(\kappa, \omega) \cos(\pi n\omega/E_0) \mathrm{d}\omega / \int S(\kappa, \omega) \mathrm{d}\omega$$
$$= P \operatorname{Re} S(\kappa, t)/S(\kappa, 0) \quad (3)$$

with $t = \pi n/E_0$.

Thus the resultant polarization in the original direction gives a direct measure of the real part of the intermediate scattering law, i.e., the time correlation function $S(\kappa, t)$ at various times t. A time scan is then obtained by varying the product of the precession field times the path length.

The above technique of measurement is applicable to all types of scattering. In the special case of paramagnetic as well as isotropic magnetic samples, such as spin glasses having no net magnetization along any preferred direction, the spin flip scattering from the sample can be directly used to achieve the spin echo. In other words the spin flipper coil at the sample is eliminated and only the magnetic scattering contributes to the measured spin echo signal. This can be seen by examining the scattered beam polarization given by [4,5]

$$\boldsymbol{P}' = -\hat{\boldsymbol{\kappa}}(\hat{\boldsymbol{\kappa}} \cdot \boldsymbol{P}) , \quad (4)$$

where $\boldsymbol{P}$ is the incident polarization.

If $\kappa \parallel x$ axis (see fig. 1)

$$P_y' = 0 = \tfrac{1}{2}P_y - \tfrac{1}{2}P_y \text{ and } P_x' = -P_x = -\tfrac{1}{2}P_x - \tfrac{1}{2}P_x ,$$

(x, y) being the Larmor precession plane. In this plane the vector $\boldsymbol{P}_1' = (\frac{1}{2}P_y, -\frac{1}{2}P_x)$ corresponds to the $n \to -n$ spin flip necessary for spin echo, while $(-\frac{1}{2}P_y, -\frac{1}{2}P_x)$ gives only a 180° phase shift in the precession.

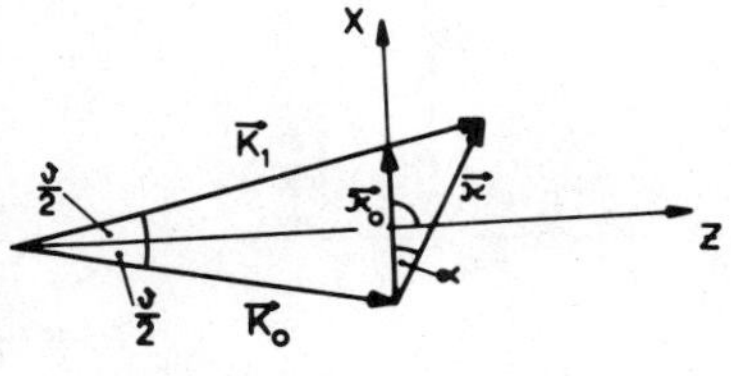

Fig. 1. The geometry of the scattering triangle for fixed scattering angle ϑ. (κ_0 and κ are the elastic and inelastic momentum transfer vectors).

This way the $\boldsymbol{P}_1'$ component of the polarization will produce a spin echo with a maximum amplitude of $|P_1'| = \frac{1}{2}$.

In addition to the spin echo scans we have also carried out simple polarization analysis with the incident neutron polarization sequentially along the three directions x, y, z. Since the scattering vector $\boldsymbol{\kappa}$ is in the x, z plane as shown in fig. 1, where α is the angle between the elastic scattering vector $\boldsymbol{\kappa}_0$ and $\boldsymbol{\kappa}$, we have

$$P_x' = -P \cos^2 \alpha \quad \text{for } \boldsymbol{P} \parallel x ,$$
$$P_z' = -P \sin^2 \alpha \quad \text{for } \boldsymbol{P} \parallel z ,$$
$$P_y' = 0 \quad \text{for } \boldsymbol{P} \parallel y ,$$

$$P_x' + P_z' = -P(\cos^2 \alpha + \sin^2 \alpha) = -P. \quad (5)$$

Thus the average polarizations $\langle P_x' \rangle$ and $\langle P_z' \rangle$ add to give the total scattering intensity. Furthermore since the angle α is directly related to the inelasticity of the scattering, the measurements of $\langle P_x' \rangle$ and $\langle P_z' \rangle$ can yield a measure of inelasticity as shown by Maleev [5] and Drabkin at al. [6]. In fact the difference $\langle P_x' \rangle - \langle P_z' \rangle$ provides, to a good approximation, a measurement of the intermediate scattering law Re $S(\kappa, t)$ at $t = \mathrm{ctg}(\frac{1}{2}\vartheta)/2E_0$.

From eq. (5) we have

$$\langle P_x' \rangle = -P \langle \cos^2 \alpha \rangle$$
$$= -P \int S(\kappa, \omega) \cos^2 \alpha \, \mathrm{d}\omega / \int S(\kappa, \omega) \mathrm{d}\omega$$
$$= -\tfrac{1}{2}P \int S(\kappa, \omega)(1 + \cos 2\alpha) \mathrm{d}\omega / S(\kappa, 0) \quad (6)$$

and

$$\langle P_z' \rangle = -\tfrac{1}{2}P \int S(\kappa, \omega)(1 - \cos 2\alpha) \mathrm{d}\omega / S(\kappa, 0) , \quad (7)$$

hence

$$\langle P_z' \rangle - \langle P_x' \rangle = P \int S(\kappa, \omega) \cos 2\alpha \, \mathrm{d}\omega / S(\kappa, 0) . \quad (8)$$

For small energy transfers ω, we find that

$$\alpha \simeq \frac{\omega}{4E_0} \mathrm{ctg}(\tfrac{1}{2}\vartheta) . \quad (9)$$

Thus

$$\langle P_z' \rangle - \langle P_x' \rangle = P \int S(\kappa, \omega) \cos\left(\frac{\omega}{2E_0} \mathrm{ctg}(\tfrac{1}{2}\vartheta)\right) \mathrm{d}\omega / S(\kappa, 0)$$
$$= P \operatorname{Re} S(\kappa, t)/S(\kappa, 0) \quad (10)$$

where $t = \mathrm{ctg}(\frac{1}{2}\vartheta)/2E_0$.
It turns out, however, that the requirement of small energy transfers (eq.(9)) is not very strict, basically because larger α's contribute little to the integrals (8) and (10) for typical quasi-elastic spectra.

We note that for isotropic spin systems $S(\kappa, t)$ is purely real, thus in this case we measure directly the time correlation function.

The measurements were made on the IN11 spin echo spectrometer at the ILL using neutrons of incident wavelength 5.9 Å with a wavelength spread of 35%, fwhm. Thus, the measured $S(\kappa, t)$ is in fact averaged over the κ-range defined by the momentum resolution of the spectrometer. This κ-averaging, however, has no influence on the results since we find in the present case that $S(\kappa, t)/S(\kappa, 0)$ is independent of κ within the accuracy of the measurement. Furthermore, it can also be shown that for relaxation type spectra eq. (2) still remains a good approximation with E_0 corresponding to the average incoming neutron wavelength. The question of momentum resolution of the spectrometer aside, the κ for the measured correlation function is given by the elastic scattering vector since only small energy transfers are involved in its measurement. This however is not so for the total scattering intensity (shown in fig. 2 for two scattering angles of 5° and 20°) obtained from $\langle P_x' \rangle + \langle P_z' \rangle$ (eq. (5)), which represents the integral over energy transfers comparable with E_0. At high temperatures the broad ω distribution results in an averaging over a large range of scattering vectors whereas at low temperatures as the energy distribution gets narrower the scattering is averaged over a narrower range of κ. Since $S(\kappa)$ is strongly peaked in the forward direction we have a relative increase of the total intensity with decreasing temperature from this effect alone.

In fig. 3 we show the results for the intermediate scattering law $S(\kappa, t)$ measured at scattering angle ϑ = 5° ($\langle \kappa \rangle = 0.092$ Å^{-1}). In the abscissa we have used the logarithmic time scale in order to represent the data over a time range varying over several orders of magnitude. The data have been normalized to the total polarization efficiency of the spectrometer, and represented as the ratio $S(\kappa, t)/S(\kappa, 0)$ where $t = 0$ corresponds in effect to times smaller than 10^{-14} s as seen from earlier neutron scattering experiments. In the diagram we have included a curve representing a simple relaxation process, given by the exponential $e^{-\gamma t}$ for $\gamma = 0.5$ meV. It should be noted that on such a diagram a simple exponential function always maintains the same shape, being displaced laterally for different values of γ.

The present practical limits of the time range over which $S(\kappa, t)$ can be measured directly by the spin-echo technique is $5 \times 10^{-11} \leqslant t \leqslant 5 \times 10^{-9}$ s. In fig. 3 the data points for $t = 3 \times 10^{-12}$ s for the two scattering vectors are obtained from polarization analysis along the x- and z-directions and using eq. (10). This limited time range prevents the determination of

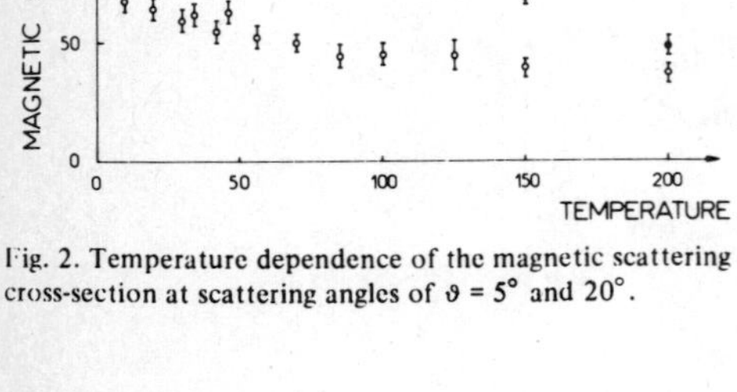

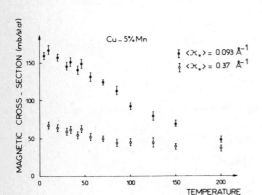

Fig. 2. Temperature dependence of the magnetic scattering cross-section at scattering angles of ϑ = 5° and 20°.

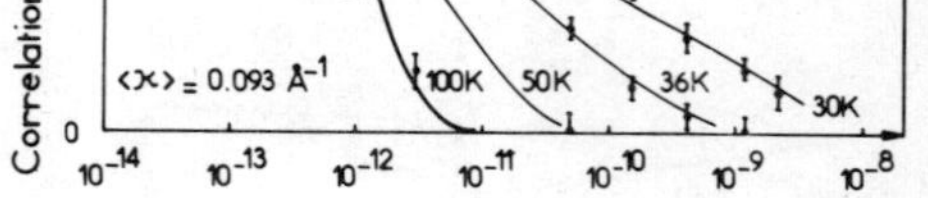

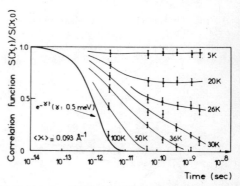

Fig. 3. The measured time dependent spin correlation function for Cu–5 at% Mn at various temperatures. The thick line corresponds to the simple exponential decay. The thin lines are guides to the eye only.

the shape of $S(\kappa, t)$ at short times and especially its measurement at higher temperatures where $S(\kappa, t)$ decreases very rapidly with t. It is interesting however that within the limits of the experimental error bars the shape of $S(\kappa, t)$ is found to be independent of κ (over the momentum interval of 0.047 – 0.37 $Å^{-1}$ covered in the present measurements). It is emphasized that this observation holds for temperatures below 45 K where non-zero values of the correlation function can be measured in the time range $3 \times 10^{-12} < t < 2 \times 10^{-9}$ s covered in the present experiment. We therefore express $S(\kappa, t)$ as

$$S(\kappa, t) = S_1(\kappa) S(t), \qquad (11)$$

where $S_1(\kappa)$ represents the spatial correlations between the spins and $S(t)$ the time-dependent correlation function which in this case must coincide with the self correlation function $\langle S_i(0) S_i(t) \rangle$. Monte Carlo computer simulation calculations of Ising spin glasses [7] yield self correlation functions which at first sight appear to be qualitatively similar to the present results. These, however, have the form

$$S(t) = \text{const} - \ln(t), \qquad (12)$$

which would be represented by straight lines in fig. 3. The marked curvature in $S(t)$ for $t \lesssim 10^{-11}$ s is therefore interesting and possibly reflects the influence of the rapid Korringa relaxation mechanism ($\tau \leqslant 10^{-11}$ s for $T \geqslant 5$ K) which is necessarily present in metallic spin glass systems with 3d magnetic atoms as suggested by earlier measurements also [8]. The model calculations as well as computer simulations, of course, do not include the Korringa mechanism but treat solute–solute exchange couplings only. Since the latter must dominate the spectrum for $t > 10^{-11}$ s the qualitative similarity of the measured $S(t)$ with the model calculations for $t > 10^{-11}$ s can be regarded as reasonably supporting the models, or simply as showing that a distribution of barrier heights may be present in real spin glasses [9].

In conclusion, we believe the present measurement of the intermediate scattering law $S(\kappa, t)$ over a time interval exceptionally large for a single experiment provide the most clear insight obtained up to now into the dynamics of spin glasses, and that the novel techniques we have used seem to be particularly well adapted for its measurement.

References

[1] A.P. Murani, J. de Phys. C6 (1978) 1517.
[2] N. Ahmed and T.J. Hicks, Solid State Commun. 15 (1974) 415.
[3] F. Mezei, Z. Phys. 255 (1972) 146.
[4] O. Halpern and H.R. Johnson, Phys. Rev. 55 (1939) 898.
[5] S.V. Maleev, Sov. Phys. JETP Lett. 2 (1966) 338.
[6] G.M. Drabkin, A.I. Okorokov, E.I. Zabidarov and Ya.A. Kasman, Sov. Phys. JETP 29 (1969) 261.
[7] K. Binder and K. Schröder, Phys. Rev. B14 (1976) 2142.
[8] A.P. Murani, Phys. Rev. Lett. 41 (1978) 1406.
[9] Julio F. Fernandez and Rodrigo Medina, Phys. Rev. B19 (1979) 3561.

Selected Issues from

Lecture Notes in Mathematics

Vol. 662: Akin, The Metric Theory of Banach Manifolds. XIX, 306 pages. 1978.

Vol. 665: Journées d'Analyse Non Linéaire. Proceedings, 1977. Edité par P. Bénilan et J. Robert. VIII, 256 pages. 1978.

Vol. 667: J. Gilewicz, Approximants de Padé. XIV, 511 pages. 1978.

Vol. 668: The Structure of Attractors in Dynamical Systems. Proceedings, 1977. Edited by J. C. Martin, N. G. Markley and W. Perrizo. VI, 264 pages. 1978.

Vol. 675: J. Galambos and S. Kotz, Characterizations of Probability Distributions. VIII, 169 pages. 1978.

Vol. 676: Differential Geometrical Methods in Mathematical Physics II, Proceedings, 1977. Edited by K. Bleuler, H. R. Petry and A. Reetz. VI, 626 pages. 1978.

Vol. 678: D. Dacunha-Castelle, H. Heyer et B. Roynette. Ecole d'Eté de Probabilités de Saint-Flour. VII-1977. Edité par P. L. Hennequin. IX, 379 pages. 1978.

Vol. 679: Numerical Treatment of Differential Equations in Applications, Proceedings, 1977. Edited by R. Ansorge and W. Törnig. IX, 163 pages. 1978.

Vol. 681: Séminaire de Théorie du Potentiel Paris, No. 3, Directeurs: M. Brelot, G. Choquet et J. Deny. Rédacteurs: F. Hirsch et G. Mokobodzki. VII, 294 pages. 1978.

Vol. 682: G. D. James, The Representation Theory of the Symmetric Groups. V, 156 pages. 1978.

Vol. 684: E. E. Rosinger, Distributions and Nonlinear Partial Differential Equations. XI, 146 pages. 1978.

Vol. 690: W. J. J. Rey, Robust Statistical Methods. VI. 128 pages. 1978.

Vol. 691: G. Viennot, Algèbres de Lie Libres et Monoïdes Libres. III, 124 pages. 1978.

Vol. 693: Hilbert Space Operators, Proceedings, 1977. Edited by J. M. Bachar Jr. and D. W. Hadwin. VIII, 184 pages. 1978.

Vol. 696: P. J. Feinsilver, Special Functions, Probability Semigroups, and Hamiltonian Flows. VI, 112 pages. 1978.

Vol. 702: Yuri N. Bibikov, Local Theory of Nonlinear Analytic Ordinary Differential Equations. IX, 147 pages. 1979.

Vol. 704: Computing Methods in Applied Sciences and Engineering, 1977, I. Proceedings, 1977. Edited by R. Glowinski and J. L. Lions. VI, 391 pages. 1979.

Vol. 710: Séminaire Bourbaki vol. 1977/78, Exposés 507–524. IV, 328 pages. 1979.

Vol. 711: Asymptotic Analysis. Edited by F. Verhulst. V, 240 pages. 1979.

Vol. 712: Equations Différentielles et Systèmes de Pfaff dans le Champ Complexe. Edité par R. Gérard et J.-P. Ramis. V, 364 pages. 1979.

Vol. 716: M. A. Scheunert, The Theory of Lie Superalgebras. X, 271 pages. 1979.

Vol. 720: E. Dubinsky, The Structure of Nuclear Fréchet Spaces. V, 187 pages. 1979.

Vol. 724: D. Griffeath, Additive and Cancellative Interacting Particle Systems. V, 108 pages. 1979.

Vol. 725: Algèbres d'Opérateurs. Proceedings, 1978. Edité par P. de la Harpe. VII, 309 pages. 1979.

Vol. 726: Y.-C. Wong, Schwartz Spaces, Nuclear Spaces and Tensor Products. VI, 418 pages. 1979.

Vol. 727: Y. Saito, Spectral Representations for Schrödinger Operators With Long-Range Potentials. V, 149 pages. 1979.

Vol. 728: Non-Commutative Harmonic Analysis. Proceedings, 1978. Edited by J. Carmona and M. Vergne. V, 244 pages. 1979.

Vol. 729: Ergodic Theory. Proceedings 1978. Edited by M. Denker and K. Jacobs. XII, 209 pages. 1979.

Vol. 730: Functional Differential Equations and Approximation of Fixed Points. Proceedings, 1978. Edited by H.-O. Peitgen and H.-O. Walther. XV, 503 pages. 1979.

Vol. 731: Y. Nakagami and M. Takesaki, Duality for Crossed Products of von Neumann Algebras. IX, 139 pages. 1979.

Vol. 733: F. Bloom, Modern Differential Geometric Techniques in the Theory of Continuous Distributions of Dislocations. XII, 206 pages. 1979.

Vol. 735: B. Aupetit, Propriétés Spectrales des Algèbres de Banach. XII, 192 pages. 1979.

Vol. 738: P. E. Conner, Differentiable Periodic Maps. 2nd edition, IV, 181 pages. 1979.

Vol. 742: K. Clancey, Seminormal Operators. VII, 125 pages. 1979.

Vol. 755: Global Analysis. Proceedings, 1978. Edited by M. Grmela and J. E. Marsden. VII, 377 pages. 1979.

Vol. 756: H. O. Cordes, Elliptic Pseudo-Differential Operators – An Abstract Theory. IX, 331 pages. 1979.

Vol. 760: H.-O. Georgii, Canonical Gibbs Measures. VIII, 190 pages. 1979.

Vol. 762: D. H. Sattinger, Group Theoretic Methods in Bifurcation Theory. V, 241 pages. 1979.

Vol. 765: Padé Approximation and its Applications. Proceedings, 1979. Edited by L. Wuytack. VI, 392 pages. 1979.

Vol. 766: T. tom Dieck, Transformation Groups and Representation Theory. VIII, 309 pages. 1979.

Vol. 771: Approximation Methods for Navier-Stokes Problems. Proceedings, 1979. Edited by R. Rautmann. XVI, 581 pages. 1980.

Vol. 773: Numerical Analysis. Proceedings, 1979. Edited by G. A. Watson. X, 184 pages. 1980.

Vol. 775: Geometric Methods in Mathematical Physics. Proceedings, 1979. Edited by G. Kaiser and J. E. Marsden. VII, 257 pages. 1980.